UCLA Symposia on Molecular and Cellular Biology, New Series

Series Editor
C. Fred Fox

RECENT TITLES

Volume 13
Molecular Biology of Host–Parasite Interactions, Nina Agabian and Harvey Eisen, *Editors*

Volume 14
Biosynthesis of the Photosynthetic Apparatus: Molecular Biology, Development and Regulation, J. Philip Thornber, L. Andrew Staehelin, and Richard B. Hallick, *Editors*

Volume 15
Protein Transport and Secretion, Dale L. Oxender, *Editor*

Volume 16
Acquired Immune Deficiency Syndrome, Michael S. Gottlieb and Jerome E. Groopman, *Editors*

Volume 17
Genes and Cancer, J. Michael Bishop, Janet D. Rowley, and Mel Greaves, *Editors*

Volume 18
Regulation of the Immune System, Harvey Cantor, Leonard Chess, and Eli Sercarz, *Editors*

Volume 19
Molecular Biology of Development, Eric H. Davidson and Richard A. Firtel, *Editors*

Volume 20
Genome Rearrangement, Melvin Simon and Ira Herskowitz, *Editors*

Volume 21
Herpesvirus, Fred Rapp, *Editor*

Volume 22
Cellular and Molecular Biology of Plant Stress, Joe L. Key and Tsune Kosuge, *Editors*

Volume 23
Membrane Receptors and Cellular Regulation, Michael P. Czech and C. Ronald Kahn, *Editors*

Volume 24
Neurobiology: Molecular Biological Approaches to Understanding Neuronal Function and Development, Paul O'Lague, *Editor*

Volume 25
Extracellular Matrix: Structure and Function, A. Hari Reddi, *Editor*

Volume 26
Nuclear Envelope Structure and RNA Maturation, Edward A. Smuckler and Gary A. Clawson, *Editors*

Volume 27
Monoclonal Antibodies and Cancer Therapy, Ralph A. Reisfeld and Stewart Sell, *Editors*

Volume 28
Leukemia: Recent Advances in Biology and Treatment, David W. Golde and Robert Peter Gale, *Editors*

Volume 29
Molecular Biology of Muscle Development, Charles Emerson, Donald Fischman, Bernardo Nadal-Ginard, and M.A.Q. Siddiqui, *Editors*

Volume 30
Sequence Specificity in Transcription and Translation, Richard Calendar and Larry Gold, *Editors*

Volume 31
Molecular Determinants of Animal Form, Gerald M. Edelman, *Editor*

Volume 32
Papillomaviruses: Molecular and Clinical Aspects, Peter M. Howley and Thomas R. Broker, *Editors*

Volume 33
Yeast Cell Biology, James Hicks, *Editor*

Volume 34
Molecular Genetics of Filamentous Fungi, William Timberlake, *Editor*

Volume 35
Plant Genetics, Michael Freeling, *Editor*

Please contact the publisher for information about previous titles in this series.

UCLA Symposia Published Prior to 1983

(Numbers refer to the publishers listed below.)

1972
Membrane Research **(2)**

1973
Membranes **(1)**
Virus Research **(2)**

1974
Molecular Mechanisms for the Repair of DNA **(4)**
Membranes **(1)**
Assembly Mechanisms **(1)**
The Immune System: Genes, Receptors, Signals **(2)**
Mechanisms of Virus Disease **(3)**

1975
Energy Transducing Mechanisms **(1)**
Cell Surface Receptors **(1)**
Developmental Biology **(3)**
DNA Synthesis and Its Regulation **(3)**

1976
Cellular Neurobiology **(1)**
Cell Shape and Surface Architecture **(1)**
Animal Virology **(2)**
Molecular Mechanisms in the Control of Gene Expression **(2)**

1977
Cell Surface Carbohydrates and Biological Recognition **(1)**
Molecular Approaches to Eucaryotic Genetic Systems **(2)**
Molecular Human Cytogenetics **(2)**
Molecular Aspects of Membrane Transport **(1)**
Immune System: Genetics and Regulation **(2)**

1978
DNA Repair Mechanisms **(2)**
Transmembrane Signaling **(1)**
Hematopoietic Cell Differentiation **(2)**
Normal and Abnormal Red Cell Membranes **(1)**
Persistent Viruses **(2)**
Cell Reproduction: Daniel Mazia Dedicatory Volume **(2)**

1979
Covalent and Non-Covalent Modulation of Protein Function **(2)**
Eucaryotic Gene Regulation **(2)**
Biological Recognition and Assembly **(1)**
Extrachromosomal DNA **(2)**
Tumor Cell Surfaces and Malignancy **(1)**
T and B Lymphocytes: Recognition and Function **(2)**

1980
Biology of Bone Marrow Transplantation **(2)**
Membrane Transport and Neuroreceptors **(1)**
Control of Cellular Division and Development **(1)**
Animal Virus Genetics **(2)**
Mechanistic Studies of DNA Replication and Genetic Recombination **(2)**

1981
Immunoglobulin Idiotypes **(2)**
Initiation of DNA Replication **(2)**
Genetic Variation Among Influenza Viruses **(2)**
Developmental Biology Using Purified Genes **(2)**
Differentiation and Function of Hematopoietic Cell Surfaces **(1)**
Mechanisms of Chemical Carcinogenesis **(1)**
Cellular Recognition **(1)**

1982
B and T Cell Tumors **(2)**
Interferon **(2)**
Rational Basis for Chemotherapy **(1)**
Gene Regulation **(2)**
Tumor Viruses and Differentiation **(1)**
Evolution of Hormone-Receptor Systems **(1)**

Publishers

(1) Alan R. Liss, Inc.
41 East 11th Street
New York, NY 10003

(2) Academic Press, Inc.
Orlando, FL 32887

(3) W.A. Benjamin, Inc.
2725 Sand Hill Road
Menlo Park, CA 94025

(4) Plenum Publishing Corp.
233 Spring Street
New York, NY 10013

MOLECULAR GENETICS OF FILAMENTOUS FUNGI

MOLECULAR GENETICS OF FILAMENTOUS FUNGI

Proceedings of a UCLA Symposium
Held in Keystone, Colorado
April 13–19, 1985

Editor

William E. Timberlake
Department of Genetics
University of California
Davis, California

Alan R. Liss, Inc. • New York

Address all Inquiries to the Publisher
Alan R. Liss, Inc., 41 East 11th Street, New York, NY 10003

Printed in the United States of America.

Library of Congress Cataloging-in-Publication Data

Main entry under title:

Molecular genetics of filamentous fungi.

Includes index.

1. Fungi—Genetics—Congresses. 2. Molecular genetics—Congresses. I. Timberlake, William E. II. Title: Filamentous fungi.
QK602.M65 1985 589.2′0415 85-23155
ISBN 0-8451-2633-4

Contents

Contributors

D.J. Ballance, Department of Microbiology, University of Bristol, Bristol BS8 1TD, UK **[15]**

James Baum, Department of Genetics, University of Georgia, Athens, GA 30602 **[95]**

J.W. Bennett, Department of Biology, Tulane University, New Orleans, LA 70118 **[345]**

R.K. Beri, Department of Microbiology, University of Bristol, Bristol BS8 1TD, UK **[15]**

Christopher R. Birkett, Biological Laboratory, University of Kent, Canterbury, Kent CT2 7NJ, England **[265]**

Robert Brambl, Department of Plant Physiology, The University of Minnesota, St. Paul, MN 55108 **[207]**

P.A. Burns, Department of Genetics, University of Edinburgh, Edinburgh EH9 3JN, UK; present address: Department of Biology, York University, Downsview, Ontario M6P 3K5, Canada **[117]**

F.P. Buxton, Department of Genetics, The University of Leeds, Leeds LS2 9JT, UK; present address: Allelix Inc., Mississauga, Ontario L4V 1P1, Canada **[127]**

Mary E. Case, Department of Genetics, University of Georgia, Athens, GA 30602 **[95]**

S.D. Clouse, Department of Plant Pathology, University of California, Davis, CA 95616; present address: Plant Molecular and Cellular Biology Lab, The Salk Institute, San Diego, CA 92138 **[405]**

Catherine M. Corrick, Department of Genetics, University of Melbourne, Parkville, Victoria 3052, Australia **[157]**

M. Crawford, Institute of Biological Chemistry, Biochemistry/Biophysics Program, Washington State University, Pullman, WA 99164 **[421]**

Ernest H. Creaser, Department of Genetics and Centre for Recombinant DNA Research, Research School of Biological Sciences, Australian National University, Canberra A.C.T. 2601, Australia **[171]**

Jim E. Cutler, Biology Department, Reed College, Portland, OR 97202; present address: Department of Microbiology, Montana State University, Bozeman, MT 59717 **[321]**

R. Wayne Davies, Department of Molecular Biology, Allelix Inc., Mississauga, Ontario L4V 1P1, Canada **[185]**

Rowland H. Davis, Department of Molecular Biology and Biochemistry, University of California at Irvine, Irvine, CA 92717 **[145]**

The number in brackets is the opening page number of the contributor's article.

Isabella Di Silvestro, Agriculture Industrial Development, SpA, The Research Center, 95100 Catania, Italy **[439]**

Colin H. Doy, Department of Genetics and Centre for Recombinant DNA Research, Research School of Biological Sciences, Australian National University, Canberra A.C.T. 2601, Australia **[171]**

J.R.S. Fincham, Department of Genetics, University of Edinburgh, Edinburgh EH9 3JN, UK; present address: Department of Genetics, University of Cambridge, Cambridge CB2 3EH, UK **[117]**

Kay E. Foster, Biological Laboratory, University of Kent, Canterbury, Kent CT2 7NJ, England **[265]**

E.H. Froeliger, Department of Medical Microbiology, University of Vermont, Burlington, VT 05405 **[39]**

John Gambino, Department of Pharmacology, University of Medicine and Dentistry of New Jersey, Rutgers Medical School, Piscataway, NJ 08854 **[239]**

Robert Geever, Department of Genetics, University of Georgia, Athens, GA 30602 **[95]**

D.G. Gilchrist, Department of Plant Pathology, University of California, Davis, CA 95616 **[405]**

Norman H. Giles, Department of Genetics, University of Georgia, Athens, GA 30602 **[95]**

J.A. Glazebrook, Department of Genetics, The University of Leeds, Leeds LS2 9JT, UK **[127]**

T. Goosen, Department of Genetics, Agricultural University, Wageningen, The Netherlands **[29]**

Keith Gull, Biological Laboratory, University of Kent, Canterbury, Kent CT2 7NJ, England **[265]**

David I. Gwynne, Department of Molecular Biology, Allelix Inc., Mississauga, Ontario L4V 1P1, Canada **[185]**

Layne Huiet, Department of Genetics, University of Georgia, Athens, GA 30602; present address: CSIRO Plant Industry, Canberra City, A.C.T.2601, Australia **[95]**

Michael J. Hynes, Department of Genetics, University of Melbourne, Parkville, Victoria 3052, Australia **[157]**

Jonathan W. Jarvik, Department of Biological Sciences, Carnegie-Mellon University, Pittsburgh, PA 15213 **[253]**

Heather J. Kane, Department of Genetics and Centre for Recombinant DNA Research, Research School of Biological Sciences, Australian National University, Canberra A.C.T. 2601, Australia **[171]**

Joan M. Kelly, Department of Genetics, University of Melbourne, Parkville, Victoria 3052, Australia; present address: Department of Genetics, University of Adelaide, South Australia, Australia **[157]**

Jane H. Kinnaird, Department of Genetics, University of Edinburgh, Edinburgh EH9 3JN, UK **[117]**

P.E. Kolattukudy, Institute of Biological Chemistry, Biochemistry/Biophysics Program, Washington State University, Pullman, WA 99164 **[421]**

Michael R. Kuchka, Department of Biological Sciences, Carnegie-Mellon University, Pittsburgh, PA 15213 **[253]**

Timothy J. Littlejohn, Department of Genetics, University of Melbourne, Parkville, Victoria 3052, Australia **[157]**

Robin Lockington, University of Essex, Colchester, UK; present address: Institut de Microbiologie, Batiment 409, Universite Paris Sud, 91405 Orsay, France **[185]**

James S. Lovett, Department of Biological Sciences, Purdue University, West Lafayette, IN 47907 **[187]**

A.N. Martensen, Department of Plant Pathology, University of California, Davis, CA 95616 **[405]**

George A. Marzluf, Department of Biochemistry, Ohio State University, Columbus, OH 43210 **[1,83]**

Gregory S. May, Department of Pharmacology, University of Medicine and Dentistry of New Jersey, Rutgers Medical School, Piscataway, NJ 08854 **[239]**

James A. McCloskey, Department of Medical Chemistry, College of Pharmacy, The University of Utah, Salt Lake City, UT 84112 **[321]**

B.L. McFarland, Department of Plant Pathology, University of California, Davis, CA 95616; present address: Chevron Biotechnology Group, Chevron Chemical Co., Richmond, CA 94804 **[405]**

Fiona M. Mellon, School of Biological Sciences, Queen Mary College, London E1 4NS, England **[69]**

Robert L. Metzenberg, Department of Physiological Chemistry, University of Wisconsin, Madison, WI 53706 **[295,309]**

N. Ronald Morris, Department of Pharmacology, University of Medicine and Dentistry of New Jersey, Rutgers Medical School, Piscataway, NJ 08854 **[239]**

Ewa Morzycka-Wroblewska, Department of Physiological Chemistry, University of Wisconsin, Madison, WI 53706; present address: Department of Molecular Biology and Genetics, Johns Hopkins University, School of Medicine, Baltimore, MD 21205 **[295]**

A.M. Muñoz-Rivas, Department of Botany, University of Vermont, Burlington, VT 05405 **[39]**

Baek H. Nahm, Department of Biochemistry, Ohio State University, Columbus, OH 43210 **[83]**

S.F. Newbury, Department of Genetics, The University of Leeds, Leeds LS2 9JT, UK **[127]**

C.P. Novotny, Department of Medical Microbiology, University of Vermont, Burlington, VT 05405 **[39]**

Berl R. Oakley, Department of Microbiology, Ohio State University, Columbus, OH 43210 **[225]**

David M. Ogrydziak, Institute of Marine Resources, University of California, Davis, CA 95616 **[367]**

Jane E. Olsen, Department of Genetics and Centre for Recombinant DNA Research, Research School of Biological Sciences, Australian National University, Canberra A.C.T. 2601, Australia **[171]**

John Paietta, Department of Biochemistry, Ohio State University, Columbus, OH 43210 **[1]**

Virginia Patel, Department of Genetics, University of Georgia, Athens, GA 30602 **[95]**

John A. Pateman, Department of Genetics and Centre for Recombinant DNA Research, Research School of Biological Sciences, Australian National University, Canberra A.C.T. 2601, Australia **[171]**

Christina Patiño, Departamento de Genética Molecular, Antibióticos, S.A., 28015 Madrid, Spain **[59]**

Miguel A. Peñalva, Departamento de Genética Molecular, Antibióticos, S.A., 28015 Madrid, Spain **[59]**

David D. Perkins, Department of Biological Sciences, Stanford University, Stanford, CA 94305 **[277]**

Kimberly G. Perrine, Department of Genetics, Ohio State University, Columbus, OH 43210 **[83]**

Mark Pickett, Department of Molecular Biology, Allelix Inc., Mississauga, Ontario L4V 1P1, Canada **[185]**

P.H. Pouwels, Department of Biochemistry, Medical Biological Laboratory TNO, 2280 AA Rijswijk, The Netherlands **[29]**

Patricia J. Pukkila, Department of Biology and Curriculum in Genetics, University of North Carolina at Chapel Hill, NC 27514 **[333]**

Eliot M. Rachlin, Department of Medical Chemistry, College of Pharmacy, The University of Utah, Salt Lake City, UT 84112 **[321]**

A. Radford, Department of Genetics, The University of Leeds, Leeds LS2 9JT, UK **[127]**

Karin D. Rodland, Biology Department, Reed College, Portland, OR 97202; present address: Department of Cell Biology and Anatomy, Oregon Health Sciences University, Portland, OR 97201 **[321]**

Franco Rollo, Department of Cell Biology, University of Camerino, 62032 Camerino, Italy **[439]**

Victor Rubio, Departamento de Genética Molecular, Antibióticos, S.A., 28015 Madrid, Spain **[59]**

Peter J. Russell, Department of Biology, Reed College, Portland, OR 97202 **[321]**

Flora Sánchez, Departamento de Genética Molecular, Antibióticos, S.A., 28015 Madrid, Spain **[59]**

Claudio Scazzocchio, University of Essex, Colchester, UK; present address: Institut de Microbiologie, Batiment 409, Universite Paris Sud, 91405 Orsay, France **[185]**

Heather Sealy-Lewis, University of Essex, Colchester, UK; present address: Department of Biochemistry, University of Hull, Hull, UK **[185]**

Eric U. Selker, Department of Physiological Chemistry, University of Wisconsin, Madison, WI 53706; present address: Department of Biology and Institute of Molecular Biology, University of Oregon, Eugene, OR 97403 **[295,309]**

C.L. Soliday, Institute of Biological Chemistry, Biochemistry/Biophysics Program, Washington State University, Pullman, WA 99164 **[421]**

José M. Fernández Sousa, Departamento de Genética Molecular, Antibióticos, S.A., 28015 Madrid, Spain **[59]**

C.A. Specht, Department of Botany, University of Vermont, Burlington, VT 05405 **[39]**

Judith N. Stevens, Department of Physiological Chemistry, University of Wisconsin, Madison, WI 53706 **[295,309]**

William E. Timberlake, Department of Plant Pathology, University of California, Davis, CA 95616 **[xix,29]**

Angeles Touriño, Departamento de Genética Molecular, Antibióticos, S.A., 28015 Madrid, Spain **[59]**

Monica L.-S. Tsang, Department of Pharmacology, University of Medicine and Dentistry of New Jersey, Rutgers Medical School, Piscataway, NJ 08854; present address: Bureau of Biological Research, The State University of New Jersey, Rutgers College, Piscataway, NJ 08854 **[239]**

B.G. Turgeon, Department of Plant Pathology, Cornell University, Ithaca, NY 14853 **[383]**

Geoffrey Turner, Department of Microbiology, University of Bristol, Bristol BS8 1TD, UK **[15]**

Brett Tyler, Department of Genetics, University of Georgia, Athens, GA 30602; present address: Department of Genetics, Research School of Biological Sciences, Australian National University, Canberra City, A.C.T. 2601, Australia **[95]**

R.C. Ullrich, Department of Botany, University of Vermont, Burlington, VT 05405 **[39]**

H.W.J. van den Broek, Department of Genetics, Agricultural University, Wageningen, The Netherlands **[29]**

C.A.M.J.J. van den Hondel, Department of Biochemistry, Medical Biological Laboratory TNO, 2280 AA Rijswijk, The Netherlands **[29]**

R.F.M. van Gorcom, Department of Biochemistry, Medical Biological Laboratory TNO, 2280 AA Rijswijk, The Netherlands **[29]**

M. Ward, Department of Microbiology, University of Bristol, Bristol BS8 1TD, UK **[15]**

James A. Weatherbee, Department of Pharmacology, University of Medicine and Dentistry of New Jersey, Rutgers Medical School, Piscataway, NJ 08854 **[239]**

J.G.H. Wessels, Department of Plant Physiology, Biological Center, University of Groningen, Haren, The Netherlands **[193]**

Rod A. Wing, Institute of Marine Resources, University of California, Davis, CA 95616 **[367]**

C.P. Woloshuk, Institute of Biological Chemistry, Biochemistry/Biophysics Program, Washington State University, Pullman, WA 99164 **[421]**

O.C. Yoder, Department of Plant Pathology, Cornell University, Ithaca, NY 14853 **[383]**

Miriam E. Zolan, Division of Biological Sciences, University of Michigan, Ann Arbor, MI 48109 **[333]**

William V. Zucker, Agriculture Industrial Development, SpA, The Research Center, 95100 Catania, Italy **[439]**

Preface

This volume represents a series of papers presented at the 1985 UCLA Symposium on "Molecular Genetics of Filamentous Fungi" held at Keystone, Colorado, April 13–19, 1985. The meeting was attended by scientists from around the world, representing both academic institutions and industrial research laboratories. It was held in conjunction with two other meetings: "Yeast Cell Biology," organized by Jim Hicks, and "Plant Genetics," organized by Mike Freeling. The interfacing of these three symposia greatly enhanced scientific interactions and exchanges.

By all accounts, the Keystone Symposium was a great success. Many new and exciting developments were described, perhaps most notably the successful transformation of plant pathogenic and industrially important fungi. In addition, it was made clear that the rapid technological advances with *Neurospora crassa* and *Aspergillus nidulans*, including procedures for directed mutagenesis and gene cloning by complementation of genetic defects, make these organisms exceedingly attractive for the molecular genetic analysis of a variety of important cellular processes.

This volume does not contain papers from all the speakers at the symposium. Submission of manuscripts was voluntary, and some speakers felt the information they presented was not appropriate for a proceedings volume. Thus, this volume does not necessarily reflect the proceedings of the meeting in its entirety.

I would like to thank Monsanto Corporation Research Laboratories, E.I. du Pont de Nemours & Company, Squibb Institute for Medical Research, Allelix, Inc., and Bristol-Meyers Company, Industrial Division for their financial support of the symposium. Without their assistance, it would not have been as successful. I also would like to thank Rohm and Haas Company for financial support of the one day joint session. Many thanks are extended to the UCLA Symposia office staff for their assistance in organizing the meetings, especially the one day when all three groups met concurrently. Finally, I am indebted to the many speakers who accepted no or minimal reimbursement for their expenses. They made it possible for me to provide funds to scientists who otherwise would have been unable to attend.

William E. Timberlake

Molecular Genetics of Filamentous Fungi, pages 1–13

DEVELOPMENT OF SHUTTLE VECTORS AND GENE MANIPULATION TECHNIQUES FOR NEUROSPORA CRASSA[1]

John Paietta and George A. Marzluf

Department of Biochemistry, The Ohio State University
Columbus, Ohio 43210

ABSTRACT Using a new approach for plasmid DNA recovery from Neurospora transformants, a screening has been carried out for a stable vector which contains an autonomously replicating sequence (*ars*). A potential shuttle vector (pJP102) has been isolated from a clone bank containing a selectable marker (*qa-2*$^{+}$) and Neurospora BamHI chromosomal fragments. The clone bank was used to transform lithium acetate-treated Neurospora spores and the plasmid DNA was extracted from the resulting transformants which were grown in liquid culture. pJP102 was isolated following transformation of *E. coli* to ampicillin resistance with the extracted plasmid DNA. In addition, we have carried out gene disruption in Neurospora using linear DNA fragments carrying the *am*$^{+}$ sequence interrupted by a selectable marker (*qa-2*$^{+}$). Only a low proportion, approximately 10%, of transformants have integration events that occur at the homologous site, whereas most integrations of transforming DNA take place in non-homologous regions. The replacement of a specific region in Neurospora is possible, but is complicated due to heterokaryon formation and to multiple and apparently non-homologous integrations.

INTRODUCTION

A major problem in applying recombinant DNA technology to Neurospora, and to other filamentous fungi, has been a lack of vectors suitable for the development of an efficient cloning system, i.e. plasmids that can replicate

[1]This work was supported by NIH grant GM-23367.

autonomously in both Neurospora and E. coli. We are interested in the development of a shuttle vector for Neurospora as well as the capability to carry out gene manipulations, such as the substitution of modified cloned sequences for resident chromosomal sequences.

Stohl and Lambowitz (1) constructed a vector (pALS-1) which consists of a mitochondrial plasmid, pBR325, and the qa-2$^+$ gene of Neurospora. This approach yielded a potentially useful shuttle vector although much of the vector DNA is frequently lost during passaging through Neurospora (1,2). Another shuttle vector, pDV1001, has been reported but it has been found to be difficult to recover from Neurospora transformants (3,4).

An alternative approach which we have used involves the isolation of putative chromosomal replication origins through the use of a new plasmid recovery procedure. The technique we have devised for plasmid recovery from Neurospora transformants is to grow spores treated with lithium acetate (5) and transformed with plasmid DNA in mass in liquid culture, followed by harvest of the mycelial growth and extraction of the plasmid DNA to transform E. coli to ampicillin resistance. This procedure serves as the basis for a direct, efficient selection of plasmids with improved properties for use as a shuttle vector.

Once a gene of interest has been cloned, the ability to perform specific manipulations on it and then to reintroduce it into the genome would be a powerful experimental tool. In yeast a cloned DNA segment can be used to replace the homologous region in the yeast chromosome (i.e., one-step gene disruption) (6,7). In Neurospora, and in other organisms, because of the varying degrees of homologous integration that occur, this type of approach is not so straightforward. In Neurospora, some transformants have integrated a re-introduced gene at the expected homologous site, but in many others the gene has integrated at an unlinked site (8,9). Despite these problems, we describe here directed gene disruption of the Neurospora am gene.

RESULTS

Recovery of plasmid DNA from Neurospora transformants.

When plasmid pSD3 (Figure 1) is used to transform the aro-9 qa-2 inl strain, the qa-2$^+$ transformants grow slowly

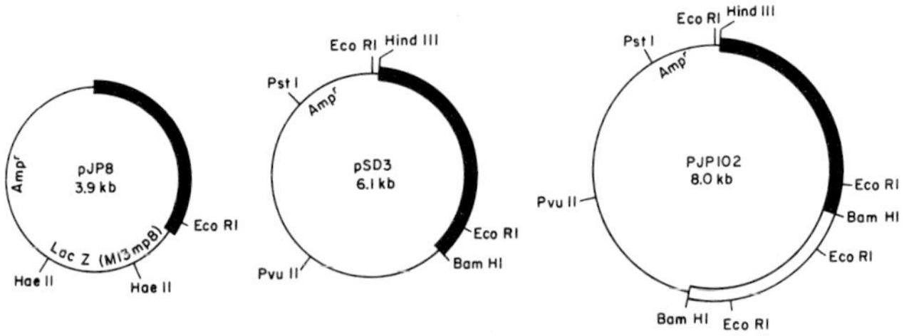

FIGURE 1. Restriction maps of plasmids pJP8, pSD3 and pJP102. Plasmid pJP8 is a derivative of pUC8 which was constructed by ligation of the BamHI + BglII *qa-2*$^{+}$ fragment from pVK57 into the NdeI site of pUC8 (overhangs were filled). pSD3 is a derivative of pVK57 which has a BamHI fragment deleted. pJP102 has a 1.9 kb Neurospora BamHI fragment inserted into pSD3. Key: solid black, Neurospora *qa-2*$^{+}$ gene; open, tentative Neurospora *ars* 102 sequence.

on agar and 4-5 days of incubation are required before colonies appear. There is a parallel lag in growth if pSD3-treated spores are instead placed in liquid culture, and three to four days are required before substantial growth is present. DNA was extracted from liquid cultures of transformants obtained with pSD3 plasmid and used to transform *E. coli* strain HB101, with selection for ampicillin resistance. Plasmid pSD3 was recovered from such cultures, and the highest number of *E. coli* colonies obtained with DNA extracted at 2 to 3 days of growth of the Neurospora transformants after which the plasmid becomes progressively harder to extract. When transformation of Neurospora was carried out with a mixture of 98% pBR322 and only 2% pSD3, pSD3 (but not pBR322) plasmid was recovered.

In contrast, plasmid pSD3 can be recovered only infrequently from mycelial cultures of individual *qa-2*$^{+}$ transformants (1 of 40 transformants). Such transformants appear to have integrated the *qa-2*$^{+}$ gene and carry little or no free plasmid.

Recovery of pSD3 in a modified form.

The pSD3 plasmid isolates recovered from Neurospora transformants were screened for possible alterations. It was found that about 10% of these isolates had been modified

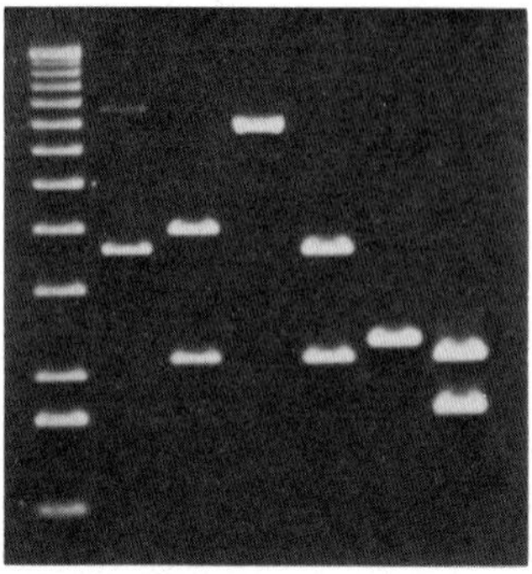

FIGURE 2. Agarose gel (0.8%) electrophoresis of pSD3, pJP3-6 and pJP3-53 DNA. Lane 1, 1 kb size markers; lanes 2, 4, 6, uncut pSD3, pJP3-6, and pJP3-53, respectively; lanes 3, 5, 7, HindIII + TthIII 1 digests of pSD3, pJP3-6, and pJP3-53, respectively.

(Figure 2). The modifications that have been detected are, in most cases, deletions of 300-500 bases that occur in the pBR322 DNA sequence in the region carrying the tet^r gene (inactive in pSD3). Larger deletions are also found but occur less frequently. Some of the modified plasmids (e.g., pJP3-6) transform Neurospora at somewhat higher frequencies than does pSD3, but do not appear to replicate better in Neurospora than pSD3.

Selection for replicating segments.

The liquid culture technique represents a solution of some practical problems in trying to select for replicating DNA segments within a clone bank. The technique allows the entire mass of transformants from an experiment to be screened easily and at a time chosen to optimize recovery of plasmids. A clone bank of BamHI fragments of Neurospora chromosomal DNA in the plasmid pSD3 was used to transform the Neurospora strain aro-9 qa-2 inl in an experiment as outlined in Figure 3. We expected that a plasmid which replicates well or at least has improved stability would be easier to recover. From a number of screenings of the clone bank, 11 different plasmids have been obtained. Four of the plasmids isolated in this way were more readily extracted from Neurospora transformants than pSD3.

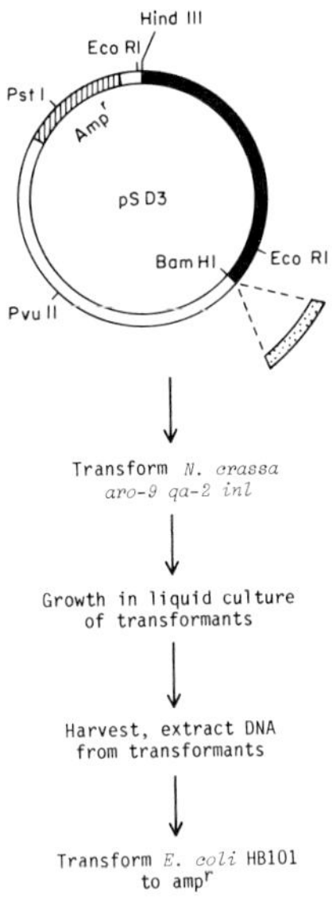

FIGURE 3. Schematic representation of the liquid culture plasmid recovery procedure. A clone bank was constructed in pSD3 of random BamHI fragments from a digest of Neurospora chromosomal DNA and used as indicated . Key: solid black, Neurospora qa-2^+ gene; stippled, random BamHI Neurospora inserts; striped, amp^r gene.

Table 1 demonstrates that none of the plasmid isolates enhance the transformation frequency substantially in comparison with the original vector, pSD3. However, the data shows that the isolate pJP102 is more readily recovered than pSD3 from liquid culture. pJP102 is also the only isolate with which free plasmid DNA can be recovered efficiently from cultures of individual Neurospora transformants (3 of 10 transformants). pJP102 contains a 1.9 kb DNA fragment which is a unique sequence in the Neurospora genome and may contain an ars sequence. We are currently subcloning portions of this fragment, carrying a putative ars sequence, into pJP8 (constructed from pUC8 and the qa-2^+ gene) (Figure 1). The pJP8 derivatives are being used as a starting point for continued development of a shuttle vector.

TABLE 1
TRANSFORMATION AND RECOVERY FREQUENCIES OF PLASMIDS[a]

Plasmid Isolate	Insert Size (kb)	No of Stable *Neurospora* transformants[b]	No. of individual *Neurospora* transformants carrying free plasmid[c]	No. of *E. coli* transformants from liquid culture extraction[d]
pSD3	--	953	0/10	6
pJP101	1.8	984	1/10	9
pJP102	1.9	1086	3/10	72
pJP107	5.1	692	0/10	11
pJP111	0.5	786	1/10	24

[a]The plasmids were isolated from screens of a bank containing *Neurospora* BamHI chromosomal fragments inserted into the BamHI site of pSD3 (see Figure 3).

[b]20 μg of plasmid DNA and 5 x 10^7 conidia were used for each transformation. Plating was at a density of 1 x 10^6/plate.

[c]Ten transformants selected at random for each case were tested for the presence of recoverable plasmid DNA. DNA was extracted from mycelial cultures and used to transform *E. coli* HB101.

[d]5 x 10^7 *Neurospora* spores were treated with 20 μg DNA and grown in 500 ml medium for 84 hours. DNA was extracted as described in Materials and Methods and used to transform *E. coli* HB101 to ampicillin resistance.

Transformation with linear DNA segments.

An efficient way to substitute an in vitro-manipulated form of a gene for its normal counterpart would be by direct replacement using a linear DNA fragment. Thus, we first examined transformation with linear plasmids and with DNA fragments that carry a selectable Neurospora gene. DNA fragments containing the Neurospora qa-2$^+$ gene (BamHI + BglII from pVK57) or qa-2$^+$ flanked by Neurospora am sequences (BamHI from pJP12) that have both ends derived from Neurospora DNA (Figure 4) gave about a 2.2-fold increase in transformation compared to circular DNA. This increase is small as compared to the 10 to 1000 fold increases found in yeast (6).

Disruption of the am gene.

pJP12, which has the am$^+$ gene interrupted by the qa-2$^+$

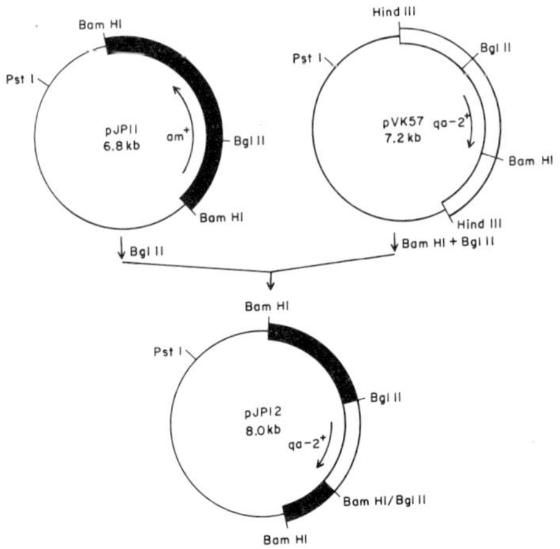

FIGURE 4. Construction of plasmids used for the gene disruption study. pJP11 has the 2.5 kb am$^+$ fragment from λc10 inserted into the BamHI site of pBR322. pJP12 was constructed by inserting a 1.2 kb BamHI + BglII qa-2$^+$ fragment isolated from pVK57 into the BglII site of pJP11. The resulting plasmid, pJP12, after BamHI digestion releases a fragment of Neurospora DNA which has am DNA flanking both sides of the qa-2$^+$ gene.

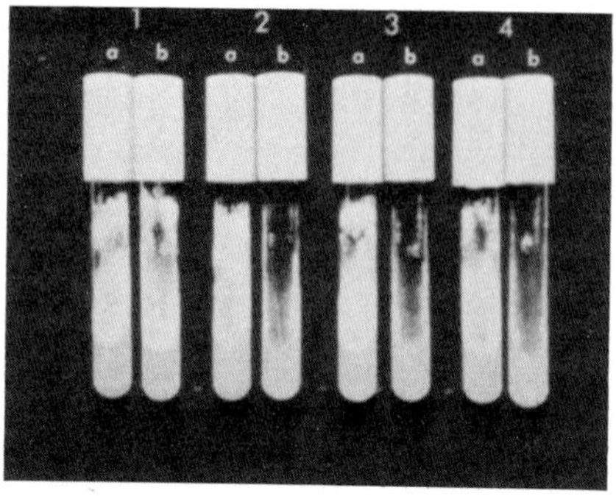

FIGURE 5. Cultures of disrupted am strains compared to am^+ and am 132 (deletion mutant) strains. The isolates were grown with ("b" series) or without ("a" series) 0.02M glycine. (1) aro-9 $qa\text{-}2^+$ am^+ inl; (2) aro-9 $qa\text{-}2^+$ am inl, carries disrupted am gene (b3-10); (3) aro-9 $qa\text{-}2^+$ am inl, carries disrupted am gene (c1-5); and (4) aro-9 $qa\text{-}2^+$ am132 inl, am deletion.

gene of Neurospora, was designed as a vector for targeting alteration in the am^+ region. Digestion of pJP12 with BamHI releases a fragment containing am DNA flanking both sides of the $qa\text{-}2^+$ gene (Figure 4). Transformation was carried out with this purified fragment with selection for $qa\text{-}2^+$ function in an aro-9 qa-2 am^+ inl strain. When the resulting $qa\text{-}2^+$ transformants were tested individually for an am^+ or am phenotype, all were am^+. This result was not unexpected since Neurospora conidia are generally multinucleate and therefore the resulting transformants are heterokaryotic, consisting of transformed as well as untransformed nuclei. In order to produce homokaryotic strains in which an am mutant phenotype could be observed, these $qa\text{-}2^+$ transformants were crossed to aro-9 qa-2 am^+ inl (untransformed genotype) and the progeny analyzed. Among 117 independent $qa\text{-}2^+$ homokaryons that were obtained, the am phenotype appeared in ten cases. The $qa\text{-}2^+$ am homokaryons demonstrate a strong am mutant phenotype (Figure 5).

Southern hybridization analysis of the $qa\text{-}2^+$ am strains confirmed that each of them had an alteration of the am^+ region. Figure 6 shows the results of analysis of one such strain, designated b3. In $qa\text{-}2^+$ am progeny derived from a backcross of b3 (e.g., b3-10), the hybridization band

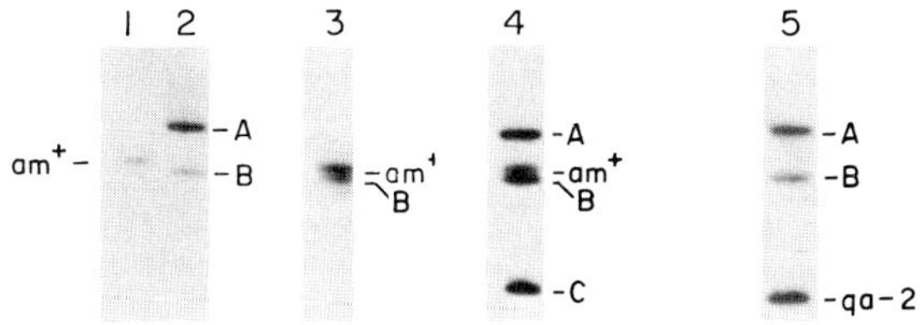

FIGURE 6. Genomic blots demonstrating a case of am$^+$ gene disruption. The Neurospora strain aro-9 qa-2 am$^+$ inl was transformed with BamHI am qa-2$^+$ fragment from pJP12, the resulting heterokaryotic transformants were crossed by an aro-9 qa-2 am$^+$ inl strain, and the progeny scored for an am phenotype. In this case, the original isolate (b3) was heterokaryotic and carried am$^+$ and am nuclei. From the cross of this strain qa-2$^+$ am and qa-2$^+$ am$^+$ homokaryotic progeny were collected. Chromosomal DNA from these strains was digested with Hind III and probed with ^{32}P-nick translated pVK57 or pJP11. The resident am$^+$ gene is at 9.1 kilobases and the resident qa-2 gene is at 3 kilobases. Lanes 1-4 were probed with pJP11 and lane 5 with pVK57. Lane 1, untransformed strain; lane 2, aro-9 qa-2$^+$ am inl homokaryon b3-10 derived from strain b3, note absence of 9.1 kilobase band in am$^+$ position and presence of disrupted am gene (band A), also an extra copy of qa-2$^+$ (band B) is present; lane 3, aro-9 qa-2$^+$ am$^+$ inl homokaryon b3-21 derived from heterokaryotic strain b3, note presence of am$^+$ band and absence of upper band corresponding to the disrupted am gene (band A), a copy of qa-2$^+$ is carried in this strain (band B); lane 4, heterokaryotic original transformant b3, b3 is am$^+$ qa-2$^+$, it carries am$^+$ and am nuclei, and has three integration events (bands A, B and C); lane 5, b3-10 probed with qa-2$^+$, it reveals the presence of qa-2$^+$ sequences in bands A and B, as well as the resident qa-2 gene site.

corresponding to the am$^+$ region (9.1 kb) is missing, while a new band at about 14 kb is present. This new hybridization band is consistent with an event in which the resident am$^+$ gene has been rendered nonfunctional by integration of the transforming fragment carrying the interrupted gene. This qa-2$^+$ am homokaryon (b3-10) has two integrated qa-2$^+$ genes, one the 14 kb band corresponding to the disrupted resident am, the second an unrelated insertion elsewhere in the genome (band B, Figure 6). When the qa-2^+ am homokaryon (b3-10) was crossed to the original parental strain, qa-2^+ am^+ progeny were obtained that have band B and the hybridization band corresponding to the normal am^+ region but lack the 14 kb disrupted am gene hybridization band.

In the sample of am disruption strains we have examined there have been no cases involving a simple replacement of sequence. In most strains am sequences in addition to the qa-2^+ gene have been added to the am^+ region. In one strain, however, a partial deletion of resident am^+ sequences has apparently occurred.

Non-homologous and multiple integration events.

When a DNA fragment carrying only the qa-2^+ gene (from pVK57) was used, only 1 of 13 transformants analyzed showed linkage to the original gene, whereas the remainder had integrated the qa-2^+ gene at unlinked sites. Similar

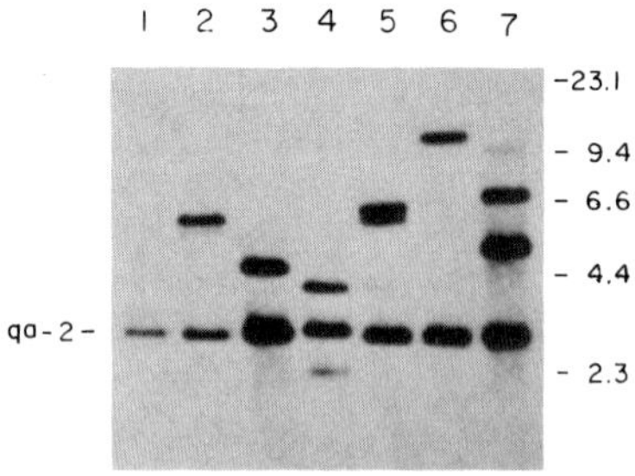

FIGURE 7. DNA structure of qa-2^+ am^+ transformants resulting from treatment of the aro-9 qa-2 am^+ inl strain with the BamHI fragment of pJP12. Each transformant is a homokaryotic isolate from a cross of a strain of untransformed genotype by the original heterokaryotic transformant. Neurospora genomic DNA was digested with HindIII and probed with ^{32}P-nick translated pVK57. The resident qa-2 gene is at 3 kilobases; integration events in the transformants have occurred at various other sites. Lane 1, untransformed strain; lanes 2-7, 6 independent homokaryotic transformants (random sample). Note double integration events in lanes 4, 5 and 7.

results were observed in transformants obtained with the am qa-2$^+$ DNA fragment from pJP12. Southern hybridizations of genomic DNA from independent am$^+$ qa-2$^+$ homokaryotic transformants revealed a number of different hybridization bands (Figure 7), suggesting that integration had occurred at various sites. Another feature which we commonly observe is evident in both Figures 6 and 7, the occurrence of multiple integrations of the transforming DNA into the genome of many transformants. When progeny from a cross involving a transformant with multiple integrations were examined, segregation of specific integrated bands from each other was observed.

DISCUSSION

Through the use of our screening procedure we have isolated plasmids, such as pJP102, which can be recovered from cultures of individual transformants to a greater extent than the starting plasmid, pSD3. The improved recovery provides a starting point for shuttle vector development. An unexpected property of the isolated plasmids is that none substantially enhance transformation frequencies. This result is in contrast to the sharp increase in transformation frequency observed in yeast for replicating vs. integrating vectors (10). At present the properties of a plasmid which carries an ars sequence in Neurospora is an open question; thus, the lack of high frequency transformation is not necessarily incompatible with the presence of an ars sequence in pJP102. The next step in vector development may be to add a Neurospora centromere to stabilize the plasmid further and improve its maintenance.

The technique described here of growing transformants in mass in liquid cultures for DNA extraction lead to at least two applications of general use. First, it should be possible to attempt direct cloning by recovery of plasmid DNA from abortive transformants. Second, screening for autonomously replicating plasmids can be accomplished rapidly and efficiently. Plasmids, derived or constructed from a variety of sources, including naturally occurring plasmids, could be screened by this procedure for improved replication properties.

The experiments reported here demonstrate that it is possible to carry out gene disruptions in Neurospora and to our knowledge represents the first deliberate gene disruption of this organism. However, the procedure is complicated by the fact that the initial transformants are heterokaryons so that the result of a disruption is not immediately obvious but only becomes evident after a cross which yields homokaryotic progeny. Gene disruption in Neurospora is further complicated because it is not possible to precisely target the manipulated sequence to the desired resident locus, the majority of integration events occurring elsewhere. Thus, we estimate that approximately 10% of the transformants studied had an integration at the am^+ locus yielding the desired gene disruption; presumably a similar pattern would be obtained with other genes. It should be possible to disrupt the resident gene in any case where a cloned gene sequence is available, including instances where the precise function of the gene and the phenotype that a mutant would cause are unknown.

An unexpected feature of Neurospora transformation is the large degree of non-homologous integration and the fact that many transformants have integrated more than one copy of the transforming DNA. Neurospora seems to represent an intermediate case, between mammalian cells and yeast, in which integrations are not entirely random, but only a low proportion of events occur at the homologous site. These findings explain the lack of a substantial increase in transformation frequency when linear DNA is used. It seems likely that multiple integrations will interfere with a genetic analysis of gene disruption experiments in Neurospora and other organisms with similar features. To overcome this problem, a transformant may have to be backcrossed to remove all but the relevant integration event prior to analysis.

REFERENCES

1. Stohl LL, Lambowitz AM (1983). Construction of a shuttle vector for the filamentous fungus, *Neurospora crassa*. Proc Natl Acad Sci USA 80:1058-1062.
2. Stohl, LL, Akins RA, Lambowitz AM (1984). Characterization of an autonomously replicating Neurospora plasmid. Nucleic Acids Res 12:6169-6178.
3. Hughes K, Case ME, Geever R, Vapnek D, Giles NH (1983). Chimeric plasmid that replicates autonomously in both *Escherichia coli* and *Neurospora crassa*. Proc Natl Acad Sci USA 80:1053-1057.

4. Hughes K, Case ME, Geever R, Vapnek D, Giles NH (1983) Retraction. Proc Natl Acad Sci USA 80:7678.
5. Dhawale SS, Paietta JV, Marzluf GA (1984). A new, rapid and efficient transformation procedure for Neurospora. Curr Genet 8:77-79.
6. Orr-Weaver TL, Szostak JW, Rothstein RJ (1981). Yeast transformation: a model system for the study of recombination. Proc Natl Acad Sci USA 78:6354-6358.
7. Rothstein RJ (1983). One-step gene disruption in yeast. Methods Enzymol 101:202-211.
8. Case, ME (1982). Transformation of *Neurospora crassa* utilizing recombinant plasmid DNA. In Hollander A, DeMoss R, Kaplan S, Konisky J, Savage D, Wolfe RS (eds): "Genetic engineering of microorganisms for chemicals," New York: Plenum Press, p 87.
9. Kinsey JA, Rambosek JA (1984). Transformation of *Neurospora crassa* with the cloned *am* (glutamate dehydronase) gene. Mol Cell Biol 4: 117-122.
10. Struhl, K, Stinchcomb DT, Scherer S, Davis RW (1979). High-frequency transformation of yeast: autonomous replication of hybrid DNA molecules. Proc Natl Acad Sci USA 76:1035-1039.

Molecular Genetics of Filamentous Fungi, pages 15–28

DEVELOPMENT OF CLONING VECTORS AND A MARKER FOR GENE REPLACEMENT TECHNIQUES IN ASPERGILLUS NIDULANS[1]

Geoffrey Turner, D.J. Ballance,
M. Ward and R.K. Beri

Department of Microbiology, University of Bristol,
Bristol BS8 1TD, U.K.

ABSTRACT Increased frequency of transformation has been achieved by incorporation of either chromosomal or mitochondrial sequences into simple transformation vectors. A vector carrying a chromosomal sequence gave the highest frequency of transformation, and has been used to isolate isocitrate lyase (*acuD*) and fawn (*fwA*) genes by complementation of *Aspergillus* mutants.

A gene coding for oligomycin resistance, *oliC31*, has been used as a selectable marker for transformation and subsequent studies on recombination and gene conversion within transformants, and has potential as a convenient selectable marker for gene replacement techniques.

INTRODUCTION

Transformation of *Aspergillus nidulans* is now relatively simple, and a variety of selectable markers have been used (1 - 3). Transforming DNA is integrated into the chromosome either by homologous recombination with the equivalent host gene or by non-homologous recombination at another site. The latter process occurs relatively frequently, unlike the case for *Saccharomyces cerevisiae*. Thus, in principle, any gene which expresses

[1]This work was supported by grants from the Science and Engineering Research Council.

in Aspergillus can be used for transformation without the requirement for additional sequences (1, 4).

This simple type of transformation occurs with a relatively low frequency (about 100 viable transformants/μg), and in order to achieve improved frequencies which would facilitate isolation of genes by complementation of Aspergillus mutants, we have experimented with both chromosomal and mitochondrial sequences included in simple vectors. Although the initial rationale behind this approach was based on attempts to produce replicating vectors, we have no unequivocal evidence that this has been achieved, though one of the vectors has proved suitable for gene isolation.

All of the selectable markers used to date for transformation of A. nidulans have required an appropriate mutant strain as recipient. During the course of work to isolate and study the gene for subunit 9 of mitochondrial ATP synthetase, we have isolated an oligomycin resistant allele, oliC31 (5), which can be used to transform wild-type strains of A. nidulans, and have taken advantage of this to examine recombination and gene conversion events within transformants so that convenient gene replacement strategies can be devised.

ANS1: A CHROMOSOMAL SEQUENCE CONFERRING HIGH FREQUENCY TRANSFORMATION

ans1 was isolated from total A. nidulans DNA as an EcoRI fragment capable of conferring replicating ability in yeast on a non-replicating plasmid pFB6 (6) using the approach of Stinchcomb and coworkers (7). pFB6 carries the pyr4 gene of Neurospora crassa, coding for orotidine 5'-phosphate decarboxylase which is able to complement a yeast ura3 mutant. ans1 was one of a number of fragments similarly isolated, and the only one to have any effect on Aspergillus transformation, increasing transformation frequency about 40-fold over that obtained with a simple vector (Table 1). Cloning vectors pDJB2 and pDJB3 were constructed using this sequence (Fig. 1). Subcloning of the EcoRI fragment has enabled us to identify a sub-fragment with unaltered activity. A further observation is that part of the ans1 sequence is repeated elsewhere in the genome of Aspergillus (Fig. 2).

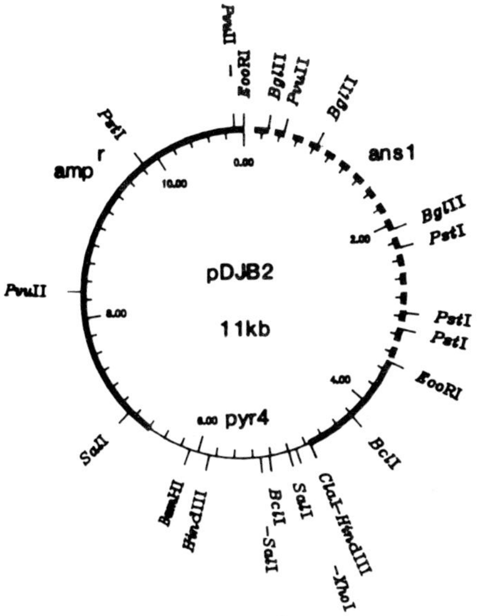

FIGURE 1. pDJB2. pDJB1 lacks the whole EcoRI ans1 fragment, and pDJB3 is a derivative of pDJB2 with the EcoRI site at 3.5kb removed. The PvuII-PstI fragment of ans-1 between 0.45kb and 2.3kb still confers high frequency transformation on simple vectors carrying pyr4.

Following transformation of Aspergillus with simple vectors, in addition to the viable transformants, a large number of very small "abortive transformants" are seen as is the case for yeast (8) and Neurospora (9). These may result from transformation without stable integration. One effect of the ans-1 sequence is to drastically reduce the proportion of these abortives. Analysis of pDJB2 transformants has suggested that the plasmid integrates into the chromosome in a number of ways, but we have not yet detected extrachromosomal plasmid by Southern hybridization against total undigested DNA.

TABLE 1
TRANSFORMATION FREQUENCY WITH DIFFERENT PLASMIDS[a]

	pDJB1	pEB3	pDJB2
Expt. 1	200	940	6250
Expt. 2	85	625	3200

[a]Transformants/μg DNA for 2 independent experiments

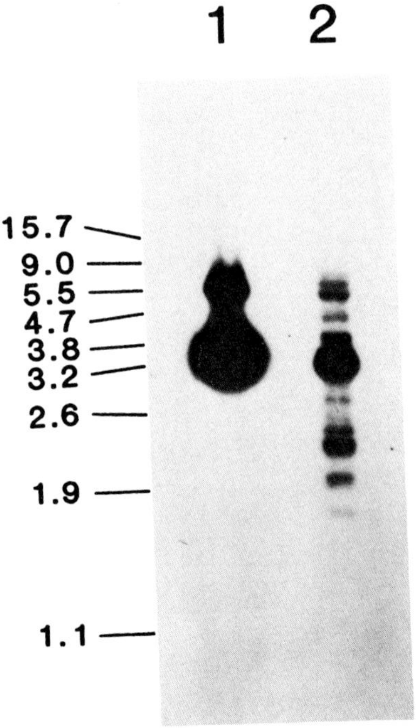

FIGURE 2. ansl-homologous sequences in total *A. nidulans* DNA. Lane 1, pDJB2 digested with *Eco*RI; Lane 2, total DNA digested with *Eco*RI. Filters were probed with labelled ansl fragment.

MITOCHONDRIAL SEQUENCES

Suppressive, cytoplasmic petite mutants of yeast contain amplified fragments of the mitochondrial genome carrying strong replication origins (10). It has been shown that such origins also function extramitochondrially in yeast, and this has led to proposals to use similar sequences from other organisms to construct autonomously replicating vectors (11). Unfortunately, it appears that sequences capable of acting as replication origins in yeast are rarely likely to exhibit similar behaviour in other organisms, and there is no rational way of finding the real mitochondrial replication origin in a fungus such as *A. nidulans*, even supposing that such an origin would function extramitochondrially. However, a phenomenon remarkably similar to the petite mutation has been reported in Aspergillus amstelodami (12), and two regions of the mitochondrial genome identified which can be excised and replicate, eventually outreplicating the complete mitochondrial genome and resulting in localized death of the mycelium (13). Neither of these regions, when incorporated into simple vectors, improved the frequency of transformation in A. nidulans.

There is close homology between the mitochondrial DNAs of these two species of Aspergillus, and the sequence of one of the replicating regions (region 2) has been compared with the equivalent region in A. nidulans (13). While the gene order was retained, a secondary structure proposed as the replication origin by comparison with yeast and human mitochondrial DNA was absent from the equivalent A. nidulans sequence. We therefore turned to region 1, which had not been analysed. This sequence (1kb) hybridizes specifically with a region of A. nidulans DNA EB3 (13) (Fig. 3). Both region 1 fragment and the equivalent A. nidulans region act as replication origins in yeast, though region 2 does

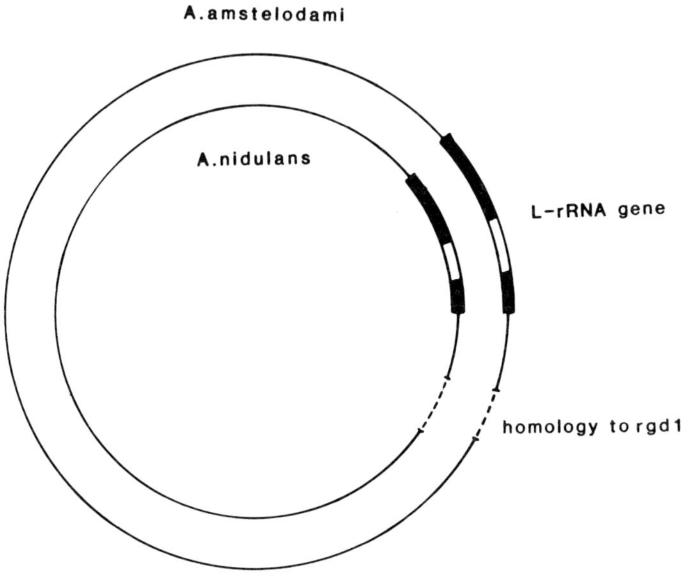

FIGURE 3. A. amstelodami region 1 (12) and homology with A. nidulans mitochondrial DNA (13).

not. When the *A. nidulans* EB3 fragment was included in the simple vector pDJB1 to form pEB3, a reproducible increase in transformation frequency of about 5-fold was observed (Table 1). Tilburn and coworkers (2) observed no frequency increase when they tested a mitochondrial fragment from the ribosomal RNA gene region isolated by the Stinchomb method (7).

AUTONOMOUSLY REPLICATING PLASMIDS?

Since chromosomal integration occurs so readily in *A. nidulans*, and since autonomously replicating vectors lacking centromeres or any form of control over copy number might be unstable, selection for stable integration might be expected at an early stage following transformation. Similar conclusions were reached by Buxton and Radford (14) when they proposed a model for the behaviour of a hypothetical replicating plasmid in a mycelial fungus. We therefore examined the stability of uridine prototrophy in the progeny of single spores taken from transformants prior to subculture (Table 2). Greater instability is observed amongst the progeny of pDJB2 and pEB3 than for pDJB1 transformants. In an attempt to detect transient extrachromosomal DNA in young transformants, we prepared total DNA from total spores harvested directly from transformation plates, and following electrophoresis of undigested DNA, probed with pDJB1. Only in the case of a pEB3 transformation was it possible to detect DNA homologous to the probe which separated from the bulk of chromosomal DNA. It is not yet clear whether this is due to autonomous replication.

It is possible to rescue plasmids from total transformant DNA into *E. coli* without cutting and ligation, but the frequency is low (e.g. 0-10/μg direct; 100/μg with cutting and ligation), and there is no significant difference between pDJB2 and pDJB1 transformants (with and without ans1). Either a small amount of free plasmid is present, possibly as a result of a low level of excision from the chromosome, or the linear DNA is recircularized by recombination events within *E. coli* (see 15).

TABLE 2
STABILITY OF URIDINE PROTOTROPHY IN YOUNG TRANSFORMANTS[a]

pDJB1	pEB3	pDJB2
100	2	38
100	3	90
100	100	100
100	20	70
1	9	15

[a]Single spores isolated from original transformant colonies were allowed to grow on minimal medium until sporulation of colonies. Spores from these were diluted and apread on complete medium, then replicated on to minimal medium, and the percentage of complete medium colonies retaining uridine prototrophy scored. For each type of transformant, progeny from 5 single spores were analysed.

ISOLATION OF GENES BY COMPLEMENTATION

A plasmid bank has been constructed in *E. coli* (13000 clones) using *TaqI* partial digest fragments (average size 7kb) of total *Aspergillus* DNA inserted at the *ClaI* site of pDJB3 following alkaline phosphatase treatment of the latter. Over 90% of the reisolated plasmids contained inserts. Plasmid DNA isolated from this bank was used to transform suitable mutants of *A. nidulans* on such a scale that 5 - 30 x 10^4 Pyr^+ transformants could be generated, and protoplasts plated on selective medium. This approach has enabled us to transform *acuD* (lacking isocitrate lyase) to Acu^+ and *fwA* (fawn conidia) to Fw^+ (green conidia).

The sequences responsible for the transformation were obtained by rescue of the ampicillin resistance carried on pDJB2 together with flanking sequences following partial digestion and religation of total DNA isolated from the transformant, then transformation of *E. coli*. A proportion of the *E. coli* clones from Acu^+ transformants and the Fw^+ transformant were able to retransform *Aspergillus* for the respective markers at a high frequency.

An *acuD* complementing clone containing a 30kb plasmid was chosen for subcloning (Fig. 4). Since the gene did not

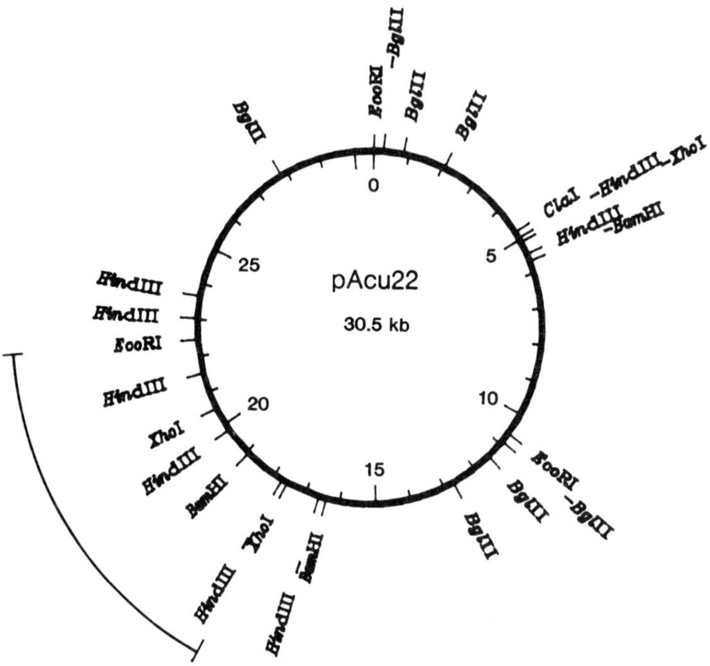

FIGURE 4. Plasmid rescue of acuD-complementing sequence in E. coli. Total A. nidulans DNA from an Acu$^+$ transformant was partially digested with EcoRI, religated, and used to transform E. coli to ampicillin resistance. This is one of the resulting plasmids which was able to retransform A. nidulans acuD to Acu$^+$ at high frequency. The region containing the complementing sequence is marked. Repetition of at least part of the ans-1 sequence can be seen between 0 - 2kb and 10.7 - 12.7kb. The active pyr4 gene lies between 16 - 17.6kb.

appear to be expressed in an equivalent E. coli mutant (aceA), sub-clones had to be tested by retransforming Aspergillus. The rescued plasmid has a complex restriction pattern, in which most of the ans-1 sequence is duplicated. Subcloning was carried out in pAT153. The results are shown in Fig. 5, and narrow down the transforming sequence to 4kb, which is capable of transforming Aspergillus acuD

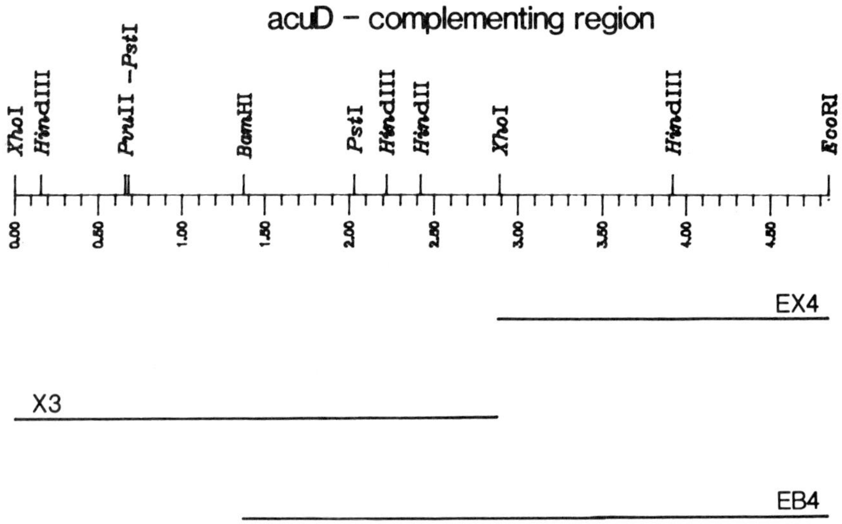

FIGURE 5. Subcloning of the acuD-complementing sequence. When subcloned in pAT153, the entire region shown transformed A. nidulans at a frequency of 200/μg. The subfragments indicated transformed at 0 - 1/μg.

to Acu^{+} at a frequency of about 100/μg. Subfragments of this region transformed Aspergillus at a very low frequency (0-1/μg), and may contain only parts of the complementing sequence such that rarer recombination events generate an intact gene.

To check that such an isolated sequence carries the gene required rather than a suppressor, a number of transformants should be analysed genetically by sexual crossing with the mutant to ascertain that integration often occurs at the desired locus.

While a high frequency of transformation is an important factor in gene isolation, it is also possible to achieve the same end by using a larger insert in a cosmid based vector (16).

OLIGOMYCIN RESISTANCE AS A SELECTABLE MARKER FOR TRANSFORMATION AND DETECTION OF GENE CONVERSION AND RECOMBINATION IN TRANSFORMANTS

Oligomycin is an inhibitor of mitochondrial ATP synthetase, acting on the Fo portion of the complex. Both nuclear and mitochondrial mutations have been reported in *A. nidulans* which confer resistance to the drug, and result from alterations to the complex. Nuclear mutations map at a single nuclear locus (*oliC*, 17, 18) on linkage group VII, and are located in the gene coding for subunit 9 of the complex, also known as the DCCD-binding protein. We isolated the *Aspergillus* gene using a heterologous probe from *Neurospora* (19), and have sequenced it to confirm its identity.

The gene has been inserted into a plasmid carrying the selectable *Neurospora pyr4* gene (Fig. 6) and this construction, pMW2, was used in transformation experiments selecting either for uridine prototrophy or oligomycin resistance.

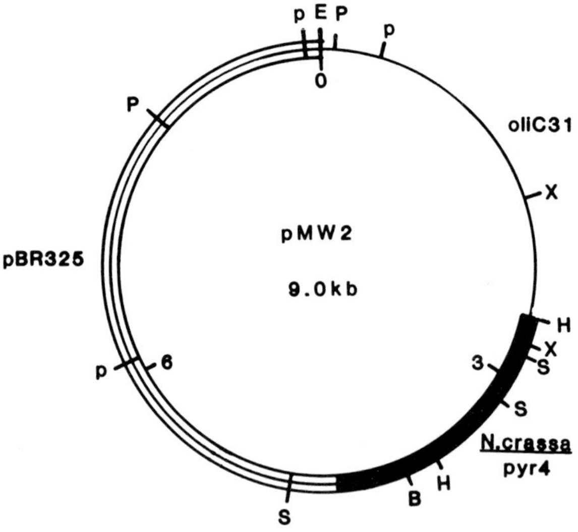

FIGURE 6. pMW2. E *EcoRI*, P *PstI*, p *PvuII*, X *XhoI*, H *HindIII*, S *SalI*, B *BamHI*.

Earlier work had shown that nuclear oligomycin resistance is semidominant in diploids (17), and led us to believe that resistance would be detectable in transformants even when both host wild-type and transforming resistant gene were present. Following transformation with pMW2, we observed both weakly resistant (Oli^{sr}) and strongly resistant (Oli^{r}) transformants, the relative frequency depending on the selection method. Oli^{r} transformants grew as well as the original oliC31 mutant on the drug, and hybridization and genetic analysis of transformants showed that the integration had occurred at the oliC locus. Oli^{sr} transformants resulted from integration either at oliC or elsewhere in the genome. While Oli^{sr} strains resulting from integration at oliC were stable if maintained on drug-free medium, growth on oligomycin for more than one day led to abundant fast growing sectors of Oli^{r} segregants. Oli^{sr} strains resulting from integration elsewhere in the genome did not show this high instability on oligomycin.

Hybridization analysis and scoring for both oligomycin resistance phenotype and uridine requirement led to the conclusions illustrated in Fig. 7. The Oli^{sr} phenotype occurred when both sensitive and resistant genes were present in the same strain, and this condition was relatively unstable only when the genes were in close proximity (at oliC). The switch from Oli^{sr} to Oli^{r} resulted from either homologous recombination, in which the non-Aspergillus sequences were excised to leave a single resistant gene, or gene conversion, when no change was detectable in the restriction map at oliC, and we assume that the sensitive gene was "converted", resulting in the formation of two resistant genes and the foreign sequences unaltered between them.

The ease by which the switch from low resistance to high resistance could be detected permitted analysis of the progeny for uridine requirement as a way of distinguishing rapidly between conversion and excision, and showed a ratio of 2 to 1. This can be compared to the equivalent situation in yeast (20), where the ratio is about 7 to 1. There is good evidence from work with yeast that two distinct mechanisms are involved, since the rad52 mutation strongly inhibits gene conversion, but not homologous recombination (20).

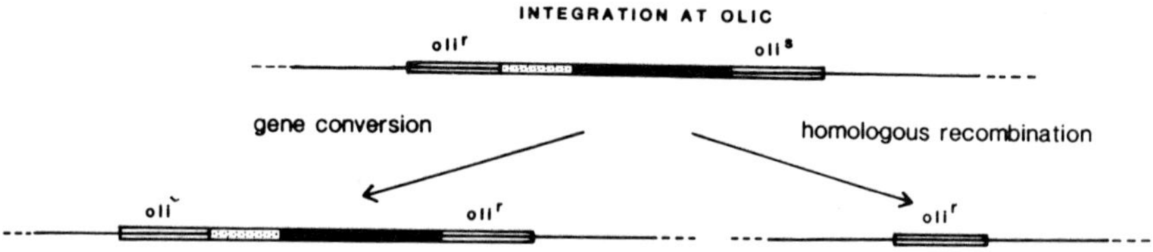

FIGURE 7. Gene conversion and homologous recombination at the *oliC* locus. Stippled region, *pyr4*; solid region, bacterial plasmid sequence. The *oliC* heterogenote (top) was phenotypically Oli^{sr} and resulted from transformation with pMW2.

When Oli^{r} strains which are still Pyr^{+} are analysed for loss of *pyr4* by replica plating, or Oli^{sr} strains for switch to Oli^{s}, where no selection is involved, the actual frequency at which the switch (homologous recombination plus gene conversion) occurs appears to be less than 10^{-3}. In yeast, the frequency is about 10^{-4} (20).

Earlier work (18) had shown that oligomycin resistant mutants tend to show increased sensitivity to triethyltin, another inhibitor of mitochondrial ATP synthetase, thus offering a counter-selection for oligomycin sensitivity. When we plated unstable Oli^{sr} strains on triethyltin medium, again fast growing sectors are seen, this time consisting of oligomycin sensitive derivatives.

Since either allele of the *oliC* gene can be positively selected, *oliC31* has a potential application as a convenient marker in vectors for gene replacement. This would be analogous to the use of cycloheximide resistance in yeast for the same purpose (21), but with the additional advantage of two way selection.

REFERENCES

1. Ballance DJ, Buxton FP, Turner G (1983). Transformation of Aspergillus nidulans by the orotidine-5'-phosphate decarboxylase gene of Neurospora crassa. Biochem Biophys Res Comm 112:284.
2. Tilburn J, Scazzocchio, C, Taylor GG, Zabicky-Zissman JH, Lockington RA, Davies RW (1983). Transformation by integration in Aspergillus. Gene 26:205.
3. Yelton MM, Hamer JE, Timberlake WE (1984). Transformation of Aspergillus nidulans by using a trpC plasmid. Proc Natl Acad Sci USA 81:1470.
4. Ballance DJ, Turner G (1985). Development of a high-frequency transforming vector for Aspergillus nidulans. Gene, in press.
5. Ward M, Turner G, unpublished results.
6. Buxton FP, Radford A (1983). Cloning of the structural gene for orotidine 5'-phosphate carboxylase of Neurospora crassa by expression in Escherichia coli. Molec Gen Genet 190:403.
7. Stinchcomb DT, Thomas M, Kelly S, Selber ER, Davis RW (1980). Eukaryotic DNA segments capable of autonomous replication in yeast. Proc Natl Acad Sci USA 77:7559.
8. Hicks JB, Hinnen A, Fink GR (1979). Properties of yeast transformation. Cold Spring Harbor Symposia in Quantitative Biology 43:1305.
9. Case ME, Schweizer M, Kushner SR, Giles NH (1979). Efficient transformation of Neurospora crassa by utilizing hybrid plasmid DNA. Proc Natl Acad Sci USA 76:5259.
10. Bernardi G (1982). The origins of replication of the mitochondrial genome of yeast. Trends in Biochem Sci 7:404.
11. Tudzynksi P, Esser K (1982). Extrachromosomal genetics of Cephalosporium acremonium II. Development of a mitochondrial DNA hybrid vector replicating in Saccharomyces cerevisiae. Curr Genet 6:153.
12. Lazarus CM, Earl AJ, Turner G, Küntzel H (1980). Amplification of a mitochondrial DNA sequence in the cytoplasmically inherited "ragged" mutant of Aspergillus amstelodami. Eur J Biochem 106:633.

13. Lazarus CM, Küntzel H (1981). Anatomy of amplified mitochondrial DNA in "ragged" mutants of Aspergillus amstelodami: excision points within protein genes and a common 215 bp segment containing a possible origin of replication. Curr Genet 4:99.

14. Buxton FP, Radford A (1984). The transformation of mycelial spheroplasts of Neurospora crassa and the attempted isolation of an autonomous replicator. Molec Gen Genet 196:339.

15. Conley EC, Saunders JR (1984). Recombination-dependent recircularization of linearized pBR322 plasmid DNA following transformation of Escherichia coli. Molec Gen Genet 194:211.

16. Timberlake WE, personal communication.

17. Rowlands RT, Turner G (1973). Nuclear and extranuclear inheritance of oligomycin resistance in Aspergillus nidulans. Molec Gen Genet 126:201.

18. Rowlands RT, Turner G (1977). Nuclear-extranuclear interactions affecting oligomycin resistance in Aspergillus nidulans. Molec Gen Genet 154:311.

19. Viebrock A, Perz A, Sebald W (1982). The imported preprotein of the proteolipid subunit of the mitochondrial ATP synthase from Neurospora crassa. Molecular cloning and sequencing of the mRNA. EMBO J 1:565.

20. Jackson JA, Fink JR (1981). Gene conversion between duplicated genetic elements in yeast. Nature 292:306.

21. Struhl K (1983). Direct selection for gene replacement events in yeast. Gene 26:231.

Molecular Genetics of Filamentous Fungi, pages 29–38

DEVELOPMENT OF A SYSTEM FOR ANALYSIS OF REGULATION SIGNALS IN ASPERGILLUS

C.A.M.J.J. van den Hondel[1], R.F.M. van Gorcom[1], T. Goosen[2], H.W.J. van den Broek[2], W.E. Timberlake[3] and P.H. Pouwels[1]

[1]Medical Biological Laboratory TNO, P.O. Box 45, 2280 AA Rijswijk, The Netherlands
[2]Dept. of Genetics, Agricultural University, Wageningen, The Netherlands
[3]Dept. of Plant Pathology, University of California, Davis, USA

ABSTRACT Gene fusion of the E.coli lacZ gene with the A.nidulans trpC gene directs the synthesis of a functional β-galactosidase fusion protein in A.nidulans. Removal of both transcription- and translation regulation sequences located upstream of the trpC-lacZ fusion gene abolishes the production of the fusion protein, indicating that its expression is controlled by these sequences. The fusion gene has been used to construct vectors for the isolation and analysis of transcription/translation signals.

INTRODUCTION

Regulation of gene expression in A.nidulans has been investigated extensively by genetic methods (1,2). Studies on the regulation of the pathways of nitrogen and carbon metabolism have shown that the expression of a number of genes is controlled by trans-acting regulatory elements (3). These studies have also revealed that regulatory proteins act on controlling regions which are located adjacent to a structural gene e.g. the amdS gene (4,5).

Little is known about the control of gene expression on the molecular level. However, the recent cloning of genes which are regulated such as the *amdS* gene (6) and the *trpC* gene (7), opens the possibility to analyse the molecular mechanisms of gene expression control. A very convenient and powerful method for such an analysis involves fusion of the promoter or regulatory region to the *lacZ* gene encoding β-galactosidase of *E.coli*. Studies in *E.coli* and *S.cerevisiae* have shown that the expression and regulation of a gene fused to *lacZ* can be assayed qualitatively *in vivo* by the formation of blue colonies containing X-gal (5-bromo-4-chloro-3-indolyl-β-d-galactopyranoside) and quantitatively by determination of the β-galactosidase activity in cellular extracts (8). The development of a transformation system for *A.nidulans* (9,10,11) makes it possible to apply this methodology also to *A.nidulans*.

In this paper we describe the construction *in vitro* of a fusion of the *E.coli lacZ* gene to the *trpC* gene of *A.nidulans*. The fusion gene was cloned into a plasmid which contains the *amdS* gene as a selection marker. *A.nidulans* strains transformed with these plasmids were analysed for *E. coli lacZ* expression. Our results show that the *E.coli lacZ* gene is expressed in *A.nidulans* as a functional fusion protein and that its expression is controlled by the *A.nidulans trpC* transcription/translation control-sequences. Based on the fusion gene a set of vectors was constructed which enable the isolation and analysis of transcription- and translation control-sequences.

RESULTS AND DISCUSSION

Construction of *A.nidulans* trpC-lacZ Gene Fusion

To fuse the *E.coli lacZ* gene to the *trpC* gene of *A.nidulans*, a *BamHI* fragment of plasmid pMC1871 (14) was used which contains the *lacZ* gene without the first seven codons. This fragment was inserted into the *BamHI* site of the *trpC* gene of *A.nidulans* present on a *XhoI* fragment of pHY101 (fig. 1). Sequence analysis of a DNA fragment which contains the connection between the *trpC* gene and the N-terminal part of the *lacZ* gene showed that the translational reading frames of both genes were in phase.

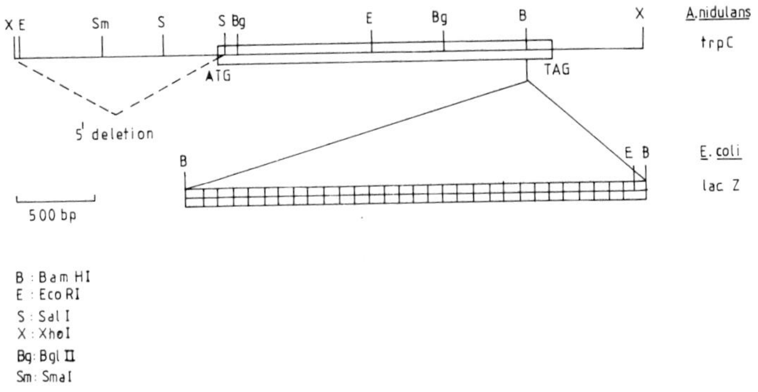

FIGURE 1. Physical map of the A.nidulans trpC gene region and the E.coli lacZ gene insertion in the former gene. The BamHI site leftward of the lacZ fragment is located upstream of the eighth codon of the lacZ gene. Deletion of the XhoI-SalI fragment, indicated as 5'-deletion, removes the transcription- and translation start sequences of the trpC gene. Only recognition sites for BamHI, EcoRI, SalI and XhoI are shown.

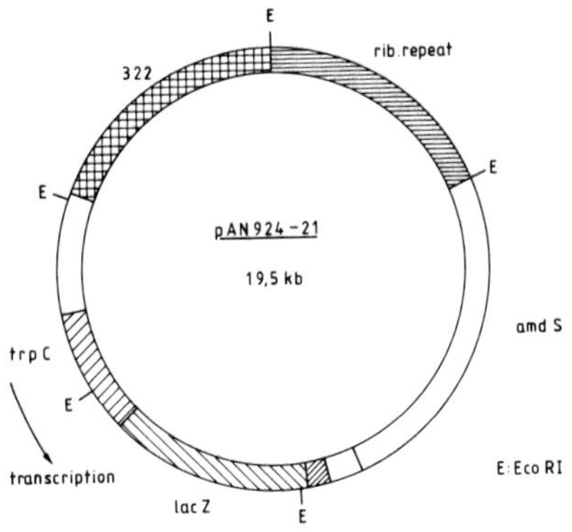

FIGURE 2. Schematic presentation of plasmid pAN924-21. amdS denotes the A.nidulans acetamidase gene, lacZ the E.coli β-galactosidase gene ( ) trpC the A.nidulans tryptophan C gene ( ), 322 pBR322 sequences ( ), rib. repeat the 3.6 kb EcoRI fragment of the A.-nidulans ribosomal repeat, containing the 18S rDNA ( ) (14). The arrow indicates the direction of transcription of the lacZ-trpC fusion gene.

The fusion gene was inserted, in two orientations, in plasmid pGW315 (12) carrying the amdS gene as a selection marker for A.nidulans, yielding plasmids pAN924-21 and pAN924-22. One of these plasmids is shown in fig. 2. In a second step we have removed the transcription- and translation start sequences of the trpC gene located upstream of the truncated fusion-gene by deleting the XhoI-SalI fragment as shown in fig. 1. The remaining SalI-XhoI fragment containing the fusion gene was inserted in two orientations in the plasmid pGW315, yielding plasmids pAN924-41 and pAN924-42.

Bèta-galactosidase Activity of A.nidulans

Studies on the expression of the ß-galactosidase (bgaA) gene of A.nidulans have shown that this gene is completely repressed by glucose or sucrose and is induced by lactose or galactose (13). This regulation can be visualized by cultivating A.nidulans agar plates containing lactose- or galactose-X-gal medium containing agar plates. Blue colonies were found under these conditions whereas cultivating on glucose or sucrose-X-gal medium containing agar plates resulted in white colonies (results not shown). This enabled us to test for the expression of the lacZ fusion gene in A.nidulans transformants, without interference by the endogeneous ß-galactosidase, namely under conditions that this gene is repressed (glucose/X-gal medium).

Expression of the lacZ Fusion Gene in A.nidulans

The A.nidulans amdS deletion strain MH1277 (6) was transformed with the plasmids pAN924-21, -22, -41 and -42. $AmdS^+$ transformants were analysed for ß-galactosidase expression under conditions that the bgaA gene is repressed. Of the transformants obtained with pAN924-21 and -22 (intact fusion-gene) 85% gave blue colonies on agar plates containing glucose-X-gal medium, whereas transformants obtained with the control plasmid pGW315 gave colourless colonies only, suggesting that the lacZ fusion results in the production of E.coli ß-galactosidase.

Of the $AmdS^+$ transformants obtained with the plasmids pAN924-41 and -42 (truncated fusion-gene) 100% gave colourless colonies on agar plates containing glucose-X-gal medium, suggesting that deletion of the transcription- and translation-start sequences in front of the lacZ fusion gene abolishes the production of β-galactosidase.

In addition, the β-galactosidase activity was measured in extracts of A. nidulans wild type and A.nidulans transformants obtained with pAN924-21. The results showed that the β-galactosidase activity of the transformants which gave blue colonies on X-gal plates is at least five hundred fold higher than that of A.nidulans wild type, cultivated under the same conditions (glucose medium; results not shown).

To verify the presence of the fusion-gene in $AmdS^+$ transformants, Southern blots of chromosomal DNA digested with BamHI or EcoRI were hybridized with labeled lacZ or trpC DNA, respectively. Analysis of the results (fig. 3) clearly shows that the intact fusion gene is present in A.nidulans transformed with pAN924-21 or -22 as well as with pAN924-41 or -42. The latter results also exclude the possibility that the absence of β-galactosidase activity in these transformants is caused by the absence of the fusion gene.

Northern blot analysis of $polyA^+$ RNA isolated from the $AmdS^+$ transformants, revealed that an RNA species of the expected size for the fusion messager (5.5 kb), is present in transformants containing the intact fusion gene and absent in those which contain the truncated fusion gene (results not shown). These results indicate that the fusion gene is properly transcribed and that the transcription of it is controlled by the signals of the trpC gene.

To prove that the β-galactosidase activity found in the extracts of the A.nidulans transformants really is a result of the presence of E.coli β-galactosidase in the fusion protein, a Western blotting experiment was carried out using antiserum against E.coli β-galactosidase. As shown in fig. 4 a protein of the expected size (190 kD) is present in the extract of A.nidulans transformed with pAN924-21, whereas no protein band can be seen in the control lanes. This result indicates that the E.coli lacZ gene is expressed in A.nidulans as a fusion protein with

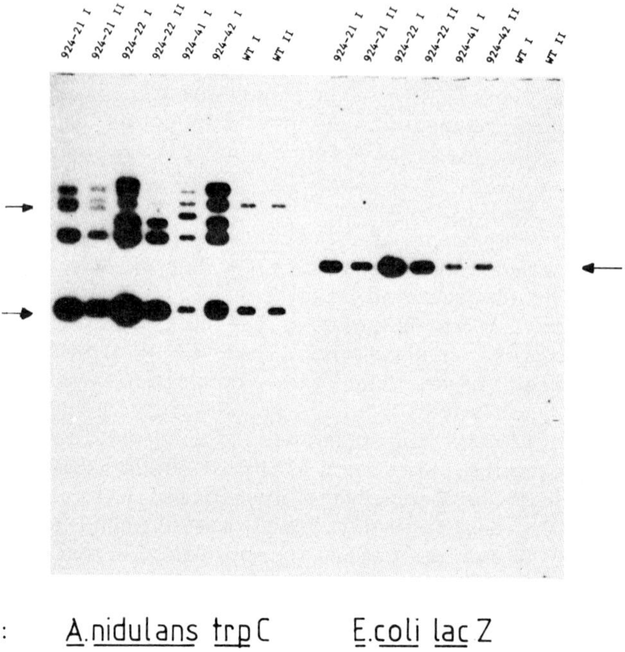

FIGURE 3. Hybridization analysis of A.nidulans transformants. The A. nidulans amdS deletion strain MH1277 was transformed with the plasmids pAN924-21, -22, -41 and -42 according to the method described by K. Wernars et al. (15). DNA of different transformants was isolated as described by Yelton et al. (13) and fractionated by electrophoresis in 1% agarose gels after digestion with BamHI or EcoRI, as indicated. The gels were blotted and the blots were hybridized with either a ^{32}P-labeled lacZ gene containing PstI fragment of pMC1871 or a trpC gene containing XhoI fragment of pHY101, as indicated. Thereafter the blots were washed at $68^{o}C$ in a buffer containing 0.04 M Na^{+} and autoradiographed. The arrows indicate the position of the chromosomal EcoRI fragments containing together the trpC gene or the BamHI fragment containing the lacZ gene.

the trpC gene. This result, together with the finding of β-galactosidase activity in extracts of A.nidulans, proves that the fusion gene expresses a functional β-galactosidase. The fact that no expression of the lacZ fusion gene is found after removal of the transcription- and translation start sequences of the trpC gene located upstream of the fusion gene, strongly suggests that its expression is controlled by these sequences.

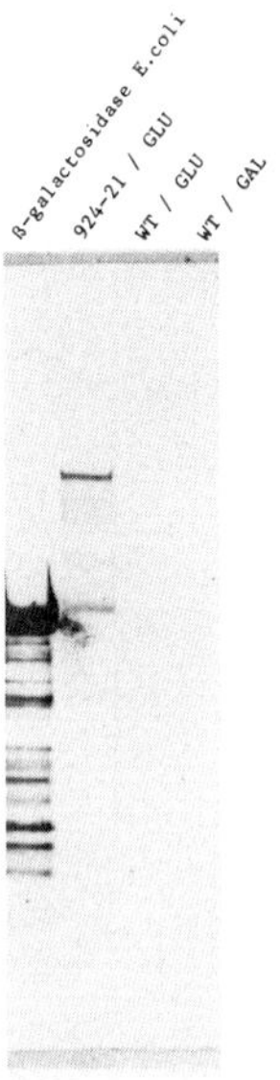

FIGURE 4. Western blots of extracts of A.nidulans wild type and A.nidulans transformed with pAN924-21. Extracts from A.nidulans wild type and A.nidulans transformed with pAN924-21 were electrophoresed on a 6% polyacrylamide gel. The gel was blotted and hybridized with E.coli β-galactosidase antiserum according to a procedure developed by J. Visser, Dep. of Genetics, Wageningen, The Netherlands (this procedure will be published elsewhere). A 20% pure preparation of β-galactosidase was used as a marker. A.nidulans wild type was grown in minimal medium with 1% glucose or galactose.

The results described above, demonstrating that the lacZ-trpC gene fusion is functional in A.nidulans, enable a further analysis of the promoter and regulatory signals of the trpC gene. Also similar fusion can be made with other Aspergillus genes to study their gene expression and/or regulation.

Vectors for Isolation and Analysis of Transcription- and Translation Signals

Based on the trpC-lacZ fusion gene, a set of vectors was constructed to isolate and analyse transcription- and translation signals. These vectors contain a unique BamHI site in three different reading frames in front of either the truncated trpC-lacZ fusion-gene (fig. 5, pAN923-31, 32 and 33) or of a fusion-gene from which the promoter and the trpC part upstream of the lacZ gene were deleted (fig. 5, pAN923-41, 42 and 43). The vectors carry the A.nidulans argB gene as a selectable marker. With these vectors DNA fragments, containing translation- and transcription-activating elements, can be isolated and their properties can be studied.

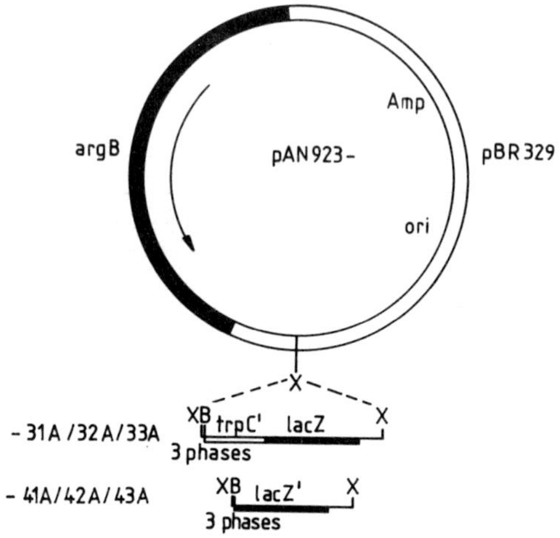

FIGURE 5. Schematic presentation of vectors to isolate transcription- and translation signals. argB denotes the A.nidulans ornithine carbamoyl transferase gene, lacZ the E.coli ß-galactosidase gene and trpC the A.nidulans tryptophan C gene. The truncated genes are indicated with a quote. The arrow indicates the direction of transcription. In the A-version of the plasmids, both the transcription direction of the argB gene and the truncated lacZ fusion-gene is counter clockwise. In the B-version the transcription directions are opposite. The unique BamHI site is located in three different reading frames in front of the truncated lacZ fusion-gene.

A system is now available to clone and analyse promoter sequences, transcription-activating sequences and other regulatory sequences in order to get more insight in the control of gene expression and gene regulation in Aspergillus.

REFERENCES

1. Arst HN Jr., Bailey CR (1977). The regulation of carbon metabolism in Aspergillus nidulans. In Smith JE, Pateman JA (eds): "Genetics and physiology of Aspergillus", New York: Academic press, p 131.
2. Timberlake WE (1980). Developmental gene regulation in Aspergillus nidulans. Dev Biol 78:497.
3. Marzluf GA (1981). Regulation of nitrogen metabolism and gene expression in fungi. Microbiol Rev 45:437.
4. Hynes MJ (1979). Fine structure mapping of the acetamidase structural gene and its controlling region in Aspergillus nidulans. Genetics 91:381.
5. Hynes MJ (1982). A cis-dominant mutation in Aspergillus nidulans affecting the expression of the amdS gene in the presence of mutations in the unlinked gene, amdA. Genetics 102:139.
6. Hynes MJ, Corrick CM, King JA (1983). Isolation of genomic clones containing the amdS gene of Aspergillus nidulans and their use in the analysis of structural and regulatory mutations. Mol Cell Biol 3:1430.
7. Yelton MM, Hamer JE de Souza ER, Mullaney EJ, Timberlake WE (1983). Developmental regulation of the Aspergillus nidulans trpC gene. Proc Natl Acad Sci USA 80:7576.
8. Cassadaban MJ, Matinez-Arias A, Shapira SK, Chou J (1983). β-Galactcsidase gene fusions for analyzing gene expression in E.coli and yeast. In Wu R, Grossman K, Moldave K (eds): "Methods in Enzymology", Vol 100, New York: Academic press, p 293.
9. Ballance DJ, Buxton FP, Turner G (1983). Transformation of Aspergillus nidulans by the orotidine-5'-phosphate decarboxylase gene of Neurospora crassa. Biochem Biophys Res Commun 112:284.
10. Tilburn J. Scozzacchio C, Taylor GG, Zabicky-Zissman JH, Lockington RA, Davies RW (1983). Transformation by integration in Aspergillus nidulans. Gene 26:205.

11. Yelton MM, Hamer JE, Timberlake WE (1984). Transformation of Aspergillus nidulans using a trpC plasmid. Proc Natl Acad Sci USA 81:1470.
12. Wernars K, Goosen T, Wennekes LMJ, Visser J, Bos CJ, van den Broek HWJ, van Gorcom RFM, van den Hondel CAMJJ, Pouwels PH (1985). Gene amplification in Aspergillus nidulans by transformation with vectors containing the amdS gene. Curr Genet (in press).
13. Fantes PA, Roberts CF (1973). ß-Galactosidase activity and lactose utilization in Aspergillus nidulans. J Gen Microbiol 77:471.
14. Lockington RA, Taylor GG, Winther M, Scazzocchio C, Davies RW (1982). A physical map of the ribosomal DNA repeat unit of Aspergillus nidulans. Gene 20:135.

Molecular Genetics of Filamentous Fungi, pages 39–57

TRANSFORMING BASIDIOMYCETES[1]

R.C. Ullrich,[*] C.P. Novotny,[§] C.A. Specht,[*] E.H. Froeliger[§] and A.M. Muñoz-Rivas[*]

Departments of Botany[*] and Medical Microbiology[§]
University of Vermont, Burlington, VT 05405

ABSTRACT This paper discusses problems associated with transforming Basidiomycetes and suggestions for how some of them may be solved. Our comments are based on our experience with *Schizophyllum commune*. In addition to general considerations, detailed remarks address four problem areas: recipient cells, vectors, presentation of DNA, and selectable markers. Results of transforming Schizophyllum to antibiotic resistance with the phosphotransferase gene of Tn5 are included. Other strategies for obtaining genes from Basidiomycetes for use as selectable markers are discussed.

INTRODUCTION

Within the last decade the development of nucleic acid technology has made it possible to answer molecular genetic questions about fungi when answers were merely wishfully anticipated before. The transformation of fungal cells with exogenously applied DNA is one of these powerful genetic techniques. It complements other techniques used in recombinant DNA methodologies and for some genes may be the only means to isolate them for molecular studies. In other studies the return of

1 This work was supported by NSF grants PCM 8203496 and 8402107, University of Vermont BSC 183-1 and NIH International Research Fellowship F05 TW0 3396-01.

exogenous DNA to the cell is a critical step for: identifying the function of a gene, observing its effects on wild-type or mutant cells, or making an *in vivo* analysis of gene function after *in vitro* gene modification.

In our laboratories, the primary subject of research is *Schizophyllum commune*. Schizophyllum is a prototypic bifactorial heterothallic Basidiomycete. Our initial interest in this experimental organism was derived from its mating-type-determined development and the extensive genetic analysis that has been performed on it (1). Briefly, the mating-type genes act as master regulators of the genes expressed in sexual development. Mating-type is determined by two genes, the *A*- and *B*-factors, each composed of two functionally redundant loci: α and β. The four loci are multiallelic. Matings between two haploid mycelia (i.e., homokaryons) are compatible only when the mates carry different alleles for at least one *A*-factor locus and for at least one *B*-factor locus. Dramatic alterations of mycelial morphology develop in compatible matings and the derived fungus, i.e., the dikaryon, is the fertile state which, in the correct environment, produces the mushroom with meiotic products termed basidiospores. Some pairings of haploid strains differ in only one of the two mating-type factors; such a mating is partially compatible and yields a heterokaryotic mycelium that is only partially differentiated along the developmental pathway that leads from homokaryon to dikaryon.

These mating-type genes are of interest to us for several reasons. One, they act as the principal regulatory genes of this intriguing developmental pathway (2). Two, Mendelian genetic studies suggest they function differently (3) from the studied mating-type genes of the ascogenous yeasts *Saccharomyces cerevisiae* (4-7) and *Schizosaccharomyces pombe* (8,9). And three, genetic studies suggest they may be analogous to another large and biologically significant group of mating-type genes found in higher plants (10-12).

Molecular and functional analysis of these genes may be possible once they have been isolated. However, because the ability to recognize mating-type genes is dependent on their biological function, it is unlikely that they will be identified and isolated in the absence of a transformation system and the ability to transform Schizophyllum cells for mating type. For these reasons we have invested considerable time and effort developing a transformation system.

Presently we are aware of reports of transformation in six genera of true fungi; five: Saccharomyces (13,14), Neurospora (15), Podospora (16,17), Schizosaccharomyces (18) and Aspergillus (19,20,21) are Ascomycetes. A single report on Ustilago maydis provides putative evidence for transformation of a Basidiomycete (22). The status of transformation in Ustilago (Dr. Geoffrey R. Banks, pers. commun.) is similar to that from our own results with Schizophyllum. The same selective marker, Neo, encoding resistance to aminoglycoside antibiotics was used with each organism. Protoplasts were exposed to plasmids containing Neo with or without other sequences from the respective Basidiomycetes. In each system positive Southerns were obtained from putative transformants; however, most putative transformants were negative in Southern blots and the transformants are unstable. Emphasis in both laboratories is shifting from the use of Neo to other selective markers.

In the remaining discussion we will present results and discuss strategies for transforming Schizophyllum. Our efforts may provide useful messages for those contemplating the development of a transformation system, particularly for a Basidiomycete. One of the foremost messages is that there is no single "magical" method to transform fungal cells. The literature illustrates a diversity of strategies and procedures that work for specific organisms, but these approaches have not been universally transferable to other organisms. The specificity may reflect unique characteristics of a given species: procedures necessary for DNA to gain entrance to the cell, sequences requisite to DNA being maintained in the cell, and sequences essential to gene expression. The second observation is that there are no guarantees that your efforts will be successful. This is especially true because most of the results are negative; consequently, there is little information to interpret and suggest how to modify experiments. A third point is that you cannot perform all of the methods and variations of them; to do so, the needs exceed the means. Consequently activities must be limited to those strategies assessed to have the best chance of success. Unfortunately, all of the information required to make these decisions is usually not available, and some of it may be presently unattainable. Fortunately, only one successful method is required initially. With a positive result, the essential features can be identified and the method can be improved.

We confine our discussion to four problem areas: recipient cells, vectors, presentation of DNA, and selectable markers.

RECIPIENT CELLS

We regularly make protoplasts of Schizophyllum from mycelium or basidiospores using modifications of the procedures of deVries and Wessels (23). Basidiospores (5×10^8) are suspended in 3.0 ml of yeast extract-peptone-glucose-salts, liquid complete medium (LCYM) and shaken at 200 rpm, at 30°C for 4 h. The spores are pelleted and resuspended in 3.0 ml magnesium osmoticum (23) containing 10 mg/ml each of Cellulase T.v. concentrate (Miles Laboratories) and Novozym 234 (Novo Laboratories). This suspension is held at 30°C for approximately 2 h to yield protoplasts (one from each germinating spore) that are uniform in size and cellular constituents. Alternatively, ca. 5.0 g mycelium grown in LCYM is suspended in 25 ml of magnesium osmoticum containing the enzymes (20 mg/ml) listed above. This suspension must be held at 30°C for variable lengths of time (ca. 2 hr) depending upon strains used and degree of hyphal degradation desired. Protoplasts from mycelium, unlike those derived from spores, are heterogeneous in size and cellular constituents, e.g. many lack nuclei. We recover concentrations (10^8-10^9/ml) of purified protoplasts from these suspensions by filtration or centrifugation in combination with various osmotica (e.g. 0.5M $MgSO_4$ or 0.5M sorbitol). Protoplasts are washed by repeated centrifugation and resuspension in 0.5M sorbitol. The viability of protoplasts from spores can be as great as 50%, whereas that of protoplasts from mycelium is much lower (e.g. 10%) due to their heterogeneity.

VECTORS

One obstacle to transformation of Basidiomycetes is that there are few known natural vectors. A useful virus is unknown, and plasmids such as yeast 2μ plasmid or Podospora sensecence plasmid are unknown for most Basidiomycetes. Plasmids of mitochondrial origin have been reported in Agaricus (24), but the utility of these as transformation vectors remains to be determined. Experience with Ascomycetes and Imperfects suggests that plasmids of mitochondrial origin may be useful transformation vectors (16,17,25). We would encourage others to screen their species of interest for possible natural vectors.

The alternative to a natural vector is a man-made one. One approach is to provide an origin of replication or autonomously replicating sequence (ars) to the vector being constructed. Ideally this will be DNA from the species of interest. We used the YIp5 SHY2 yeast system (26) to recover Schizophyllum ars in yeast. Twenty ars containing plasmids (including seven different ars) were recovered. Nineteen were of mitochondrial origin despite the fact that only 3-4% of the input DNA was mitochondrial. At least six out of 12 EcoR1 restriction fragments of mitochondrial DNA contain ars sequences. The ars sequences were added to derivatives of plasmid pBR322 containing various selectable markers. None of these constructions yielded transformants. In an effort to construct an integration vector, we inserted various restriction fragments from the ribosomal DNA unit repeat or other Schizophyllum DNA sequences into many plasmids. One of these constructions proved noteworthy, and this is discussed below under selectable markers.

PRESENTATION OF DNA

We have attempted to transform Schizophyllum protoplasts, hyphal cells, and basidiospores in a variety of protocols using many different buffers, reagents, temperatures, etc. Most commonly we use polyethylene glycol to fuse protoplasts after treating the cells in a variety of conditions or solutions. Early in our experiments we determined that our most thoroughly washed protoplast preparations contained nuclease activity (4-5 double-stranded cuts per 20 kb of linear DNA incubated in the transformation suspension at 20°C for 30 min). This activity may derive from the protoplasting enzymes, the protoplasts themselves, or both. Consequently, we frequently protect the exogenous DNA by packaging it in synthetic liposomes (27). Electrophoresis of DNA extracted from the transformation suspensions immediately before plating cells for regeneration, reveals most of the encapsulated plasmid vector is protected from nuclease activity even when exogenous DNase is added to the liposome preparation.

One unknown that looms large when transformation fails is the question of whether or not DNA enters the cells. Recently we devised experiments that may permit us to assess if exogenous DNA penetrates the recipient cells. Although we do not have results at writing, the approach is worthy of consideration and we present it here. The problem with experiments designed to answer this question is to distinguish between DNA taken into the cells and contaminating exogenous DNA. This complexity is avoided by asking if foreign sequences in the vector are transcribed in regenerating protoplasts. Northern blots of RNA extracted from regenerating protoplasts would be probed with labelled vector DNA (the vector should contain no sequences complementary to recipient species DNA). Controls with RNase and DNase eliminate the possibility that a positive signal is derived from contaminating vector DNA from the transformation procedure. This experiment has the potential of detecting uptake and transcription without requiring translation and expression of a selectable phenotype. In the face of negative transformation results the experiment may discern which of several complex requirements are not being met.

SELECTABLE MARKERS

For transformation to be a useful tool, the transformant must be identifiable. In most cases identification depends upon a selectable marker. Most selections involve transformation of a mutant and identification of the transformant under conditions in which transformed cells are readily distinguished from nontransformed cells. In these cases the transforming DNA must include a gene that endows the mutant to grow in an identifiable or selective manner under specified environmental conditions. In most instances the transforming DNA is a wild-type gene corresponding to that gene for which the mutant is genetically deficient. Obtaining a functional gene that corresponds to the genetic lesion of the mutant may be difficult for Basidiomycetes.

An assured source of a wild-type gene corresponding to a mutant gene in a Basidiomycete of interest is the wild-type strain from which the mutant was derived. If a mutant is being transformed with a complete gene library of this DNA, then in theory, the correct sequences should be present. Of course, there is no way of knowing a priori if the required sequence is intact in the library; it may have been disrupted or lost during construction of the library. The construction of complete gene libraries is not a trivial matter. It requires time and skill, and should not be taken lightly, even for fungi with their relatively small genomes. Even if a functional gene is present, it is unlikely that the initial transformations of a fungus will be efficient enough to yield any transformants when DNA from a complete gene library is used. The low efficiency associated with initial transformations of a species demand that the transforming DNA be greatly enriched for the sequence of interest; usually this means that the sequence has been cloned.

We are primarily interested in the mating-type genes for which there should be powerful selective methods to identify transformants (28-31). Nevertheless, we chose not to use mating type as a selectable marker because little is known of the structure of the mating-type genes, e.g. size may preclude the likelihood of cloning them in a functional manner. For these reasons we think it preferable to use an isolated gene encoding a selectable phenotype.

This logic leads to the next problem encountered by those attempting to transform Basidiomycetes. As discussed above, the most sensible source of genes for transformation would be from the species of interest. But relatively few genes have been isolated from

Basidiomycetes. For example, from Schizophyllum these include: two cellulase genes and a β-glucosidase gene (Dr. Vern Seligy, pers. commun.), a gene implicated in fruiting (32), ribosomal DNA genes (33, B. Buckner, unpub. results), and the mitochondrial cytochrome oxidase subunit III and cytochrome b genes (C. A. Specht, and L. Phelps, unpubl. results). None of these are useful selective markers; so we must identify and isolate a suitable gene from Schizophyllum or use a foreign gene.

We have tried to isolate Schizophyllum genes for use as selectable markers by several methods. One method used Schizophyllum complete gene libraries to transform and complement mutants of organisms for which efficient transformation systems have been established (E. coli: hisB, leuB, pyrF, argH, aroD, purC, D, E, F, G, H, and trpA, B, C and F; and S. cerevisiae: his3, leu2, ura3 and trp1). No complementation by Schizophyllum sequences was observed. A second method involved an attempt to isolate a gene for heavy metal resistance; resistance to heavy metals could be a selective marker. We isolated a series of strains increasingly resistant to heavy metals and characterized the patterns of restriction fragments derived from repetitive DNA sequences in the sensitive parent and the resistant derivatives. Under heavy metal toxicity, eukaryotes frequently become resistant by gene amplification of a sequence encoding the metallothionine gene. In theory, an amplified gene could be identified as a new band of repetitive sequence appearing in restriction digests of the resistant mutant. This fragment could be isolated and used to recover the intact gene sequence. We were unable to detect such a band in DNA from the resistant strains.

The alternative to a Schizophyllum gene as a selectable marker is a foreign gene. Although the number of available foreign genes continues to increase, their utility is not assured. One problem is, for many isolated foreign genes, mutants of the corresponding genes in the Basidiomycete of interest have not been identified. A harsh reality is that little biochemical genetics has been done on most Basidiomycetes. This lack of knowledge hinders progress by making a guessing game out of the requirement for gene/mutant match. For several mutants of Schizophyllum we have knowledge of the enzymatic deficiencies, and in some cases the corresponding foreign genes are available. We have used the following genes from S. cerevisiae (URA3, LEU2 and ADE1,3,4,5-7,8) in various plasmid constructions to determine if they would complement Schizophyllum mutants either known or thought to be mutant in the corresponding gene. No transformants were detected. This is not entirely surprising for many features of a gene could render it nonfunctional in a new species. These include promoter sequences, ribosome-binding sites, introns, sequences essential to splicing, polyadenylation sites and possibly transcriptional termination sites. It may be that some of these characteristics are unique and that fundamental differences between species make it improbable that foreign genes, would function, even if they were from other classes of fungi. Despite these liabilities, the availability of isolated foreign genes requires their consideration. Even genes from evolutionarily distant prokaryotes may be useful, particularly if they are engineered for expression in eukaryotes.

Prokaryotic genes that are uniquely attractive as selective markers in eukaryotes are those coding for antibiotic resistance. The attraction is that you need not match the transforming gene to a mutation of a corresponding gene in the recipient. Some antibiotic-resistance genes are poor candidates: those that convey resistance by virtue of coding for parts of membrane complexes, or those specific to antibiotics for which eukaryotes are insensitive. A bacterial gene with demonstrated utility is the neomycin-resistance gene (Neo) from either Tn5 or Tn601. This gene encodes resistance to aminoglycoside antibiotics by virtue of phosphotransferase activity and has been used in transformation to convey resistance to animal tissue culture cells (34,35), plant protoplasts (36), plant leaf-disk explants (37), yeast (38), and Dictyostelium (39,40). Schizophyllum is relatively insensitive to the commonly used kanamycin aminoglycoside, but is sensitive to geneticin (G418). For our most successful constructions we used derivatives of plasmid pAG60 (34). pAG60 contains Neo linked to a herpes simplex promoter and a polyadenylation sequence. We shotgunned EcoR1 helper fragments of Schizophyllum DNA into the unique EcoR1 site of pAG60. DNA from a library of 3×10^4 clones was used for transformation experiments.

Using the transformation procedures described above, we obtained from 1-20 resistant colonies per 10^8 protoplasts treated with 6-15ug of liposome-encapsulated plasmid. Some resistant colonies were unstable (i.e., transiently resistant) and could not be maintained even with continued selection for G418 resistance. Among the colonies that grew sufficiently to provide DNA for probing, a small number from ca.100 tested gave positive signals in dot blots. One colony that gave a strong signal in dot blots, grew sufficiently to provide DNA for two independent Southern analyses. Each was positive, but this colony, like all of those that gave positive signals, became unstable and lost its resistance. The molecular basis of these results remains unclear. Our attempts to improve these results have been unsuccessful. Due to the low efficiency, the high background of spontaneously resistant colonies, the instability of the resistance, and the difficulties attending the analysis of unstably resistant colonies by Southerns, we have directed our attentions to isolating other Schizophyllum genes for use as selectable markers. These new approaches are technically demanding, but they avoid the problems of species specificity associated with gene expression discussed above.

The Schizophyllum gene coding orotidine -5'-phosphate decarboxylase (OMP decase, orotidine-5'-phosphate carboxy-lyase EC 4.1.1.23) is of interest to us because we have Schizophyllum mutants lacking this activity (41, E. Froeliger unpub. results). Our plan is to purify Schizophyllum OMP decase and to produce monospecific polyclonal antibody to it. This antibody would be used to screen a Schizophyllum library made in an expression vector (e.g. λgt11) and to isolate from this library clones producing antigen for this antibody. The sequence coding for the antigen would be used to isolate the entire gene from one of our Schizophyllum genomic libraries (e.g., one in λEMBL 4). The complete gene would be used to isolate OMP decase mRNA and this could be translated _in vitro_ to produce OMP decase or a precursor if it is modified _in vivo_. Identification of the protein as OMP decase by activity or immunoassay would constitute proof that the gene was isolated. This gene will be used to transform OMP decase mutants of Schizophyllum.

There are several reasons to consider this approach. There is a sensitive radioassay for OMP decase activity (42). A chromatographic procedure for rapid and extensive purification of the enzyme from S. cerevisiae (43,44) and E. coli (45) has been published, and the required materials are readily available. We have purified small amounts of Schizophyllum OMP decase and are now preparing larger amounts to immunize rabbits. Purified yeast enzyme is commercially available, and antibodies can be raised to it in mice; antibodies to yeast OMP decase react weakly with the Schizophyllum enzyme (E. Froeliger, unpubl. results). While this is too weak to isolate clones from a Schizophyllum library made in an expression vector, it suggests that Schizophyllum OMP decase will elicit an adequate immune response in experimental animals. There is also a positive selective procedure for OMP decase mutants (46), should the need for additional Schizophyllum mutants arise. Schizophyllum cells with wild-type OMP decase are sensitive to the drug, 5-fluoro-orotic acid, used in the selection of OMP decase mutants (E. Seidel, unpub. results), and other Basidiomycetes may be as well. These features make this an attractive strategy for obtaining a suitable selective marker from Schizophyllum, and we recommend it for consideration by others for the Basidiomycete of interest to them.

The second method to isolate a selectable Schizophyllum gene may be easier, but less foolproof. Although we have not completed the procedure, we can report some progress. The method assumes that foreign genes, for which there are mutations of the corresponding genes in Schizophyllum, can be used as heterologous probes in Southern blots to detect Schizophyllum restriction fragments that contain complementary sequences. Probes that give positive signals can be used to isolate the complementary Schizophyllum genomic sequences from our gene libraries. Because the positive signals from heterologous probes could be due to complementarity with other sequences, the presence of the gene sequence of interest must be verified by transformation or sequencing.

The choice of which foreign gene to use as the probe is limited by several considerations. First, there should be mutants of the corresponding gene in the Basidomycete for which transformation is being developed. Second, a restriction map of the foreign DNA to be used as a probe should be available, and the location and orientation of the foreign gene within this sequence should be known. This information is useful to simplify the interpretation of Southern blots as discussed below.

We are probing Southern blots of Schizophyllum DNA with restriction fragments containing the following foreign genes: S. cerevisiae URA3 (42), ADE2 (Dr. Jeffery F. Lamont, pers. commun.), TRP1 (47), TRP3 (48), and TRP5 (49); S. pombe ADE6 (Dr. David Beach, pers. commun.); N. crassa PYR4 (50) and TRP1 (51); A. nidulans TRPC (52); and Cochliobolus heterostrophus TRPC (Dr. Olin C. Yoder, pers. commun). There are mutants for each of the corresponding genes in Schizophyllum. For each probe we make duplicate Southern blots and demand that each positive signal appear in each lane; this detects many artifacts. The plasmid containing the probe sequence is purified as covalently closed circle on two CsCl gradients and restricted with appropriate enzymes. The restricted DNA is labelled by replacement synthesis using T4 DNA polymerase (53). We find this is the most reliable method of labelling restriction fragments. The labelled restriction fragments are electrophoresed in low melting point agarose. The fragment of interest is recovered by cutting a piece of agarose containing the fragment from the gel, and the DNA and agarose are melted together. Once thoroughly melted the agarose has no detrimental effect on the hybridization which will be conducted at $60^{o}C$. Nitrocellulose Southern blots are rinsed in prehybridization buffer (8x SSC, 5x Denhardt's, 0.1% SDS, 10mM Tris pH 7.5, 1mM EDTA) and then hybridized with probe ($1-2x10^{6}$ cpm ^{32}P-labelled DNA/ml of the above buffer) for 20 hr at $60^{o}C$. No carrier DNA is used to prevent nonspecific binding to the nitrocellulose because it may contain sequence complementary to the probe, and therefore, increase background. After hybridization, one 5 min rinse in 2 x SSC, 0.1% SDS at room temperature and two 60 min rinses in 8 x SSC, 0.1% SDS at $60^{o}C$ are used. These conditions allow retention of probes containing sequences with 65% or greater base pair complementarity. Kodak XAR X-ray film with intensifying screen is used to make 36h to 8d exposures at $-70^{o}C$. Figure 1 illustrates results.

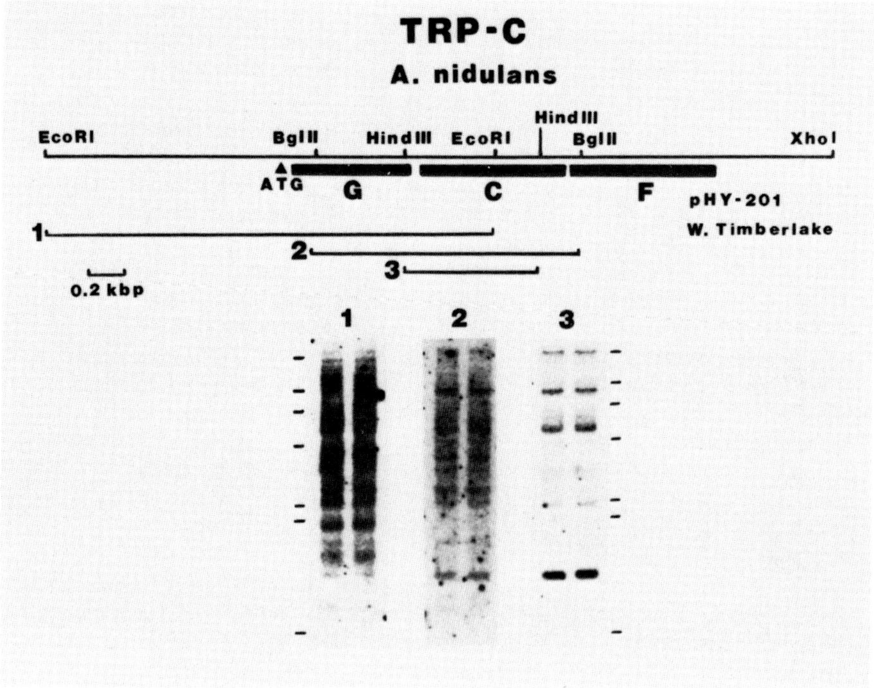

FIGURE 1. Southern blots of Pstl cut Schizophyllum DNA probed with various restriction fragments of A. nidulans TRPC sequence. A. nidulans TRPC gene in plasmid pHY-201 and restriction map courtesy of Dr. W. E. Timberlake (52). GCF are the three functional domains of the TRPC trifunctional polypeptide. ATG, translational initiation codon; restriction sites as indicated. Restriction fragments (1,2,3) were ^{32}P-labelled by replacement synthesis using T4 polymerase and used to probe duplicate lanes (1,2,3) of Pstl cut Schizophyllum DNA, respectively. Lowered criterion conditions for hybridization and rinsing are specified in the text. Hind III phage λ restriction fragments (length standards) indicated as dashes to the left and right of autoradiograms.

Although no results can be generalized, probes containing the 5' flanking sequence of a foreign gene frequently bind to multiple fragments of the blot, whereas sequences internal or 3' in the gene usually bind to fewer fragments. Some probes such as S. cerevisiae URA3 and TRP1 do not bind to Schizophyllum DNA. We are currently completing the last of our Southern analyses. Whenever possible we have probed blots with corresponding genes from two or more organisms (e.g. A. nidulans TRPC, N. crassa TRP1, and S. cerevisiae TRP1; S. cerevisiae ADE2 and S. pombe ADE6). If two different foreign probes for corresponding genes bind to the same Schizophyllum fragment, this provides great encouragement to isolate this fragment and proceed with its use. We are now screening our Schizophyllum λ library for genomic sequences that hybridize to these probes. The genomic sequences retrieved will be subcloned in various plasmid vectors or used directly for transformation of Schizophyllum mutants.

CONCLUDING REMARKS

It should be obvious from our experiences that developing a transformation system can require a prodigous effort. Negative results in the absence of further information are not interpretable and confound the process. For any particular strategy, critical information is frequently missing and this precludes assessment of the likelihood for success. Despite the difficulty of the studies, there are many positive aspects. Our efforts at transformation have stimulated the initiation of other projects related to transformation. Studies on OMP decase (41, E. Froeliger unpub. results), mitochondrial DNA (54, C.A. Specht, L. Phelps, and Dr J. Burke, unpub. results), ribosomal DNA (33, C.A. Specht and B. Buckner, unpub. results), and tryptophan metabolism (Dr. A. Muñoz-Rivas and B. Drummond, unpub. results) are now on going in our laboratories.

In this paper we have shared our experiences developing transformation in Schizophyllum. The lessons we have learned may be helpful to those developing transformation in other Basidiomycetes. At the same time many of our comments may also apply in the broader context of transforming other organisms. Many of our remarks are tempered by the vision of hindsight; a few concern untested approachs that now seem worthwhile. The former have been informative to us; we hope the latter will be more so.

ACKNOWLEDGMENTS

Our thanks to the many scientists who have generously shared DNA sequences from other organisms with us.

REFERENCES

1. Raper JR (1970). "Genetics of Sexuality in Higher Fungi." New York: Ronald, p 283.
2. Ullrich RC (1978). On the regulation of gene expression: Incompatibility in Schizophyllum. Genet 88:709.
3. Ullrich RC, Novotny CP (1985). Nucleic acid studies in Schizophyllum. In Moore D, Casselton LA, Wood DA, Frankland JC (eds): "Developmental Biology of Agarics," Cambridge, UK: Cambridge University Press. (In press).
4. Hicks JB, Strathern JN, Herskowitz I (1977). The cassette model of mating-type interconversion. In Bukhari A, Shapiro J, Adhaya S (eds): "DNA Insertion Elements, Plasmids and Episomes." Cold Spring Harbor: Cold Spring Harbor Laboratory, p 457.
5. Hicks JB, Strathern JN, Klar AJS (1979). Transposable mating type genes in Saccharomyces cerevisiae. Nature 282:478.
6. Klar AJS, Strathern JN, Broach JR, Hicks JB (1981). Regulation of transcription in expressed and unexpressed mating type cassettes of yeast. Nature 289:239.
7. Nasmyth KA, Tatchell K, Hall BD, Astell C, Smith M (1981). A position effect in the control of transcription at mating-type loci. Nature 289:244.
8. Egel R, Kohli J, Thuriaux P, Wolf K (1980). Genetics of the fission yeast Schizosaccharomyces pombe. Ann Rev Genet 14:77.
9. Beach DH (1983). Cell type switching by DNA transposition in fission yeast. Nature 305:682.
10. East EN, Mangelsdorf AJ (1925). A new interpretation of the hereditary behavior of self-sterile plants. Proc Natl Acad Sci USA 11:116.

11. Bateman AJ (1947). Number of S-alleles in a population. Nature 160:337.
12. Bateman AJ (1952). Self-incompatibility systems in angiosperms. Heredity 6:285.
13. Hinnen AJ, Hicks JB, Fink G (1978). Transformation of yeast. Proc Natl Acad Sci USA 75:1929.
14. Beggs, JD (1978). Transformation of yeast by a replicating hybrid plasmid. Nature 275:104.
15. Case ME, Schweizen M, Kushner SR, Giles NH (1979). Efficient transformation of Neurospora crassa by utilizing hybrid plasmid DNA. Proc Natl Acad Sci USA 76:5259.
16. Tudzynski P, Stahl U, Esser K (1980). Transformation to senescence with plasmid-like DNA in the Ascomycete Podospora anserina. Curr Genet 2:181.
17. Stahl U, Tudzynski P, Kück U, Esser K (1982). Replication and expression of a bacterial-mitochondrial hybrid plasmid in the fungus Podospora anserina. Proc Natl Acad Sci USA 79:3641.
18. Beach D, Nurse P (1981). High-frequency transformation of the fission yeast Schizosaccharomyces pombe. Nature 290:140.
19. Ballance DJ, Buxton FP, Turner G (1983). Transformation of Aspergillus nidulans by the orotidine-5'-phosphate decarboxylase gene of Neurospora crassa. Biochem Biophys Res Commun 112:284.
20. Tilburn J, Scazzocchio C, Taylor GG, Zabicky-Zissman JH, Lockington RA, Davies RW (1984). Transformation by integration in Aspergillus nidulans. Gene 26:205.
21. Yelton MM, Hamer JE, Timberlake WE (1984). Transformation of Aspergillus nidulans by using a trpC plasmid. Proc Natl Acad Sci USA 81:1470.
22. Banks GR (1983). Transformation of Ustilago maydis by a plasmid containing yeast 2-micron DNA. Curr Genet 7:73.
23. deVries OMH, Wessels JGH (1972). Release of protoplasts from Schizophyllum commune by a lytic enzyme preparation from Trichoderma viride. J Gen Microbiol 73:13.
24. Mohan M, Meyer R, Anderson J, Horgen P (1984). Plasmid-like DNAs in the commercially important mushroom genus Agaricus. Curr Genet 8:607.
25. Stohl LL, Akins RA, Lambowitz AM (1984). Characterization of deletion derivatives of an

autonomously replicating *Neurospora* plasmid. Nucleic Acids Res 12:6169.

26. Struhl K, Stinchcomb DT, Scherer S, Davies RW (1979). High-frequency transformation of yeast: autonomous replication of hybrid DNA molecules. Proc Natl Acad Sci USA 76:1035.
27. Dellaporta SL, Fraley RT (1981). Delivery of liposome-encapsulated nucleic acids into plant protoplasts. Plant Mol Biol Newsletter 2:59.
28. Raper JR, Boyd DH, Raper CA (1965). Primary and secondary mutations at the incompatibility loci in *Schizophyllum*. Proc Natl Acad Sci USA 53:1324.
29. Parag Y (1962). Mutations in the B incompatibility factor in *Schizophyllum commune*. Proc Natl Acad Sci USA 48:743.
30. Koltin Y (1968). The genetic structure of the incompatibility factors of *Schizophyllum commune*: comparative studies of primary mutations in the *B* factor. Mol Gen Genet 102:196.
31. Raudaskoski M, Stamberg J, Bawnik N, Koltin Y (1976). Mutational analysis of natural alleles at the *B* incompatibility factor of *Schizophyllum commune* α2 and β6. Genetics 83:507.
32. Dons JJM, Springer J, deVries SC, Wessels JGH (1984). Molecular cloning of a gene abundantly expressed during fruiting-body initiation in *Schizophyllum commune*. J Bacteriol 157:802.
33. Specht CA, Novotny CP, Ullrich RC (1984). Strain differences in ribosomal DNA from the fungus *Schizophyllum commune*. Curr Genet 8:219.
34. Colbere-Garapin F, Horodniceanu F, Kourilsky P, Garapin A (1981). A new dominant hybrid selective marker for higher eukaryotic cells. J Mol Biol 150:1.
35. Southern PJ, Berg P (1982). Transformation of mammalian cells to antibiotic resistance with a bacterial gene under control of the SV40 early promoter region. J Mol Appl Genet 1:327.
36. Bevan MW, Flavell RB, Chilton MD (1983). A chimaeric antibiotic resistance gene as a selectable marker for plant cell transformation. Nature 304:184.
37. Horsch RB, Fry JE, Hoffman NL, Eichholtz D, Rogers SG, Fraley RT (1985). A simple and general method for transferring genes into plants. Science 227:1229.

38. Jimenez A, Davies J (1980). Expression of a transposable antibiotic resistance element in Saccharomyces. Nature 287:869.
39. Barclay SL, Meller E (1983). Efficient transformation of Dictyostelium discoideum amoebae. Mol Cell Biol 3:2117.
40. Nellen W, Silan C, Firtel RA (1984). DNA-mediated transformation in Dictyostelium discoideum: regulated expression of an actin gene fusion. Mol Cell Biol 4:2890.
41. DiRusso C, Novotny CP, Ullrich RC (1983). Orotidylate decarboxylase activity in Schizophyllum commune. Exptl Mycol 7:90.
42. Rose M, Grisafi P, Botstein D (1984). Structure and function of the yeast URA3 gene: expression in Escherichia coli. Gene 29:113.
43. Brody RS, Westheimer FH (1979). The purification of orotidine-5'-phosphate decarboxylase from yeast by affinity chromatography. J Bio Chem 254:4238.
44. Levine HL, Brody RS, Westheimer FH (1980). Inhibition of orotidine-5'-phosphate decarboxylase by 1-(5'-phospho-β-D-ribofuranosyl) barbituric acid, 6-azauridine 5'-phosphate, and uridine 5'-phosphate. Biochem 19:4993.
45. Donovan WP, Kushner SR (1983). Purification and characterization of orotidine-5'-phosphate decarboxylase from Escherichia coli K-12. J Bacteriol 156:620.
46. Boeke JD, LaCroute F, Fink GF (1984). A positive selection for mutants lacking orotidine-5'-phosphate decarboxylase activity in yeast: 5-fluoro-orotic acid resistance. Mol Gen Genet 197:345.
47. Tschumper G, Carbon J (1980). Sequence of a yeast DNA fragment containing a chromosomal replicator and the TRP1 gene. Gene 10:157.
48. Zalkin H, Paluh JL, van Cleemput M, Moye WS, Yanofsky C (1984). Nucleotide sequence of Saccharomyces cerevisiae genes TRP2 and TRP3 encoding bifunctional anthranilate synthase: indole-3-glycerol phosphate synthase. J Biol Chem 259:3985.
49. Zalkin H, Yanofsky C (1982). Yeast gene TRP5: structure, function, regulation. J Biol Chem 257:1491.
50. Buxton FP, Radford A (1983). Cloning of the structural gene for orotidine 5'-phosphate

carboxylase of Neurospora crassa by expression in Escherichia coli. Mol Gen Genet 190:403.

51. Schectman MG, Yanofsky C (1983). Structure of the trifunctional trp-1 gene from Neurospora crassa and its aberrant expression in Escherichia coli. J Mol Appl Genet 2:83.
52. Mullaney EJ, Hamer JE, Roberti KA, Yelton MM, Timberlake WE (1985). Primary structure of the trpC gene from Aspergillus nidulans. Mol Gen Genet (In press).
53. Maniatis T, Fritsch EF, Sambrook J (1982). "Molecular cloning, a laboratory manual." Cold Spring Harbor: Cold Spring Harbor Laboratory, p 545.
54. Specht CA, Novotny CP, Ullrich RC (1983). Isolation and characterization of mitochondrial DNA from the basidiomycete Schizophyllum commune. Exptl Mycol 7:336.

Molecular Genetics of Filamentous Fungi, pages 59–68

STUDIES ON TRANSFORMATION OF CEPHALOSPORIUM ACREMONIUM

Miguel A. Peñalva, Angeles Touriño, Cristina Patiño, Flora Sánchez, José M. Fernández Sousa and Victor Rubio

Departamento de Genética Molecular, Antibióticos, S.A.
Bravo Murillo 38, 28015 - Madrid - SPAIN

ABSTRACT Spheroplasts of the filamentous, antibiotic-producing fungus *Cephalosporium acremonium* have been transformed with plasmids containing the Tn903 aminoglycoside 3'-phosphotransferase (I) (APH(3')I gene, selecting for resistance to G418. Transformation was of the integrative type with a frequency of 0.2-1 transformants per ug of DNA. Selection of transformants was improved when the APH(3')I gene was governed by the yeast ADCI promoter. Subfragments of a 1.94 Kb mitochondrial DNA segment of *Cephalosporium acremonium* with ARS activity in yeast were subcloned. All but one of the subfragments had ARS activity. The nucleotide sequence of the 1.94 Kb segment has been deduced. A consensus ARS sequence has been found in the subfragment with the highest ARS activity. Other sequences related to the consensus one could account for the ability of the subfragments in promoting autonomous replication.

INTRODUCTION

The ascomycete *Cephalosporium acremonium* is a filamentous fungus that produces the B-lactam antibiotic cephalosporin C. It belongs to the group of fungi imperfecti (i.e.: it is known only in an asexual state) and genetic recombination can only be obtained via hyphal anastomosis or protoplast fusion, processes that are rather inefficient (1). Therefore, application of recombinant DNA techniques to this microorganism represents an alternative for the genetic improvement of antibiotic-producing strains. A transformation method for a fungus was first established in the yeast *Saccharomyces cerevisiae* (2,3) and was based in polyethylene glycol (PEG)-induced DNA uptake by spheroplasts. The general procedure has been successfully applied to filamentous fungi such as Neurospora crassa (4) and Aspergillus nidulans (5). C. acremonium cells are sensitive to low concentrations of the aminoglycoside G418. The APH(3')I encoded by the E. coli transposon Tn903 is known to phosphorylate and inactivate this antibiotic, as well as other 2-deoxystreptamine aminoglycosides such as neomycin or kanamycin (6). Furthermore, it has been shown that the Tn903 APH(3')I gene, whose complete nucleotide sequence is known (7), can be expressed in yeast conferring resistance to G418 (6). Therefore, we decided to use this gene as a selectable marker in C. acremonium.

A 1.94 Kb mitochondrial DNA fragment of C. acremonium has been shown to have ARS activity in yeast (8,9,10). This fragment could be useful for the construction of autonomously replicating vectors and therefore we have performed a detailed examination of the structural characteristics that might be relevant for its ARS properties.

METHODS

a) Transformation of C. acremonium.

C. acremonium ATCC 11550 was grown in defined medium at 28ºC (11); mycelium was recovered by filtration and washed with water. One gram (wet weight) was resuspended in 10 ml of 0.2M McIlvaine's buffer, pH 7.3, containing 0.5M B-mercaptoethanol, and was incubated for 15 min at 30ºC. Mycelium was washed again and resuspended (0.1 g/ml) in lytic mixture - zymolase 20T (Miles) 20 mg/ml, lytic enzyme L1 (BDH) 20 mg/ml in 1M sorbitol - and incubated for 6 hours at 30ºC. Spheroplasts were purified from mycelial debris, washed and finally taken in a minimal volume of 1M sorbitol, 10mM $CaCl_2$, 10mM Tris-HCl, pH 7.5 (STC). 0.1 ml samples containing $1x10^8$ spheroplasts were mixed with 5-10 ug of DNA and incubated for 15 min at room temperature. Then, 1 ml of 40% (w/v) PEG 4000, 10mM $CaCl_2$, 10mM Tris-HCl, pH 7.5 was added and the mixture was incubated for 15 min further at room temperature. Spheroplasts were then recovered by centrifugation resuspended in STC, mixed with 5 ml of regeneration medium (PRM) at 50ºC (PRM is 1% glucose, 0.5% yeast extract, 0.5% casamino acids, 0.4M sucrose and 2% agar) and plated on PRM plates. Spheroplasts were regenerated at 28ºC for 18 hours and then overlaid with PRM containing G418 to give a final concentration of 50 ug/ml. Plates were incubated at 28ºC.

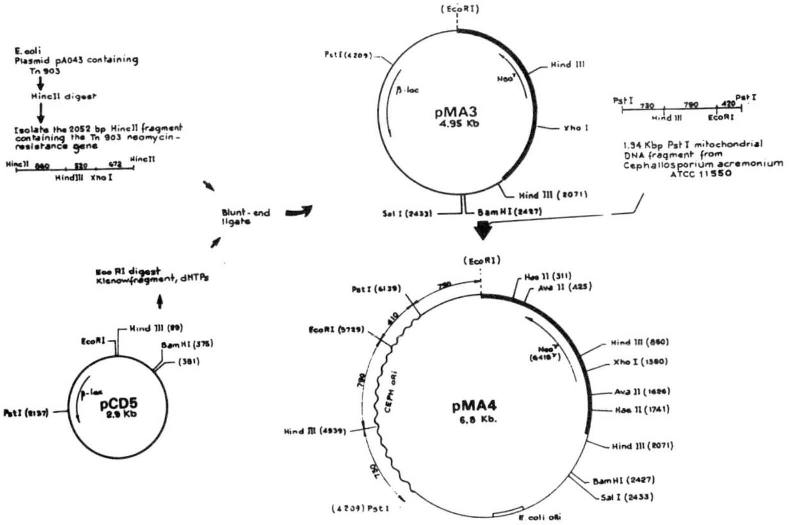

Figure 1 - Construction of plasmids containing the Tn903 APH(3')I gene.

b) Miscellaneous Procedures.

E. coli was transformed according to Hanahan (12). Plasmid DNA was obtained from E. coli by alkaline lysis (13) and purified by CsCl-EtBr gradient centrifugation. Oligonucleotides were synthesized using an Applied Biosystems apparatus and purified as described by the manufacturer. Yeast transformation was according to Beggs (2) and yeast DNA preparations enriched in plasmid DNA were made as described by Zakian (14). The nucleotide sequence of the mitochondrial ARS was determined by the method of Sanger (15) using M13 mp8 or M13 mp9 (16). Probes used for Southern blot hybridization (17) were labelled by nick-translation (18).

RESULTS

a) Transformation of C. acremonium.

To transform C. acremonium a high yield of spheroplasts must be obtained from the mycelial mass. The conditions given in Methods allowed a yield of $5x10^8$-$1x10^9$ spheroplasts per

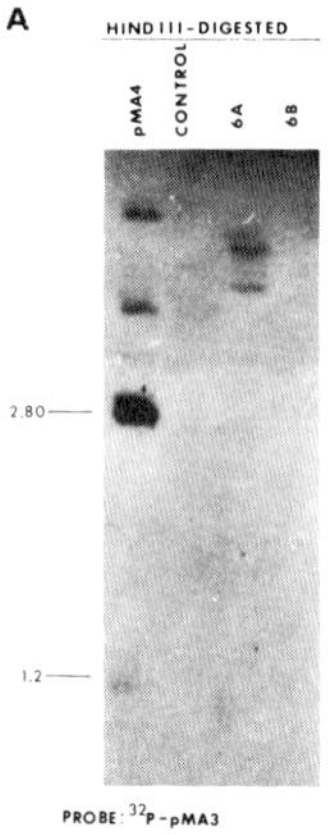

Figure 2 - Southern analysis of transformants. C. acremonium G418 resistant clones were grown in liquid medium plus 50 ug/ml of G418. Mycelium from 25 ml cultures was collected and used to isolate spheroplasts, which were lysed in the presence of 1% SDS. Total DNA was purified from the lysates by phenol extraction and repeated isopropanol precipitations. After digestion with HindIII, DNA was analyzed by Southern blot hybridization using ^{32}P-pMA3 as a probe.

0.5 g of mycelial wet weight. 5-10% of spheroplasts regenerated in PRM medium after 3 days at 28ºC and the contamination of non-spheroplast cells is usually less than 0.1% of the regenerated spheroplasts. The construction of plasmids used in transformation experiments is described in fig. 1. We subcloned a 2058 bp HincII fragment containing the APH(3')I gene in plasmid pCD5 (19). The resulting plasmid, pMA3, confers neomycin and ampicillin resistance to E. coli and contains a single PstI site in which we have cloned the 1.94 Kb DNA fragment that presumably contains the replication origin of mitochondrial DNA, and the resulting plasmid, pMA4 (fig. 1) has been used in transformation experiments. We have obtained 1-2 G418 resistant colonies per ug of DNA after 7-10 days of incubation. This frequency was independent on the DNA used, either pMA3 or pMA4. Colonies resistant to G418 were twice subcultured in liquid medium containing this antibiotic. DNA was isolated from these cultures as described in fig. 2, restriction enzyme digested and analyzed by Southern blot hybridization for the presence of sequences containing the Tn903. Fig. 2 shows the analysis of G418-resistant colo-

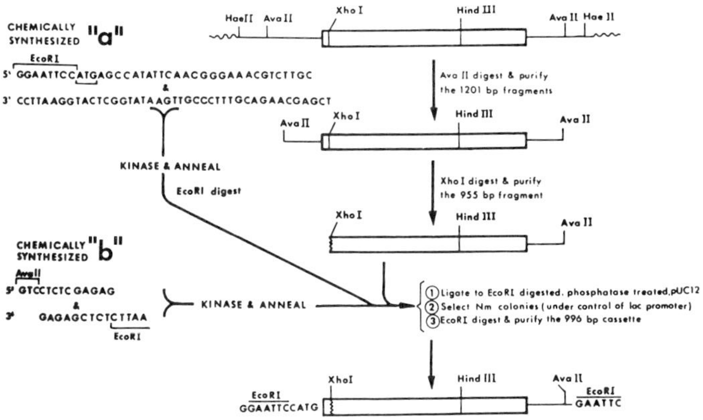

Figure 3 - Construction of a cassette fragment containing the Tn903 APH(3')I gene (boxed structure in the scheme). The different constructions are explained in the text.

nies obtained with pMA4. Usually, only 20-50% of the resistant colonies contained sequences that hybridize to the transforming vector. So, the transformation frequency would be 0.2-1 transformants per ug of DNA. In all cases, transformation ocurred by integration of the plasmid into the host chromosome.

b) Construction of a Plasmid with the APH(3')I Gene under the Control of a Yeast Promoter.

To improve the expression of the *E. coli* APH(3')I gene in *C. acremonium*, we decided to place it under the control of a strong fungal promoter. To eliminate several ATG triplets which are present before the initiation codon of the APH(3')I gene (6) and to provide it with appropriate cohesive ends to be cloned in an expression vector, we constructed a cassette fragment flanked by EcoRI sites as described in fig. 3. A purified AvaII fragment obtained from pMA3 was digested with XhoI which cuts it at a single site 30bp downstream of the initial ATG of the gene. To restore this part of the gene, an adapter "a" that provides an EcoRI site just before the initial ATG codon was synthesized (fig. 3). A second adapter "b" which converts the AvaII end in a second EcoRI cohesive end was also synthesized. The mixture of EcoRI-

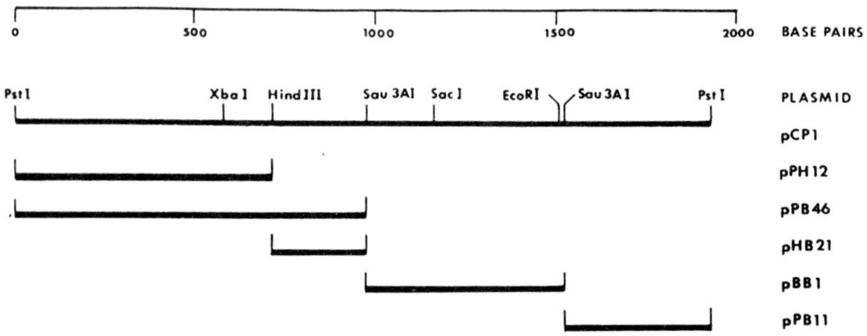

PLASMID	mt DNA INSERT(Kb)	TRANSFORMATION FREQUENCY (Transformants/μg $\times 10^{-4}$)
pCP1	1.94	0.47
pPH12	0.72	0.3
pPB46	0.98	0.32
pHB21	0.26	0.28
pBB1	0.56	1.07
pPB11	0.42	<0.01
pMA2	—	<0.01

Figure 4 - ARS activity of subfragments from the 1.94 Kb mitochondrial DNA fragment. Subfragments were cloned in the yeast integrative vector pMA2 and the resulting recombinant plasmids were used to transform S. cerevisiae D483 (Mat a, leu2-112, his4, can1). The relevant inserts and the names of the plasmids carrying them are shown in the figure. Transformation frequencies are given in the table.

digested adapter "a", adapter "b" and XhoI-AvaII fragment was ligated to EcoRI-digested and phosphatase-treated pUC12 (16) and, after transforming E. coli cells, neomycin resistant colonies were recovered. Plasmid from one of these clones was purified and digested with EcoRI to obtain a 996 bp cassette fragment that contains the APH(3')I structural region flanked by EcoRI cohesive ends. The fragment was then inserted in the single EcoRI site of the yeast expression plasmid pAAR6 (20). This site is located between the promoter and terminator sequences of the yeast ADCI gene. A hybrid plasmid with the fragment in the right orientation was purified and used to transform C. acremonium.G418-resistant colonies appeared on the plates at low frequency (0.3/ug of DNA in 5 days) and

Subfragment	PstI-HindIII (1) (728)	HindIII-Sau3A (728) (986)	Sau3A-Sau3A (986)(1534)	Sau3A-PstI (1534)(1943)
Transformation frequency/$\mu g x 10^{-4}$	0.3	0.28	1.07	< 0.01
A+T%	74.3	68.7	74.67	68.05
ARS consensus - like sequences	TcTTATATaTT TTaTATATTgT AcTgATATTTT TTgTAaGTTTA TTTTAaGTTTA ATTTtTGgTTT TTTTtgGTTTA AaaTATATTTT ATaTATtTTTT TTTTATAaTaA AaTTATGaTTT ATTTcTATTcT TcTTATATTaA TaTTATGgTTT cTTTtTATTTT ATTTtaGTTTA cTTTAaATTTT AaaTATATTTA ATaTtTATTTT ATTTcTATTaA ATTaAgGTTTA cTTTATtTTTT TTTTtTATTTA AcTTgTATTTA	TaTTcTATTTT gTTTATcTTTA ATTTATGTTag TTTcATcTTTA	TaTTATATTgA gTTTAaATTTA ATTTAaATTTA ATTTAaATgTT TaTTATGTaTA ATgTATATTaA AaTaATATTTT TTTTATAggTT TgTTATATTTg ATaTtTGTTTT TTTTcTtTTTT ATgTgTATTTT TaTTtTATTTT TTTTATtTTTA ATTcATATTTA TTaTATAaTTA ATTTATATTTA TTTTATAccTA ATTTAaATTTA ATTTAaATTTA	TTcTATGTTTT ATTTATATaTT TTaTATATTcT TTaTtTATTTT ATTTATtTTTA ATTTgTcTTTA

Table I - ARS-like sequences found in the subfragments described in fig. 4. Numbers in parenthesis represent the nucleotide position of the restriction site in the 1941 bp fragment. A sequence identical to the consensus one is boxed.

grew vigorously when subcultured in G418-containing medium. DNA was isolated from these clones and analyzed by Southern blot hybridization. Most of the clones tested contained the Tn903 APH(3')I gene in an integrated form (not shown).

c) Subcloning and Sequence Analysis of the 1.94 Kb Fragment of *C. acremonium* Mitochondrial DNA with ARS Activity.

To analyze the DNA fragment from *C. acremonium* that confers ARS phenotype in yeast (9,10), we subcloned different restriction fragments in the yeast integrative vector pMA2. This plasmid was obtained by insertion of a 2.2 Kb DNA fragment containing the yeast LEU2 gene (21) in the SalI site of pBR325.Plasmid DNA was purified from the different clones and used to transform yeast. The results of this experiment are shown in fig. 4. All but one of the subfragments conferred ARS phenotype and, interestingly, the central 560 bp Sau3A fragment allows even a higher transformation efficiency than the complete 1.94 Kb fragment. To confirm that transformants contained autonomously replicating plasmids, DNA was isolated from yeast leu^+ clones and analyzed by Southern blot hybridization. All the yeast leu^+ clones contained plasmids of the expected size (not shown).

We have determined the complete nucleotide sequence of the 1.94 Kb fragment. The fragment is very rich in A:T pairs (72.5%). A minimal consensus sequence for ARS has been published (22) and it seems to represent a structural requirement for autonomous replication. Table I shows the sequences more closely related to the consensus one and their location in the 1.94 Kb fragment. It is worth noting that only the 560 bp Sau3A subfragment with the highest ARS activity contains a sequence identical to the consensus (A/TTTTATPuTTTA/T). Consensus-related sequences are found in all the subfragments (see table I) which might be responsible for promoting autonomous replication.However, the role of the high A:T content of the sequences can not be ruled out.

DISCUSSION

We have developed a transformation procedure for C. acremonium selecting for G418 resistance. We have used for transformation plasmid pMA3 which contains the Tn903 APH(3')I gene. This plasmid should transform in an integrative way C. acremonium, and, to get a replicating plasmid we cloned in it the 1.94 Kb DNA fragment that presumably contains the mitochondrial replication origin. In all transformants so far studied, however, Southern analysis shows that integrative transformation is taking place, in agreement with the low frequency of transformation obtained.

The amount of G418 used for selection of transformants and the long incubation time required allowed the appearance of spontaneous resistant mutants, which had to be distinguished from real transformants by hybridization. A similar situation has been described in yeast (23). In this case, an increase in the G418 concentration gave rise to a loss of transformants (23). To improve the expression of the E. coli APH(3')I gene in C. acremonium we placed it under the control of the strong yeast ADCI promoter. When this construction was used for transformation G418-resistant colonies appeared in a shorter time and most of them showed to have integrated the transforming plasmid, suggesting that a high expression of the gene allowed a correct selection of transformants.

We have performed an analysis of the ARS activity of the 1.94 Kb mitochondrial DNA fragment finding that several subfragments confer ARS activity in yeast. To correlate this ARS phenotype with primary structure, we have sequenced the 1.94 Kb mitochondrial DNA fragment and found a canonical ARS sequence in the subfragment with the highest ARS activity. Other ARS-related sequences might be responsible for the activity of the subfragments, although some of these sequences are found in a subfragment which does not present ARS phenotype. Whether or not mitochondrial ARS sequences represent true replication origins is still unclear and we are at present carrying *in vivo* experiments to identify the *C. acremonium* mitochondrial DNA replication origin.

ACKNOWLEDGMENTS

Thanks are due to Dr. A. Jiménez for critical reading of the manuscript and to Mrs. D. Fernández for typing it.

REFERENCES

1. Minuth W, Tudzyski P, Esser K (1982). Extrachromosomal genetics of *Cephalosporium acremonium*. Curr Genet 5:227.
2. Beggs JD (1978). Transformation of yeast by a replicating hybrid plasmid. Nature 275:108.
3. Hinnen A, Hicks JB, Fink GR (1978). Transformation of yeast. Proc Natl Acad Sci USA 75:1929.
4. Case ME, Scheweizer M, Kushner SR, Giles NH (1979). Efficient transformation of *Neurospora crassa* utilizing hybrid plasmid DNA. Proc Natl Acad Sci USA 76:5259.
5. Yelton MM, Hamer JE, Timberlake WE (1984). Transformation of *Aspergillus nidulans* by using a trpC plasmid. Proc Natl Acad Sci USA 81:1470.
6. Jiménez A, Davies J (1980). Expression of a transposable antibiotic resistant element in *Saccharomyces*. Nature 287:869.
7. Oka A, Sugisaki H, Takanami M (1981). Nucleotide sequence of the kanamycin resistance transposon Tn903. J Mol Biol 147:217.
8. Stinchcomb D, Thomas M, Kelley J, Selker E, Davis RW (1980). Eukaryotic DNA segments capable of autonomous replication in yeast. Proc Natl Acad Sci USA 77:4559.

9. Tudzynski P, Esser K (1982). Extrachromosomal genetics of Cephalosporium acremonium. Curr Genet 6:152.
10. Skatrud PL, Queener SW (1984). Cloning of a DNA fragment from Cephalosporium acremonium which functions as an autonomous replication sequence in yeast. Curr Genet 8:155.
11. Hamlyn PF, Bradshaw RE, Mellon FM, Santiago CM, Wilson JM, Peberdy JF (1981). Efficient protoplast isolation from fungi using commercial enzymes. Enzyme Microb Technol 3:321.
12. Hanahan D (1983). Studies on transformation of Escherichia coli with plasmids. J Mol Biol 166:557.
13. Birnboim HC, Dolly J (1979). A rapid alkaline extraction procedure for screening recombinant plasmid DNA. Nucl Acids Res 7:1513.
14. Zakian VA (1981). Origin of replication from Xenopus laevis mitochondrial DNA promotes high-frequency transformation of yeast. Proc Natl Acad Sci USA 78:3128.
15. Sanger F, Nickens S, Coulson AR (1977). DNA sequencing with chain-terminating inhibitors. Proc Natl Acad Sci USA 74:5463.
16. Messing J (1983). New M13 vectors for cloning. Meth Enzymol 101:20.
17. Southern EM (1975). Detection of specific sequences among DNA fragments separated by gel electrophoresis. J Mol Biol 98:503.
18. Rigby PWJ, Deickmann M, Rhodes C, Berg P (1977). Labeling deoxyribonucleic acid to high specific activity in vitro by nick-translation with DNA polimerase I. J Mol Biol 113:237.
19. Scott MRD, Westphal KH, Rigby PWJ (1983). Activation of mouse genes in transformed cells. Cell 34:557.
20. Ammerer G (1983). Expression of genes in yeast using the ADCI promoter. Meth Enzymol 101:192.
21. Andreadis A, Hsu YP, Kohlhaw GB,Schimmel P (1982). Nucleotide sequence of yeast LEU2 shows 5'-noncoding region has sequences cognate to leucine. Cell 31:319.
22. Broach JR, Li YY, Fieldman J, Jayaram M, Abraham J, Nasmyth KA, Hicks JB (1982). Localization and sequence analysis of yeast origins of DNA replication. Cold Spring Harbor Symp Quant Biol 47:1165.
23. Hollenberg CP (1982). Cloning with 2-um DNA vectors and the expression of foreign genes in Saccharomyces cerevisiae. In Hofschneider PH, Goebel W (eds): "Gene cloning in organisms other than E. coli". Berlin: Springer Verlag, p 119.

Molecular Genetics of Filamentous Fungi, pages 69–82

PROTOPLAST FUSION AND HYBRIDIZATION IN PENICILLIUM[1]

Fiona M. Mellon.

School of Biological Sciences,
Queen Mary College, Mile End Road.
London. E1 4NS.

ABSTRACT Fungal protoplasts have proved to be very useful in genetic studies over the last decade. The initial interest was in interspecific protoplast fusion, particularly in the *Aspergilli* and *Penicillia*. Fusion of protoplasts was found to be induced by polyethylene glycol (PEG) and fusion products were first isolated following nutritional complementation of auxotrophic markers. More recently selection techniques involving the use of dead donors and drug resistance have been developed. Fusion products can be characterised genetically by haploidization analysis and biochemically by analysis of the mitochondrial genomes, isoenzyme banding patterns and antibiotic production. In recent years protoplasts have formed the basis of fungal transformation systems in a variety of species. An important feature of these systems is the need to have a reproducible protoplasting procedure in terms of the numbers of protoplasts isolated and their regenerative capacity.

INTRODUCTION

Traditionally, commercially important species, such as *Penicillium chrysogenum* have been modified by mutation and selection (1,2). The lack of sexuality as a utilizable breeding system in this fungus has been a mitigating factor in the highly successful exploitation of this technique.

[1]This work was supported by an S.E.R.C. C.A.S.E. award funded by the Science and Engineering Research Council and the Ministry of Defence.

The classical random mutation and selection techniques, which were important in the development of highly productive mutants in the absence of fundamental knowledge, have now been largely replaced by more rational (directed) selection techniques grounded on more scientific bases, for example, the use of fermentation prescreening of mutagenized cells prior to laboratory fermentation studies (3). However there is a limit to the improvements which can be achieved by repeated mutagenesis of a single strain (4,5). Moreover, deleterious gene mutations and chromosome aberrations are often induced concomitantly with the desired changes (6,7, 8). For these reasons it would be advantageous to use recombination methods, in addition to mutation, for strain improvement. For example, the potential performance of a high yielding strain might be limited by a mutation which could be replaced by the introduction of the missing genetic material from another strain. Recombination could also be useful to introduce several benefical factors simultaneously into any one strain and the cumulative effect of these could be much greater than the effect of any one alone.

The history of recombination studies with P. chrysogenum started with Pontecorvo and Sermonti (9), who discovered the parasexual cycle in this fungus. Early work designed to enhance penicillin yields was summarized by Sermonti (10) and a number of possible reasons for the lack of success were put forward. Subsequently, a need was recognised for a more fundamental approach to deal with the problems highlighted which included strain instability, selection against alleles and parental genome segregation due to translocations in divergent line crosses (11). Working with sister strains or strains of low divergence a genetic map was constructed (12) and exploited in later recombination studies.

A successful investigation, referred to by Ball(12) was carried out of the possibility of recombining genes for penicillin titer improvement between strains much more highly divergent than those used by others in earlier studies. This study adapted elements of the biometrical approach advocated by Caten and Jinks. Furthermore, Queener and Baltz have claimed success in improving penicillin yield by recombining strains of P. chrysogenum (13) and, in an empirically based study, Calam et al (14) were successful with a combination of diploidization and mutation and Elander and co-workers improved titer, albeit rarely by diploidization of sister strains (2, 15). These reports indicate that the parasexual cycle has been used in the industrial laboratories as a routine technique directed at improving penicillin productivity of *P. chrysogenum*.

Why Protoplast Fusion?

In *P. chrysogenum* and *Cephalosporium acremonium* the onset of parasexual phenomena via hyphal fusion is relatively rare, protoplast fusion is a means of producing heterokaryons at a high frequency. This is intraspecies fusion, examples of the successful application of this technique include 1) the fusion of heterokaryon incompatible strains of *Aspergillus nidulans* which enabled Dales and Croft (16,17) to investigate the genetic basis of incompatibility in this species; 2) Hamlyn and Ball (18) used protoplast fusion to produce heterokaryons in *C. acremonium*. They crossed a high yielding, slow growing strain to a low yielding, fast growing strain. They isolated one prototroph and analysed 600 of its segregants. One of these gave 40% more Cephalosporin C and grew faster and sporulated better than the prototroph.

Protoplast fusion is the only means of producing inter-species hybrids and therefore allows the study of the genetic homology between species. It also allows the possibility of hybrids with combined properties of the parents such as the introduction of a sexual cycle or the production of novel metabolites. The use of protoplast fusion for industrial strain improvement has been advocated by numerous authors (19,20,21,22,23,24,25,26,27,28,29) because of the wider scope for genetic recombination afforded by this technique (24,29). One of the factors contributing to this reserve is the necessity of introducing selective markers, e.g. auxotrophic mutations, into parental strains. Mutations which affect basic nutritional processes in an organism often have pleiotrophic effects on the production of commercially important products (30,31,32). However, recently various other selection strategies have been developed (Table 1).

Protoplast fusion techniques, first described in filamentous fungi in 1975 by Anné and Peberdy (33) and almost simultaneously by Ferenczy (34), have proved to be a valuable method for interspecific hybridization. So far only two genera of filamentous fungi, the *Penicillia* and the *Aspergilli*, have been studied in detail.

Kevei and Peberdy (35,36) have investigated the hybridization of *A. nidulans* and *A. rugulosus*, two closely-related species, and found that protoplast fusion led to heterokaryosis, hybrid formation following nuclear fusion

TABLE 1
SELECTION STRATEGIES

Strategy	Comments and References
Nutritional complementation	This is the most commonly used method, it involves two auxotrophic mutants (35,36,37)
Drug resistance	This method allows one of the parental strains to be a wild-type strain. The heterokary ons are selected on stabilised minimal medium plus the appropriate inhibitor at concentrations which prevent growth of protoplasts of the sensitive parent (38).
Dead donor	Also allows the use of wild-types, protoplasts can be inactivated by UV irradiation (39,40) or by heat.
Metabolic poisons	First developed for mammalian cells this method has recently been adapted for fungi.
Protoplast: nucleus fusion	This method uses isolated nuclei from wild-type strains fused with auxotrophic mutants (41)

and subsequent haploidization, all of which are key steps in the parasexual cycle. The wide range of haploid types obtained in the segregant progeny suggested some functional similarities in genome organization between these closely related species. Hybrids of A. nidulans and A. rugulosus resembled, in morphology and behaviour, intraspecific diploids of heterokaryon incompatible strains of A. nidulans also obtained by protoplast fusion. Bradshaw and Peberdy hav extended this work (42) by producing hybrids between strains of A. rugulosus and mitotic master strains of A. nidulans with a genetic marker on each linkage group. Analysis of segregants induced by growth on medium containing benomyl revealed recombination between every pair of unlinked markers. Parental combinations of markers were often recovered at significantly higher frequences than expected. This aberrant segregation was not correlated with any particular pair of linkage groups and was attributed to inter-species incompatibility. The segregation of genetic markers of A. rugulosus from hybrids suggested that A. nidulans and A. rugulosus may differ in haploid chromosome number and chromosome size. In sexual crosses between A. nidulans and strains containing chromosomes of mixed parental origin recombinants were recovered. Ferenczy et al (43) have fused protoplasts of diverse species such as A. nidulans and A. fumigatus, unstable complementation products were recovered at low frequencies and there was no evidence of a regular parasexual cycle. In the light of these various data Kevei and Peberdy (44) carried out an exhaustive study on protoplast fusion and interspecific hybridization within the A. nidulans species group. A total of 28 possible crosses with 8 closely related species were attempted; 8 gave fusion products. The primary fusion products proved to be interspecific heterokaryons. Prolonged cultivation under selective conditions, in some cases, resulted in interspecific hybrid formation as a consequence of somatic nuclear fusion. The haploidization experiments confirmed that the hybrids were the phenotypic expression of a nucleus with both parental nuclear genomes and so corresponding to an allodiploid.

The situation found in the Penicillia is somewhat different. Anné and Peberdy studied crosses between closely related (45, 46, 47, 48) and less related species of Penicillium (49, 50, 51) and have found differences between these two types of cross. Heterokaryons produced between rather closely related species e.g. P. chrysogenum and P. notatum, P. chrysogenum and P. cyaneo-fulvum, P. chrysogenum and P. citrinum or P. cyaneo-fulvum and P. citrinum behaved almost as intraspecific heterokaryons. They developed well, but with irregular colony morphology and at somewhat slower growth rate than intraspecies heterokaryons. Like intraspecies heterokaryons they produced spores of both the complementing parents on non-selective media (Figure 1).

In contrast to heterokaryons and hybrids obtained from crosses between closely related species, progeny recovered from protoplast fusion between less related Penicillium species were quite different in morphology and behaviour.

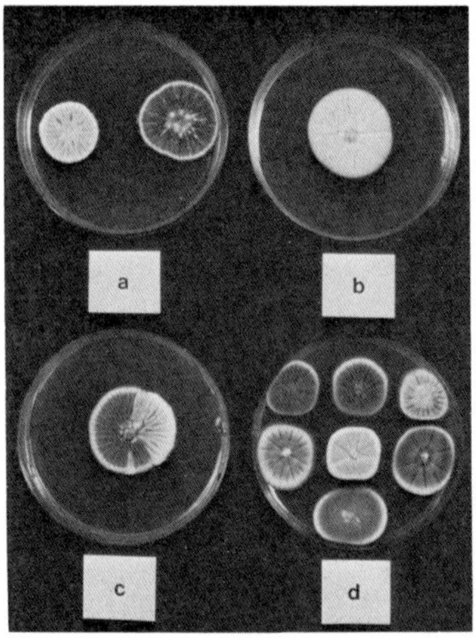

FIGURE 1. Parents, hybrids and segregants from a cross between two closely related Penicillia. (a) The parents, P.chrysogenum (left) and P. cyaneo-fulvum. (b) The first order hybrid which developed from the heterokaryon. (c) Segregation of the first order hybrid. (d) Isolated segregants.

Heterokaryons grew very slowly in atypical sparse colonies and they did not sporulate on minimal medium. From these heterokaryons hybrid progeny arose as faster growing more compact colonies, they produced low yields of white prototrophic conidia. The colonies which arose from these conidia varied phenotypically in morphology and stability. In the case of P. chrysogenum and P. roqueforti (45) and P.patulum and P. chrysogenum (51) fully stable hybrids were isolated after repeated subcylture, but from their phenotypes it was concluded that they were not haploid in their chromosomal constitution. Figure 2 outlines the taxonomic relationships of the Penicillia used in these crosses.

I will now give a brief outline of some more recent work which involved two species which are even more distantly related, P. chrysogenum and P. baarnense. P. baarnense is an unusual member of the genus because it has a sexual cycle. The aim of this study was to establish that hybridization between distantly related species was possible and therefore to attempt to isolate a recombinant which had a sexual stage and produced penicillin, such a hybrid could then form the basis of a rational breeding programme because genetic

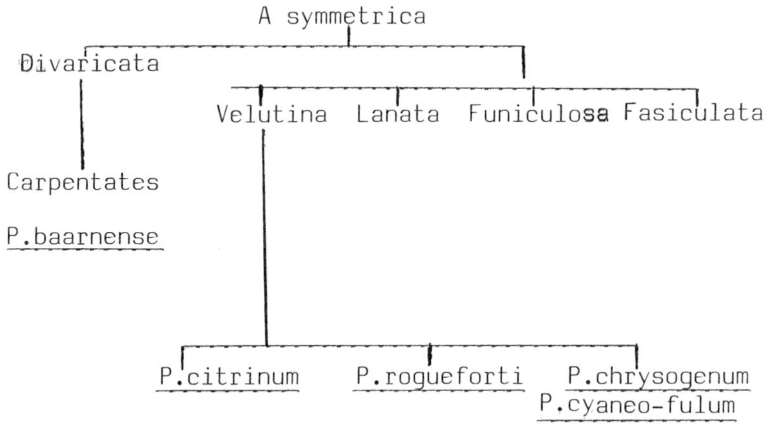

FIGURE 2. Taxonomic relationships of *Penicillia* used in crosses.

manipulations of such a strain could produce many more realistic recombinant than classic methods presently available.

RESULTS

Isolation of Protoplasts.

Conditions had already been established for the isolation of protoplasts from *P. chrysogenum* (46), however Novozym 234 was substituted for the "homebrew" enzyme used previously (52). The same conditions were found to be effective for the isolation of protoplasts from *P. baarnense* (Figure 3).

From Figure 3b it can be seen that most of the protoplasts are nucleated, however, as had been found previously the presence of a nucleus does not necessarily mean that the protoplasts will regenerate (54). *P. baarnense* has an average regeneration frequency of 15% while *P. chrysogenum* has one of 47%.

The method used for protoplast fusion is basically that of Anné and Peberdy (45) and is outlined in Figure 4.

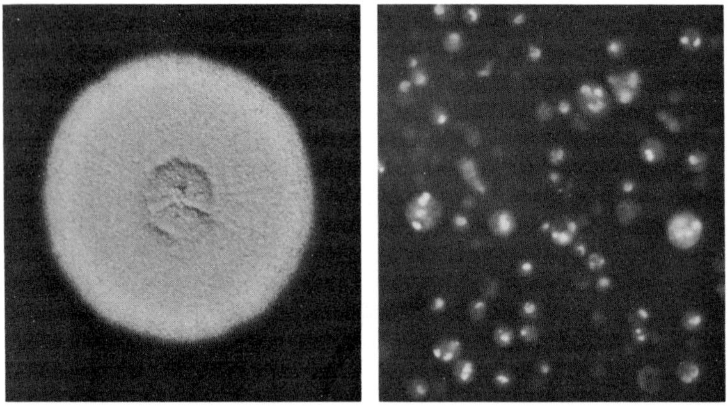

FIGURE 3. a)P. baarnense, the sexual stage. b)protoplasts from P. baarnense fixed in gluteraldehyde and stained with chromomycin (53).

Nutritional complementation was used as the selection system. Protoplasts were isolated using Novozym 234 in 0.7M NaCl (52). Equal numbers of protoplasts from each strain were mixed with 30% PEG and fusion products were selected on stabilised minimal medium, and later subcultured on minimal medium. The fusion products displayed a wide variety of morphological types (Figure 5). The fusion products also varied in their stability on minimal and complete medium, unstable fusion products segregated spontaneously on minimal medium whilst stable products only segregated after growth in the presence of a haploidising agent (paraflurophenylalanine [PFA] or benomyl). Analysis of the segregants showed that there were both parental and non-parental types. The non-parentals were of two types prototrophs and auxotrophs. The percentage of non-parentals varied between crosses. Another variation between crosses was the isolation of parental types, it was not always possible to reisolate both parentals following segregation, however, unlike previous reports (49) it was not always the same parent which was reisolated i.e. in some crosses P. chrysogenum was reisolated and in others P. baarnense was reisolated

In the absence of mitotic master strains in either of these species biochemical means of assessing hybridity were investigated. There are two ways of looking for differences

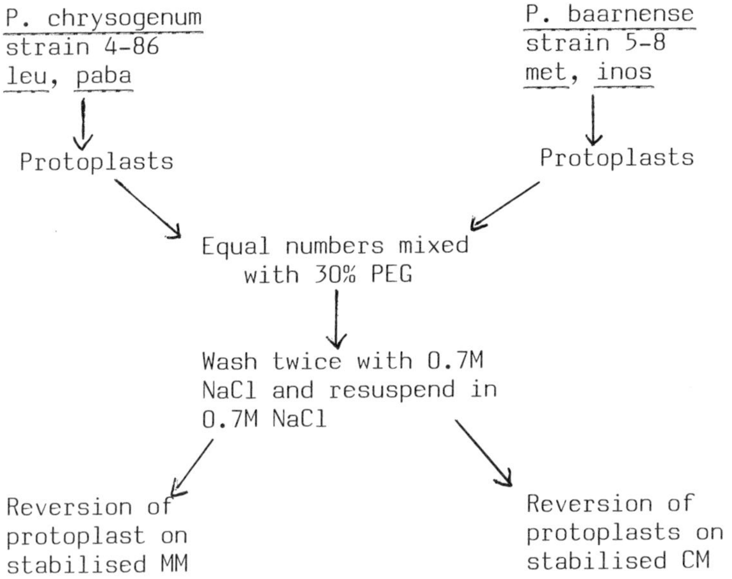

FIGURE 4. Schematic diagram showing protoplast fusion protocol.

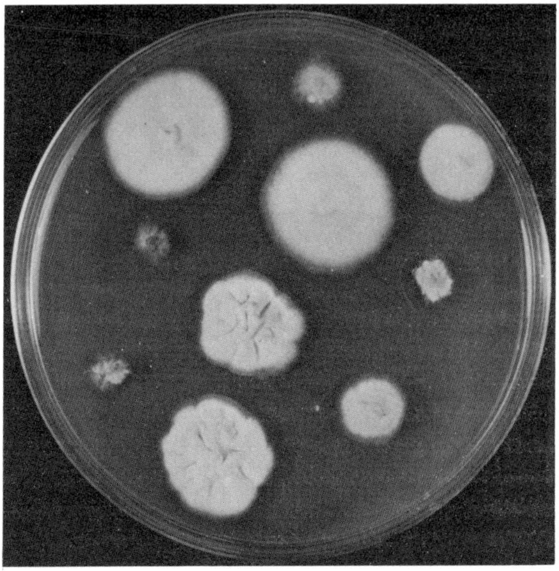

FIGURE 5. Fusion products showing morphological variety.

at the level of proteins and at the DNA level. For these reasons the isoenzyme banding patterns and the restriction endonuclease fragment banding patterns were investigated.

DISCUSSION

Hybridization between distantly related members of the genus Penicillium appears to be possible. Following protoplast fusion nutritional complementation occurs and prototrophic fusion products were isolated, these are slow growing sparse colonies. With continued cultivation faster more compact colonies with regular morphology are found. These colonies were stable or unstable on minimal (MM) and complete medium (CM) and segregate either spontaneously of by induction. The segregants may be parental or non-parental. Analysis of the segrgants showed a range of phenotypes from prototrophic to multi-auxotrophic. It is possible that the prototrophic colonies are stable aneuploids, however, the multi-auxotrophic isolates are more likely to be allohaploid since the markers involved in these crosses are all recessive in diploids.

ACKNOWLEDGMENTS

I am indebted to Professor John Peberdy and Dr. Ken Macdonald for their help and encouragement during this work.

REFERENCES

1. Backus MP, Stauffer JF (1955). The production and selection of a family of strains in Penicillium chrysogenum. Mycologia 47:429.
2. Elander RP (1967). Enhanced Penicillin biosynthesis in mutant and recombinant strains of Penicillium chrysogenum. Abh Dtsch Akad Wiss Berlin 4:403.
3. Elander RP (1979). Mutations affecting Antibiotic Synthesis in Fungi Producing B-Lactam Antibiotics. In Sebeck OK, Laskin AI (eds): "Genetics of Industrial Microorganisms," Washington DC: American Society of Microbiology, p21.
4. Demain AL (1973). Mutation and the production of secondary metabolites. Advances in Appl. Mic 16: 177.

5. Sermonti G (1979). Mutation and microbial breeding. In Sebeck OK, Laskin AI (eds): "Third International Symposium on the Genetics of Industrial Microorganisms," Washington DC: American Society of Microbiology, p 10.
6. Stauffer JF, Backus MP (1961). Induced mutations in penicillin biosynthesis Recent Advs in Bot 1:645.
7. Macdonald KD (1968) The persistance of parental genome segregation in Penicillium chrysogenum after nitrogen mustard treatment. Mut Res 5:302.
8. Elander RP, Corum CJ, De Valeria H, Wilgus RM (1976). Ultraviolet mutagenesis and cephalosporin acremonium. In Macdonald KD(ed): "Second International Symposium on the Genetics of Industrial Microorganisms," London: Academic Press, p 253.
9. Pontecorvo G, Sermonti G (1954). Parasexual recombination in Penicillim chrysogenum JGM 11:94.
10. Sermonti G (1969). "Genetics of Antibiotic Producing Microorganisms," London: Wiley Interscience,
11. Ball C (1971). Haploidization analysis in Penicillium chrysogenum. JGM 66:63.
12. Ball C (1982). Genetic approaches to overproduction of B-Lactam antibiotics in eukaryotes. In Krumphanz I, V, Sibyta B, Vanek A (eds): "Overproduction of Microbial Products," London: Academic Press, p 515.
13. Queener SW, Baltz RH (1979). Genetics of industrial microbiology. In Perlman D (Ed) "Annual Reports of Fermentation Processes "Volume 3", New York: Academic Press, p 5.
14. Calam CT, Daglish CB, McCann EP (1976) Penicillin - tactics in strain improvement. In Macdonald KD(ed): "Genetics of industrial microbiology, " London: Academic Press p 273.
15. Elander RP, Espenshade MA, Pathak SG, Pan CH (1973). The use of parasexual genetics in an industrial improvement programme with Penicillium chrysogenum. In Vanek Z, Hostalek Z, Cudlin (eds): "Genetics of industrial microorganisms volume 2", Amsterdam: Elsevier, p 239.
16. Dales RBG, Croft JH (1979). Protoplast fusion and the isolation of heterokaryons and diploids from vegetative incompatible strains of Aspergillus nidulans. FEMS Microbiology Letters 1:201.
17. Dales RBG, Croft JH (1979). Protoplast fusion and the genetic analysis of vegetative incompatibility in Aspergillus nidulans. In Ferenczy L, Farkas GL (eds): "Advances in Protoplast Research," Oxford: Pergamon Press, p 73.

18. Hamlyn PF, Ball C (1979). Recombination studies with cephalosporium acremonium In Sebeck OK, Laskin AI (eds): "Genetics of Industrial Microorganisms," Washington DC: American Society for Microbiology, p 185.
19. Pontecorvo G (1976). Presidential address in Macdonald KD (ed): "Second International Symposium on the Genetics of Industrial Microorganisms," London: Academic Press, p 1.
20. Peberdy JF (1979) . New approaches to gene transfer in fungi. In Sebek OK, Laskin AI (eds): "Genetics of Industrial Microorganisms,". Washington DC: American society for Microbiology, p 192.
21. Peberdy JF (1980). Protoplast fusion - a tool for genetic manipulation and breeding in industrial microorganisms. Enz and Mic Tech 2:23.
22. Peberdy JF (1980). Protoplast fusion - a new approach to interspecies genetic manipulation and breeding in fungi. In Ferenczy L, Farkas GL (eds): "Advances in Protoplast Research" Budapest: Academiai Kiado, Oxford: Pergamon Press, p 63.
23. Hopwood DA, Chater KF (1980). Fresh approaches to antibiotic production. Phil. Trans. R. Soc. Lond. B 290:313.
24. Ball C (1980). A discussion of microbial genetics in fermentation process development with particular reference to fungi producing high yields of antibiotics. Folia Microbiologica 25:524.
25. Hopwood DA (1981) Future possibilities for the discovering of new antibiotics by genetic engineering. In Salton MRJ, Shockman GD (eds): "Mode of Action, New Developments and Future Prospects," New York: Academic Press, p 585.
26. Hopwood DA (1981) Genetic studies of antibiotics and other secondary metabolites. Symp. Soc. Gen.. Microbiol 31:187.
27. Hopwood DA (1981) Possible application of genetic recombination in the discovery of new antibiotics in actinomycetes. In Ninet L, Bost PE, Bouanchaud DH, Florent J (eds): "The Future of Antibio-therapy and Antibiotic Research," London: Academic Press,p 407.
28. Cape RE, Gelfand DH, Innis MA Neidleman SL (1982). An introduction to the present state and future role of genetic manipulation in the development of over-producing microorganisms. FEMS Symp 13: 327.

29. Elander RP (1982) Traditional vs current approaches to the genetic improvement of microbial strains FEMS Symp 13:353.
30. Fantini AA (1962). Genetics and antibiotic production of Emericellopsis species. Genetics 47: 161.
31. Macdonald KD,Hutchinson JM, Gillett WA (1963). Isolation of auxotrophs of Penicillium chrysogenum and their penicillin yields. JGM 33:365.
32. Chang LT, Terry CA (1973). Intergenic complementation of glucoamylase and citric acid production in two species in Aspergillus. Appl. Microbiol 25:890.
33. Anné J, Peberdy JF (1975). Conditions for induced fusion of fungal protoplasts in polyethylene glycol solutions. Arch Mic 105:201.
34. Ferenczy L, Kevei F, Zsolt J (1974). Fusion of fungal protoplasts. Nature 248:793.
35. Kevei F, Peberdy JF (1977). Interspecific Hybridization between Aspergillus nidulans and Aspergillus rugulosus by Fusion of somatic Protoplasts. JGM 102:255.
36. Kevei, F, Peberdy JF (1979). Induced Segregation in Interspecific hybrids of Aspergillus nidulans and Aspergillus rugulosus obtained by Protoplast fusion. Molec Gen Genet. 170:213.
37. Peberdy JF, Eyssen H, Anne J (1977). Interspecific hybridization between Penicillium chrysogenum and Pencillium cyaneo-fulvum following protoplast fusion. Molec gen Genet. 157:281.
38. Bradshaw RE, Peberdy JF (1984). Protoplast fusion in Aspergillus: selection of interspecific heterokaryons using antifungal inhibitons. J. Mic Methods 3:27.
39. Hopwood DA, Wright HM (1981). Protoplast fusion in Streptomyces: Fusions involving ultraviolet iiradiated protoplasts. JGM 126:21.
40. Ferenczy L. (1984) Fungal Protoplast Fusion: Basic and Applied Aspects. In Beers Jr RF, Bassett EG (eds): "Cell Fusions: Gene Transfer and Transformation," New York: Raven Press, p 145.
41. Ferenczy L, Perti M (1982) Transfer of Isolated Nuclei into Protoplasts of Saccharomyces cerevisiae. Curr Mic 7: 157.
42. Bradshaw RE, Lee K, Peberdy JF (1983). Aspects of Genetic interaction in hybrids of Aspergillus nidulans and Aspergillus rugulosus obtained by Protoplast fusion. JGM 129: 3525.

43. Ferenczy L (1976) Some characteristic of intra - and inter-specific protoplast fusion products of Aspergillus nidulans and Aspergillus fumigatus. In Dudits D, Farkas GL, Maliga P (eds): "Cell Genetics in Higher Plants," Budapest: Akademiai Kiado, Oxford: Pergamon Press, p 171.
44. Kevei F, Peberdy JF (1984). Further studies on protoplast fusion and interspecific hybridization within the Aspergillus nidulans group. JGM 130: 2229.
45. Anné J, Peberdy JF (1976). Induced fusion of fungal protoplasts following treatment with polyethylene glycol. JGM 92:413.
46. Anne J (1977) Somatic hybridization between Penicillium species after induced fusion of their protoplasts. Agricultura 25:1.
47. See 37.
48. Anné J, Eyssen H (1978). Isolation of inter-species hybrids of Penicillium citrinum and Penicillium cyaneo-fulvum following protoplast fusion. FEMS Microbial Lett 4:87.
49. Anné J, Eyseen H, De Somer P (1976). Somatic hybridization of Penicillium roqueforti and P. chrysogenum after protoplast fusion. Nature 262: 719.
50. Anné J (1982) Genetic evidence for selective chromasomal loss in interspecies hybrids from P. chrysogenum and P. stoliniferum. Fems Mic Lett 14: 191.
51. Anné J (1982) Comparison of penicillins produced by interspecies hybrids from P.corysogenum. Eur J. Appl Microbiol Biotechnol 15:41.
52. Hamlyn PF, Bradshaw RE, Mellon FM, Santiago CM, Wilson JM, Peberdy JF (1981). Efficient protoplast isolation from fungi using commercial enzymes. Enz Mic Tech. 3:321.
53. Hamlyn PF (1982) Protoplast fusion and genetic analysis in Cephalosporium acremonium PhD Thesis, University of Nottingham
54. Garcia Acha I, Lopez-Belmonte F, Villanueva JR (1966). Regeneration of mycelial protoplasts of Fusarium culmorum. JGM 45:515.

Molecular Genetics of Filamentous Fungi, pages 83–94

GENETIC REGULATION OF NITROGEN METABOLISM IN *NEUROSPORA CRASSA*[1]

George A. Marzluf, Kimberly G. Perrine and Baek H. Nahm

Departments of Biochemistry and Genetics
Ohio State University, Columbus, Ohio 43210

ABSTRACT The *nit-2* and *nit-4* genes of Neurospora appear to be *major* and *minor* control genes, respectively, of the nitrogen regulatory circuit. A *nit-2* amber nonsense mutation was identified by virtue of its suppression by *Ssu-1*. Neither this *nit-2* mutant nor several *nit-3* amber nonsense mutants could be suppressed by *Ssu-1* when it was present in different nuclei of a heterokaryon. The *nit-2* amber mutant completely lacks nitrate reductase activity but contains substantial amounts of two other nitrogen-regulated enzymes suggesting that the mutant encodes a truncated protein with altered regulatory properties. Uricase is synthesized at a basal level even during nitrogen repression but can be induced to about 6-fold greater levels. Induction requires *de novo* enzyme synthesis. A single uricase species is present in uninduced cells and its synthesis rate increases upon induction. *In vitro* translation of Poly(A)$^+$ RNA yields uricase of mature size, although the amount of uricase synthesized *in vitro* was approximately the same whether the template RNA was from uninduced or induced cells.

INTRODUCTION

Molecular studies of gene expression and its regulation in the filamentous fungi can be particularly informative

[1]This research was supported by grant GM-23367 from the National Institutes of Health. KGP was supported by a NSF predoctoral fellowship.

when applied to systems that have been well defined genetically and biochemically. The nitrogen regulatory circuit of Neurospora includes a set of unlinked structural genes which specify various enzymes of nitrogen catabolism (1). This complex control network coordinates the effects of major and minor control genes plus metabolic inducers and repressors to regulate the expression of the unlinked structural genes. A major regulatory gene, nit-2, mediates nitrogen catabolite repression and turns on the expression of the structural genes in a positive manner when the cells become limited for nitrogen. Glutamine appears to be the metabolite responsible for nitrogen repression (2,3).

Synthesis of the enzymes of a specific nitrogen catabolic pathway requires induction by a substrate or intermediate of the pathway, which appears to be mediated by a minor control gene specific for each pathway. Thus, nitrate and nitrite reductase synthesis requires a functional nit-2 gene product, nitrogen limitation and simultaneous induction by nitrate, which may be mediated by the minor control gene nit-4 (4,5,6). Similarly, synthesis of amino acid catabolic enzymes and of the purine catabolic enzymes each requires pathway-specific induction (7,8,9).

Although the general features of the nitrogen circuit of Neurospora are known, there still exist significant gaps in our understanding of this regulatory system. We present here new findings concerning the nit-2 and nit-4 genes that provides additional evidence that they are control genes and encode regulatory proteins, and describe the isolation and study of amber nonsense mutants of the nit-2 and nit-3 loci. We also present new results concerning uricase, a nitrogen regulated enzyme, which is dually expressed as a constitutive and a regulated enzyme.

RESULTS

Suppressible Nit Mutations.

The nit-2 gene is a major nitrogen regulatory gene and nit-4 has been suspected to be a pathway-specific control gene which mediates nitrate induction of nitrate and nitrite reductase. Both of these control genes appear to act in a positive manner and may encode regulatory proteins (1,10). Suppressors have been known for many years in Neurospora (11), but have seldom been employed to study gene structure or expression, an important exception being definitive studies with the am gene (12). Since Ssu-1 is known to suppress amber nonsense mutations by insertion of tyrosine in response to the UAG codon (13), isolation of nit-2 or nit-4 mutants which are suppressible by Ssu-1 would argue strongly that the respective gene encoded a protein, presumably one with a regulatory function.

Mutation in at least 7 different genes yield nit mutants which lack nitrate reductase and cannot grow on nitrate; such mutants are readily selected as chlorate resistant (Table 1).

TABLE 1
GENES RELEVANT TO NITRATE METABOLISM[a]

Gene	Function
nit-1	Mo-Cofactor
nit-2	Major N Regulatory Gene
nit-3	Nitrate Reductase
nit-4	Nitrate Induction
nit-6	Nitrite Reductase
nit-7	Mo-Cofactor
nit-8	Mo-Cofactor
nit-9	Mo-Cofactor

[a]Mutation in any one of these gene except nit-6 leads to chlorate resistance. Mo-Cofactor, molybdenum cofactor contained in nitrate reductase.

Complementation tests were utilized to efficently examine a large number of new chlorate-resistant mutants to identify those which were altered in the nit-2, nit-3 or nit-4 genes. In each case the identification was confirmed by growth tests (and in certain cases by crossing to the appropriate mutant). Of 67 new chlorate-resistant strains tested we identified 10 nit-2, 32 nit-3, and 6 nit-4 mutants. The remaining nit mutants, which complemented nit-2, nit-3 and nit-4, were presumed to be mutants affecting the molybdenum cofactor and were not studied further.

Since Ssu-1 encodes a suppressor tRNA, which should be freely diffusible in the cytoplasm, one might expect that suppression by Ssu-1 would be evident in a heterokaryon. Thus, we tested each new nit-2, nit-3, and nit-4 mutant for suppression in a heterokaryon of the form Ssu-1 nit-2^u am + inl nit-2 where nit-2^u is a non-suppressible allele and nit-2 is the new mutant being tested for suppression. No case was observed where Ssu-1 in one nucleus was able to suppress a nit mutant in the other nuclear type of a heterokaryon. Each new nit-2, nit-3 and nit-4 mutant was also crossed with the corresponding Ssu-1 strain, which resulted in the identification of one suppressible nit-2 and four suppressible nit-3 (but no nit-4) mutants. The resulting Ssu-1 nit-2 strain, when crossed with wild-type, segregated approximately 25% nit progeny, which were shown to be nit-2 by several tests; the Ssu-1 nit-3 strains similarly segregated authentic nit-3 mutant progeny. These results imply that the new suppressible nit-2 and nit-3 mutants are amber nonsense mutations, based upon the known specificity of Ssu-1 (13). When these suppressible nit-2 and nit-3 mutants were re-tested in heterokaryons with Ssu-1, they consistently failed to show suppression. Thus, we have concluded that suppression in heterokaryons is not efficient or reliable and cannot be employed to identify new suppressible mutants, which is in agreement with results of Davis and Weiss (14).

Biochemical Analysis of Suppressible Strains.

We examined the growth rate and enzyme profile of the suppressible nit-2 mutant (allele KGP 0220), Ssu-1 nit-2, and the parental nit-2^+ strain. This nit-2 mutant cannot grow at all on nitrate or phenylalanine but grows at approximately 50% of the wild-type rate on uric acid. The Ssu-1 nit-2 strain grows at the wild-type rate on both nitrate and uric acid and at about 45% of the wild-type rate on phenylalanine. The nit-2 mutant lacks any detectable nitrate reductase activity, whereas Ssu-1 nit-2 possess 48% of the wild-type level (Fig. 1). Both the nit-2 and Ssu-1 nit-2 strains have about 28% of the wild-type amount of L-amino acid oxidase and possess a normal level of uricase (Fig. 1). It is obvious from these results that the inability of nit-2 to grow on phenylalanine is not due to its decreased content of L-amino acid oxidase since Ssu-1 nit-2 has a similar amount of this enzyme and yet grows at a rate approaching normal. Thus, other activities required

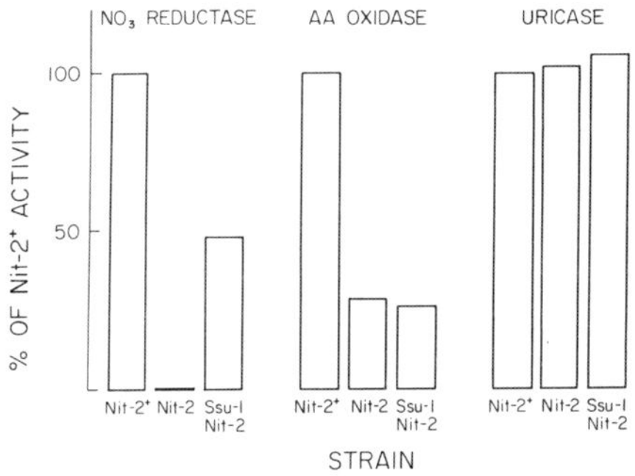

FIGURE 1. Enzyme activities present in nit-2^+, nit-2, and Ssu-1 nit-2 strains grown in each case under optimal conditions for derepression and induction. AA Oxidase, L-amino acid oxidase. The specific activity for each enzyme in nit-2^+ is assigned the value of 100% and all others are relative to it.

for amino acid catabolism, such as amino acid permeases, must be deficient in nit-2 but restored in the suppressed strain, Ssu-1 nit-2.

One suppressible nit-3 mutant (allele KGP1222) has been examined to date and possesses no detectable nitrate reductase. The corresponding Ssu-1 nit-3 strain has only about 26% of the amount of enzyme present in the parental strain. Moreover, the nitrate reductase present in Ssu-1 nit-3 is qualitatively different from the wild-type enzyme as judged by heat inactivation studies, and appears to be more thermostable than the wild-type enzyme (Fig. 2). This result suggests that the tyrosine inserted by Ssu-1 in response to the amber nonsense codon constitutes a new amino acid residue at that position in the nitrate reductase. Fig. 2 also reveals an unexpected result: a one-minute incubation of the extract at 43°C resulted in an activation, yielding approximately a 50% increase in nitrate reductase activity. Longer incubations at 43°C lead to the loss of enzyme activity. Preincubation of the enzyme for 10 min at 30°C does not give a similar activation.

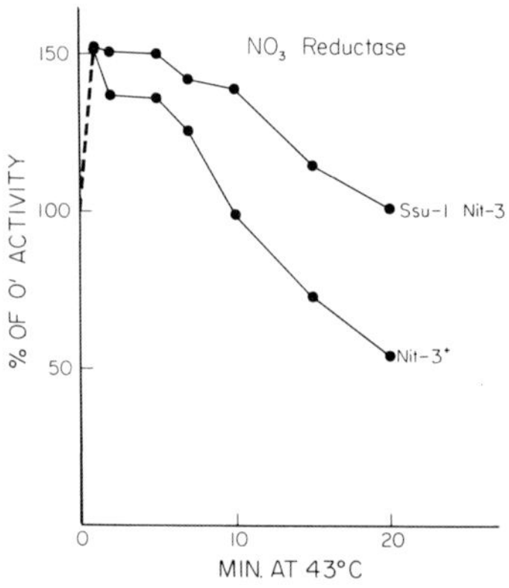

FIGURE 2. Heat inactivation of nitrate reductase from nit-3+ and Ssu-1 nit-3 strains. After incubation of aliquots of the extracts at 43°C for the indicated times, the samples were cooled in a ice bath and then assayed for remaining nitrate reductase activity.

Revertant Analysis.

We have isolated over 200 revertants each for nit-2 and nit-4 mutants in order to further define the characteristics of these putative control genes. It is particularly informative that two nit-2 revertants and one nit-4 revertant display a high degree of constitutive nitrate reductase activity; for example nit-2 (allele KGP027) revertant 40 displayed approximately 30% of the fully induced wild-type level of nitrate reductase during growth on medium containing nitrate and ammonium ion, a condition where wild-type lacks any detectable enzyme activity. Similarly, nit-4 (allele KGP 25) revertant 27 has approximately 16% of the fully induced wild-type level when grown under the same repressing conditions. The two nit-2 revertants with constitutive behavior require nitrate induction for expression of nitrate reductase but are insensitive to nitrogen repression. In contrast, the nit-4 revertant requires neither induction nor nitrogen derepression for enzyme synthesis. These results strongly reinforce previous suggestions that the nit-2 and nit-4 genes are regulatory. The non-repressible phenotype of the two nit-2 revertants is fully consistent with the proposed role for nit-2 in mediating nitrogen catabolite repression

(1). The fact that the constitutive *nit-4* revertant does not require induction agrees with its postulated role in nitrate induction (4,5,6), although its insensitivity to N-repression was unexpected.

Biochemical Genetics of Uricase.

Uricase, a purine catabolic enzyme composed of 4 identical subunits of molecular weight approximately 33,000 each, is another member of the nitrogen control circuit whose synthesis is regulated by the *nit-2* gene. However, unlike nitrate reductase, which is virtually undetectable in non-induced, repressed cultures, a significant basal level of uricase is found in cells grown under fully repressing conditions. Upon induction and derepression, uricase activity increases approximately 6-fold in *nit-2*$^+$ but declines from the basal level to an undetectable amount in *nit-2* mutant cells. Thus, we wanted to determine whether multiple, differentially regulated forms of uricase existed in Neurospora. It also seemed important to determine whether the increase in uricase activity "induced" by uric acid represented only an activation of the pre-existing enzyme or whether it required *de novo* protein synthesis.

The results displayed in Fig. 3 clearly demonstrate that Neurospora possesses only a single uricase species, and that upon induction, the increase in enzyme activity was paralleled by a corresponding increase in the synthesis of uricase. It appears conclusive that uricase induction reflects *de novo* enzyme synthesis. The uricase structural gene represents a clear case where a considerable basal level of constitutive enzyme expression can be further elevated by the appropriate inductive signals. It is also noteworthy that the overall pattern of proteins synthesized by wild-type cells grown under repressed and induced conditions was virtually identical (Fig. 3).

In Vitro Translation of Uricase.

Poly(A)$^+$ RNA was isolated from wild-type cells repressed and from cells induced for uricase synthesis. These RNAs were translated *in vitro* as described in Fig. 4. The spectrum of total protein synthesized *in vitro* with poly(A)$^+$ RNA from control and induced cells as template was nearly identical. Immunoprecipitation of uricase from these *in vitro* translations showed that the same single uricase

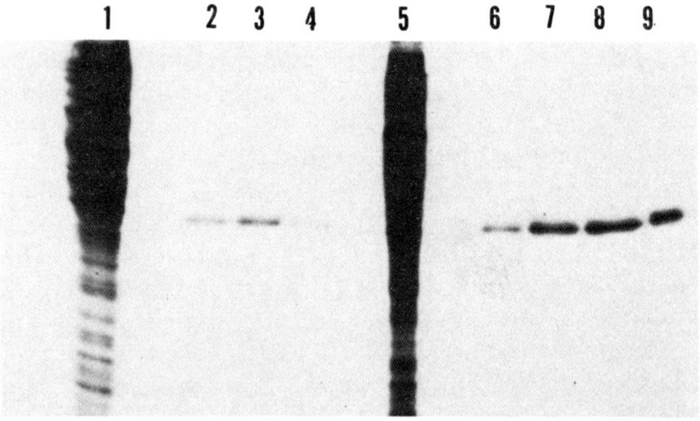

FIGURE 3. *De novo* synthesis of uricase. Mycelia from an overnight *wild-type* Neurospora culture was harvested and one-half was then incubated in repressing (high ammonium ion) medium and the other half in inducing (uric acid) medium. In each case, any new proteins being synthesized were labeled with 5 uCi/ml $^{35}SO_4$ for 2, 4, and 6 hours. Cellular proteins were extracted and electrophoresed in a 12.5% polyacrylamide/0.1% SDS gel. Lanes 1,5 show total proteins synthesized at 2 hours (50,000 cpm of incorporated ^{35}S in each case) by cells grown in control and induced medium, respectively. Lanes 2, 3 and 4 reveal uricase immunoprecipitated from control cells incubated for 2, 4, and 6 hours. Lanes 6, 7 and 8 show uricase immunoprecipitated from induced cells for these same times. In each case uricase was immunoprecipitated from 500,000 cpm of incorporated ^{35}S. Lane 9, uricase standard.

subunit species was synthesized from RNA isolated from control and induced cells, and corresponds to a subunit of mature uricase (Fig. 4). Surprisingly, when uricase was immunoprecipitated from proteins translated *in vitro* (using equal amounts of ^{35}S-labeled proteins), no increase in the amount of uricase was evident in translations with Poly(A)$^+$ RNA from induced cells as compared to that from control cells. Similar results were obtained in several replicates of this experiment. It is of obvious significance that uricase is almost completely missing in the translation products of Poly(A)$^+$ RNA isolated from induced *nit-2* mutant cells; in contrast, uricase is readily translated from Poly(A)$^+$ RNA of repressed, uninduced *nit-2* cells (Results not shown).

Uricase is one of several peroxide-producing enzymes which are localized within peroxisomes. Accordingly, it might be anticipated that its synthesis or entry into peroxisomes might involve a processing step. However, the addition of microsomal membranes, which were active in the processing of pre-lactogen to its mature form, did not cause any alteration in uricase (Results not shown). We tentatively conclude that uricase is synthesized in its mature form and is not subject to any processing, although more information in this context would be useful.

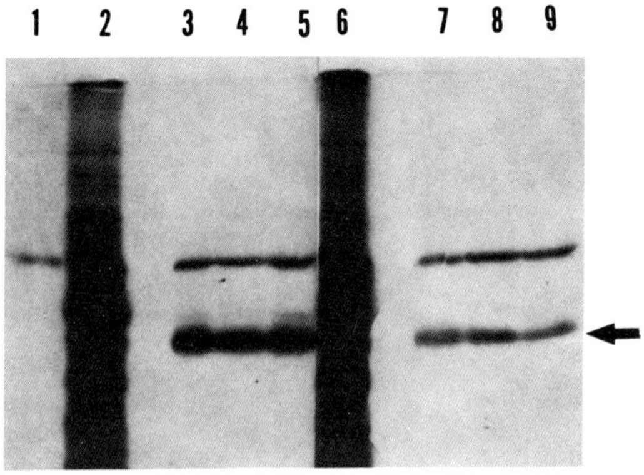

FIGURE 4. In vitro translation of uricase. Poly(A)$^+$RNA was isolated from wild-type cells either repressed or derepressed and induced for uricase for 2, 4, and 6 hours using cultures handled similarly to those described in Fig. 3. Each translation reaction mixture (25 ul total volume) contained 50 uCi of ^{35}S-methionine, 5 ug of Poly(A)$^+$RNA and the components of a rabbit reticulocyte system (New England Nuclear). Lane 1, no mRNA added; lanes 2,6 show total proteins synthesized from 2 hour RNA isolated from control and induced cells, respectively. Lanes 3, 4, and 5 show immunoprecipitated uricase from 500,000 cpm of incorporated ^{35}S-methionine translated from RNA of control cells at 2, 4, and 6 hours. Lanes 7, 8 and 9 show uricase similarly precipitated from translations with RNA from induced cells at 2, 4, and 6 hours. The arrow identifies uricase.

DISCUSSION

We describe here the isolation and characterization of an amber nonsense mutation in nit-2, a major regulatory gene of Neurospora. This nit-2 mutant (allele KGP0220) was identified as a nonsense mutation by virtue of its suppression by Ssu-1, known to insert a tyrosine residue in response to the amber UAG codon. This finding implies that the nit-2 regulatory gene product is a protein, in agreement with a previous suggestion (10). The availablity of this nonsense mutant, which should lack a full length nit-2 coded protein, and the Ssu-1 nit-2 strain which should regain the normal sized protein, should be invaluable in any attempt to directly identify this important regulatory protein. The nit-2 amber mutant was selected for chlorate resistance and, not unexpectedly, completely lacks nitrate reductase, which is largely restored by Ssu-1. However, in other properties, this nit-2 nonsense mutant does not act at all as a null mutation, e.g., it possesses a normal level of uricase and approximately 30% of the wild-type level of L-amino acid oxidase, whereas other nit-2 mutant strains have only a basal level of uricase and completely lack the oxidase (7,10). These features suggest that the suppressible mutant contains the amber codon near the end of the nit-2 coding region, yielding a truncated protein with altered regulatory properties. This truncated nit-2 protein appears to be able to "turn on" the expression of uricase and L-amino acid oxidase but not nitrate reductase; this outcome implies that the control region of each of the structural genes differs in its recognition by the nit-2 protein, and that the presumed loss of residues near the carboxy terminus somehow prevents a productive interaction between the mutant nit-2 protein and control sequences of the nit-3 gene.

Recent studies have suggested the possible existence of a second major nitrogen control gene, designated MS5, which might be responsible for recognizing the nitrogen status of the cell and controlling the expression or the activity of the nit-2 gene product (15,16). A question of considerable importance is the nature of the possible interaction between MS5 and nit-2. Very little is known about the nit-4 gene, other than it is suspected to have control function. Although 6 new nit-4 mutants were examined, none of them were suppressible by Ssu-1. The fact that a nit-4 revertant was obtained which displays constitutive nitrate reductase synthesis in the absence of nitrate adds support to suggestions (4,5,6) that the *nit-4* gene product mediates nitrate induction of nitrate and nitrite reductase.

Uricase is present at a basal level in wild-type and in *nit-2* cells growing under nitrogen-repressed conditions. Upon derepression and induction, uricase content increases approximately 6-fold in wild-type cells and clearly represents *de novo* synthesis of the enzyme. In contrast, the uricase level declines radically from the basal level when *nit-2* cells are placed under derepressed, induced conditions. Only the same single uricase species is present in repressed and in induced cells. After immunoprecipitation from *in vitro* translations with Poly(A)$^+$RNA, uricase can be readily detected in an apparently mature form of correct size. It was surprising that the amount of uricase translated *in vitro* was approximately the same, regardless of whether the Poly(A)$^+$ RNA came from induced or uninduced wild-type cells (or from uninduced *nit-2* cells). Yet, translation of Poly(A)$^+$ RNA from induced *nit-2* cells gave only a greatly decreased, barely detectable uricase band. It should be noted that if instead total RNA or Poly(A)$^-$ RNA were employed, the same type of results were obtained.

Although various explanations are obviously possible, one interpretation which could account for these results is that the uricase gene can be transcribed from one of two alternative promoters. Accordingly, under repressing conditions, the uricase gene might be transcribed only from a low level promoter, which would be active in both wild-type and *nit-2* cells. Under derepressing, inducing conditions, uricase gene expression might originate only at a second, regulated promoter to yield a different sized mRNA but with an identical uricase coding region. Under inducing conditions, both of these promoters would presumably be inactive in the *nit-2* mutant. If the mRNA species transcribed from the putative regulated promoter were poorly translated *in vitro* (in the rabbit reticulocyte system used for these experiments), all of the results would be compatible with this possible explanation. A considerable amount of experimental evidence will be required to determine whether this explanation is plausible. A molecular analysis of the uricase gene and its RNA transcript(s) should prove to be highly informative.

REFERENCES

1. Marzluf GA (1981). Microbiol. Rev. 45:437.
2. Premakumar R, Sorger GJ, Gooden D (1979). J. Bacteriol. 137:1119.
3. Wang LC, Marzluf GA (1979). Mol. Gen. Genet. 176:385.
4. Bahns M, Garrett RH (1980). J. Biol. Chem. 255:690.
5. Premakumar R, Sorger GJ, Gooden D (1978). Biochim. Biophys. Acta 519:275.
6. Tomsett AB, Garrett RH (1980) Genetics 95:6490.
7. Sikora L, Marzluf GA (1982). Mol. Gen. Genet. 186:33.
8. Sikora L, Marzluf GA (1982). J. Bacteriol. 150:1827.
9. Reinert WR, Marzluf GA (1975). Mol. Gen. Genet. 139:39.
10. Grove G, Marzluf GA (1981). J. Biol. Chem. 256:463.
11. Seale TW (1967). Genetics 56:586.
12. Seale TW (1968). Genetics 58:85.
13. Fincham, JRS (1984). Neurospora Newslett. 31:12.
14. Davis, RH, Weiss, RL (1983). Molec. Gen. Genet. 192:46.
15. Chambers, JAA, Griffon SM, Marzluf GA (1983). Current Genetics 7:51.
16. DeBusk RM, Ogilvie S (1984). J. Bacteriol. 160:656.

Molecular Genetics of Filamentous Fungi, pages 95–115

GENETIC REGULATORY MECHANISMS IN THE QUINIC ACID (QA) GENE CLUSTER OF NEUROSPORA CRASSA[1,2]

Norman H. Giles, Mary E. Case, James Baum, Robert Geever, Layne Huiet[3], Virginia Patel, and Brett Tyler[4]

Department of Genetics, University of Georgia
Athens, Georgia 30602

ABSTRACT The quinic acid gene cluster of *Neurospora crassa*, which controls the inducible utilization of quinic acid as a carbon source, has been cloned on a 17.6 kb segment of linkage group VII. Combined transformation and transcription experiments indicate that this cluster comprises seven genes transcribed as distinct mRNAs. Three of the genes (*qa-2*, *qa-3*, and *qa-4*) encode inducible *qa* enzymes while two other presumptive structural genes (*qa-x* and *qa-y*) have unknown functions. Genetic and molecular evidence indicates that there are two *qa* regulatory genes which encode, respectively, a repressor protein (*qa-1S*) and an activator protein (*qa-1F*). Two types of mutants occur in the *qa-1S* gene, semi-dominant, non-inducible (*qa-1S*$^-$) and recessive, constitutive (*qa-1S*^{c}), while in the *qa-1F* gene only recessive, non-inducible mutants (*qa-1F*$^-$) have been detected. Transcriptional evidence utilizing *qa-1S* and *qa-1F* mutants suggests that *qa-1F* acts positively in controlling transcription of itself (autogenous regulation) and of the

[1]This work was supported by NIH Grant GM28777 to N.H.G. and M.E.C.

[2]The names of authors, other than the first two, are in alphabetical order.

[3]Present address: CSIRO Plant Industry, P.O. Box 1600, Canberra City, A.C.T. 2601 AUSTRALIA.

[4]Present address: Department of Genetics, Research School of Biological Sciences, Australian National University, P.O. Box 475, Canberra City, A.C.T. 2601 AUSTRALIA.

other qa genes. The entire qa cluster has been sequenced and open reading frames identified for each of the seven gene products. The two regulatory genes encode proteins of ca. 100,650 (repressor) and ca. 88,960 (activator) daltons. S1 mapping data have identified major mRNA initiation sites and the direction of transcription for each qa gene. Two cloned and sequenced $qa\text{-}1S^-$ mutants contain missense mutations in the C-terminus of the qa-1S coding region, presumably implicating the carboxy terminal region of the repressor in inducer binding. The function of the qa-1F gene product (activator) has been studied in several qa-2 activator-independent ($qa\text{-}2^{ai}$) mutants and three temperature-sensitive activator mutants ($qa\text{-}1F^{ts}$). Based on sequence data, the $qa\text{-}2^{ai}$ mutants possess one of a variety of mutations, all in the 5' untranscribed region of qa-2. Several of these mutations have been shown to exhibit certain characteristics associated with enhancer elements. These studies have distinguished two types of requirements for transcription initiation of the qa genes -- one type apparently requires RNA polymerase II access while the second type apparently requires direct binding of activator. Furthermore, studies of DNase I hypersensitive sites show that the $qa\text{-}1F^+$ genotype is directly associated with changes in chromatin structure in the upstream region of qa-2. Data obtained in transformation experiments utilizing truncated qa-1F donor DNA lacking all 5' untranscribed, presumptive regulatory sequences eliminate an operator site 5' to the +1 mRNA initiation site for the $qa\text{-}1F^+$ gene as the sole target for the repressor and raises the possibility (but do not yet establish) that regulation in the qa system involves, in part, interactions between the two regulatory proteins. The recent development of a homologous *in vitro* transcription system for polymerase II should aid substantially in the further dissection of regulation in the qa-system at the molecular level, especially if current efforts to obtain significant production of the two regulatory proteins in expression vectors are successful.

INTRODUCTION

The quinic acid (qa) gene cluster of Neurospora crassa represents an unusual, perhaps unique, situation in eukaryotes in that apparently all the regulatory and structural genes concerned with the utilization of a single metabolite, quinic acid, as a carbon source are located in a tightly-linked cluster. The utilization of quinic acid as a sole carbon source was first studied genetically by the selection of mutants unable to utilize this substrate. The resulting mutants were initially classified into four groups by complementation and genetic crossing studies (1-3). Three of these mutant groups (qa-2, qa-3, qa-4) were shown to represent single structural genes, each encoding one of three separate, quinic acid-inducible enzymes involved in the conversion of quinic acid to protocatechuic acid (Fig. 1). The fourth group (qa-1) consisted of pleiotropic mutants, non-inducible for all three enzymes and was originally hypothesized to represent a single, positively-acting regulatory gene, presumably encoding an activator protein, which acted to turn on transcription of

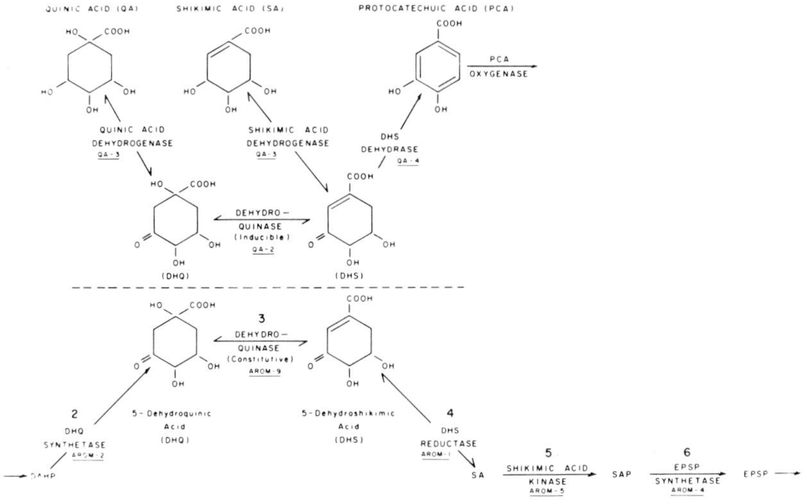

Figure 1. Diagram of gene-enzyme relationships in the inducible quinate-shikimate catabolic pathway (shown above the dotted line) in N. crassa. Also indicated (below the dotted line) are reactions in the related constitutive common aromatic synthetic pathway catalyzed by enzymes encoded by the arom cluster-gene.

the three structural genes. However, these initial studies also indicated that *qa-1* mutants comprised two quite distinct groups, mapping genetically in two non-overlapping regions of the presumptive single gene. *Qa-1*$^{F-}$ mutants were non-inducible and recessive in heterokaryons, while *qa-1*^{S} mutants consisted of two types: recessive, constitutive mutants (*qa-1*$^{S^c}$), and semi-dominant, non-inducible mutants (*qa-1*$^{S^-}$) (4). Crossing analyses indicated that the four genes were arranged in the order *qa-1*, *qa-3*, *qa-4*, *qa-2*, as a tightly-linked cluster very close to the centromere on the right arm of linkage group VII (5). Subsequent biochemical studies resulted in the purification of the three enzymes [quinate dehydrogenase (shikimate dehydrogenase) - *qa-3*; catabolic dehydroquinase - *qa-2*; and dehydroshikimate dehydratase - *qa-4*] encoded by the three *qa* structural genes, and the partial determination of amino acid sequences for the three proteins (6-8).

Definitive studies of the *qa* gene cluster at the molecular level became possible with the cloning of the entire cluster in *E. coli*. Initially, the *qa-2*$^{+}$ gene encoding catabolic dehydroquinase was cloned on a recombinant plasmid in *E. coli* by selecting for complementation with an *E. coli* auxotroph (*aroD*$^{-}$) lacking biosynthetic dehydroquinase. This experiment was attempted because earlier studies had shown that in *N. crassa* catabolic dehydroquinase could substitute for biosynthetic dehydroquinase (3) (Fig. 1). Physical and immunological tests of the dehydroquinase produced in *E. coli* demonstrated unequivocally that these hybrid plasmids contained the *N. crassa* *qa-2*$^{+}$ gene expressed with fidelity in *E. coli* (9).

The cosmid technique, combined with selection for *qa-2*$^{+}$ expression in *E. coli*, resulted in the cloning of ca. 42 kb of contiguous DNA sequence flanking the *qa-2*$^{+}$ gene (10). Although neither the *qa-3*$^{+}$ gene nor the *qa-4*$^{+}$ genes were expressed in *E. coli* (no test was available for *qa-1*$^{+}$), these additional *qa* genes were localized by transforming appropriate *N. crassa* double mutants using subclones containing the *qa-2*$^{+}$ gene and various flanking segments adjacent to *qa-2*$^{+}$ according to a procedure developed by Case *et al*. (11). Subsequent experiments showed that the two distinct groups of *qa-1* mutants were transformed by different and adjacent segments of DNA, suggesting that the *qa-1* region comprised two distinct genes, rather than distinct regions of a single gene (12). These results were confirmed when divergently transcribed mRNAs were identified, and the two genes have been designated qa-1F and qa-1S (12).

Analysis of qa mRNAs provided evidence that regulation of gene expression in this gene cluster occurs primarily at the level of transcription (13). Furthermore, the analysis of various qa pleiotropic mutants established that both the product of the qa-1F$^+$ gene and the inducer quinic acid are required for transcription, and that the qa-1S gene product is also involved in controlling this process.

QA mRNA STUDIES

Initial DNA-RNA hybridization studies and subsequent S1 nuclease mapping studies utilizing cloned qa genes as probes, established that all the genes in this cluster are coordinately transcribed as separate mRNAs rather than as a polycistronic mRNA, despite the close clustering of these genes (13-15). Surprisingly, the qa mRNA studies also identified two additional, previously undetected, genes in the qa cluster (designated qa-x and qa-y), based on the production by specific DNA segments within the qa cluster of two distinct additional major mRNA species under quinic acid and qa regulatory gene control (Fig. 2). The functions of these two genes are not yet known, but are being sought on the basis of gene inactivation studies which hopefully will produce mutants whose phenotypes can be characterized. Our present conjecture is that one of the genes may encode a permease involved in quinic acid (and possibly chlorogenic acid) uptake and the other an enzyme involved in the conversion of chlorogenic acid to quinic acid and caffeic acid.

Extensive mRNA mapping studies (15; Tyler *et al*., unpublished data) have identified the major transcription initiation sites and direction of transcription for each qa gene, and have demonstrated that transcription initiation is typically heterogeneous. In addition to this heterogeneity, most of the qa genes also promote transcription from other upstream regions giving rise to minor mRNA species, several of which appear to involve a mechanism of control quite distinct from that of the major inducible transcripts, as will be discussed subsequently.

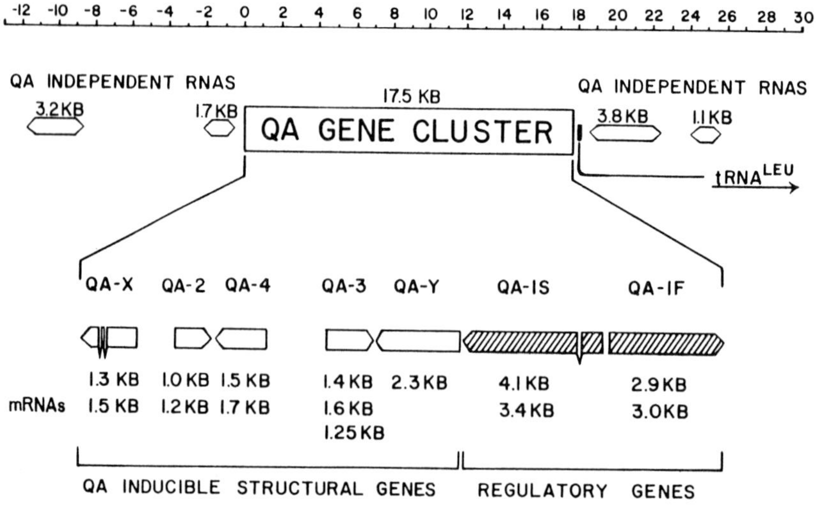

Figure 2. Transcriptional map of the 42 kb region containing the _qa_-gene cluster and adjacent regions. Seven adjacent mRNA transcripts show induction by quinic acid and define the 17.5 kb region of the _qa_-gene cluster. Other mRNAs, including the immediately adjacent $tRNA^{leu}$, show no induction by quinic acid and therefore are not part of the cluster. The direction of transcription for each of the _qa_ genes is indicated, as well as the size of the principal mRNA transcript (immediately below each gene), plus additional minor mRNA transcripts, detected by DNA-RNA blot hybridization. The positions of short introns in _qa-x_ (two) and _qa-1S_ (one) are indicated.

DNA SEQUENCE OF THE _QA_ CLUSTER

The nucleotide sequence of the entire _qa_ gene cluster has been determined (14,16-19; Geever _et al._, unpublished). The cluster comprises a segment of ca. 17.6 kb containing seven closely adjacent genes.

A summary of the data compiled from DNA sequencing and mRNA analysis is presented in Table 1. Proteins are presumed to begin at the first in-frame AUG initiation codon. These assignments for _qa-2_, _qa-3_ and _qa-4_ are confirmed by previously determined amino acid sequence data cited in the Introduction. The locations of the small introns in _qa-x_

Table 1. Characteristics of Neurospora crassa Qa Genes

Gene	mRNA species[a] (kb)	Transcription initiation regions (bp), Type[b]	Untranslated sequence (bp)[c] 5'	3'	Coding region[d] length	(MW)	Additional Characteristics
qa-x	1.3 1.45	+1 II[e] -26 II[e]	86	30	340	(37,330)	69 and 74 bp introns, heterogeneous processing
qa-2	1.0 1.2	+1 I -45 II -400 II	86	210	173	(18,270)	Native enzyme is a homododecamer, protein lacks tryptophan
qa-4	1.5 1.7	+1 II -40 II -174 II	206	71	359	(40,500)	
qa-3	1.25 1.4 1.6	+126 I +1 I -335 II -487 II	91	244	321	(35,210)	
qa-y	2.1	+94 I +1 I	137	227	537	(60,110)	Hydrophobic protein, lacks histidine
qa-1S	3.4 4.1	+1 II -51 II	408	47 730	918	(100,650)	Negative regulatory protein, 66 bp intron, GC donor splice junction
qa-1F	2.9	+1 I -65 I	329	40	816	(88,960)	Positive regulatory protein

[a] Poly(A)$^{+}$RNAs detected by DNA/RNA blot hybridizations. Estimated length in kilobases.

[b] Approximate regions of transcription initiation are given in base pairs relative to the major initiation site designated as +1. Nuclease mapping studies suggest that several distinct transcripts are initiated from each region. In addition, each transcription initiation region is distinguished by Type based on the analysis of certain transcriptional mutants (see text).

[c] 5' untranslated sequence lengths are shown only for the major +1 initiation site. Lengths of 3' untranslated sequences are approximate determinations from S1 end-mapping. Most 3' termini exhibit considerable heterogeneity.

[d] Coding regions are predicted from open reading frames presumed to begin at the first in-frame AUG codon.

[e] Qa-x transcription appears to be subject to distinct controlling mechanisms governed by qa-1S and glucose repression.

and qa-1S were determined by sequencing their mRNAs (Geever and Huiet, unpublished).

Particularly noteworthy is the evidence that the predicted molecular weights of the two presumptive regulatory proteins -- the repressor (ca. 100,650) and the activator (ca. 88,960) are both quite large. In this respect, the qa-1F activator protein resembles in size the 99,350-dalton Saccharomyces cerevisiae GAL4 activator protein (20) rather than the typically smaller activator proteins in prokaryotic systems. However, the qa-1S repressor protein is twice the size of the comparable GAL80 repressor protein (48,300 daltons) (21).

GENE REGULATION IN THE QA CLUSTER

Regulation of mRNAs for the Qa Regulatory Genes.

Studies at the mRNA level have established that the positively-acting regulatory gene qa-1F^{+} is autoregulated (13,22). In wild-type this gene makes a low basal level of qa-1F-specific mRNA which increases 50-fold upon induction, an increase which does not occur in non-inducible qa-1F^{-} mutants. The qa-1S gene is also involved in qa-1F^{+} expression since qa-1S^{c} (constitutive) mutants make high levels of qa-1F^{+} mRNA in the absence of inducer, while there is no increase in the basal level of qa-1F^{+} mRNA produced by qa-1S^{-} (non-inducible) mutants in either the presence or absence of inducer (22). Thc qa-1S gene also controls, probably indirectly via the qa-1F^{+} gene, the synthesis of its own mRNA, since the low basal level of qa-1S-specific mRNA present in non-induced wild-type cultures increases ca. 40-fold upon induction, whereas this level remains unchanged in qa-1S^{-} (non-inducible) mutants grown in the presence of inducer. Also, the level of qa-1S mRNA in a qa-1S^{c} (constitutive) mutant is much higher than in uninduced wild-type even when this mutant is grown under non-inducing conditions (12). It is also clear that the qa-1F gene is involved in the expression of the qa-1S gene since qa-1F^{-} (non-inducible) mutants make only low basal levels of qa-1S^{+} mRNA (12). The current interpretation of the interrelationships of these two proteins in regulating qa gene expression will be discussed in the following section.

General Model for Gene Regulation in the Qa Gene Cluster.

The current model hypothesizes that there are two different regulatory molecules acting in this system -- a negatively-acting repressor protein produced by the qa-1S^{+} gene, and a positively-acting activator protein produced by the qa-1F^{+} gene. The actions of these two molecules, plus that of the inducer, quinic acid, serve to regulate the utilization of quinic acid as a carbon source by *N. crassa*. The model in its simplest form postulates that the qa-1S product, the repressor, controls the expression of the qa-1F^{+} gene. In wild-type (qa-1S^{+} qa-1F^{+}) in the absence of inducer, the repressor prevents the expression of qa-1F^{+} and qa gene products are produced only at low (basal) levels. The addition of inducer releases the inhibition of qa-1F^{+} expression by repressor, whereupon the activator initiates its own synthesis (autoregulation), as well as the syntheses of all other qa mRNAs (transcriptional control). Surprisingly, the activator also stimulates the synthesis of repressor. Presumably, this serves as an economizing mechanism to ensure that ample repressor is present to rapidly turn off the system once inducer levels decline.

Several lines of evidence bear on this interpretation. Two types of qa1S mutants have been identified, the frequent class of recessive, constitutive (qa-1S^{c}), and the infrequent class of semi-dominant, non-inducible mutants (qa-1S^{-}). In qa-1S^{c} mutants, constitutive transcription of all qa genes occurs in the absence of inducer, presumably due to repressor inactivation. In qa-1S^{-} mutants, only basal levels of qa transcription are observed even in certain double mutants (qa-1S^{-}, arom-9^{-}) where the internal inducer, DHQ, accumulates as a result of the mutational block in the arom pathway (Fig. 1). This suggests that qa-1S^{-} mutants act as superrepressors insensitive to inducer inactivation. Furthermore, all qa-1S^{-} mutants revert spontaneously to constitutivity. Only one type of qa-1F mutant has been identified, the class of recessive, non-inducible mutants (qa-1F^{-}). In these mutants, only basal levels of qa transcription are observed both in internally induced double mutants (qa-1F^{-}, arom-9^{-}) and in the triple mutant qa-1F^{-}, qa-1S^{c}, arom-9^{-}. Genetically, recessive qa-1F^{-} mutants are epistatic to both qa-1S^{c} and qa-1S^{+}, suggesting that a functional activator protein encoded by qa-1F^{+} is required for qa gene transcription.

A secondary feature of qa gene regulation involves the competition for carbon source. Neurospora cultures grown in the presence of both a good carbon source, such as sucrose, and a poor carbon source, quinic acid, show low levels of qa structural gene expression. Preliminary evidence suggests that this suppression of induction by sucrose is indirect, perhaps as a consequence of catabolite exclusion. Studies in part derived from the analysis of certain constitutive qa-2 transcriptional mutants (see below) show that the relative abundance of qa-2 mRNA (as well as the other qa mRNAs) does not depend on the choice of carbon source. However, in this respect, regulation of the qa-x gene appears exceptional. The relative abundance of qa-x mRNA is elevated approximately 20-fold when cultures are shifted to carbon-limiting growth conditions, suggesting a direct effect of sucrose repression on qa-x transcription.

Mechanism of *Qa* Repressor Action.

What is the target of the repressor protein? Initially, on the simplest hypothesis, the most probable target for the repressor appeared to be an operator sequence presumably located in the controlling sequence 5' to the *qa-1F*$^+$ gene. Accordingly, in uninduced wild-type cultures the repressor would directly block *qa-1F*$^+$ transcription and thereby indirectly repress expression of the *qa* genes. Upon induction, the interaction of the inducer with the repressor would release repressor binding to the *qa-1F*$^+$ operator permitting autogeneous synthesis of activator. The increased levels of activator would then stimulate transcription of the other *qa* genes. However, recent experimental evidence has led to an important revision of this initial view concerning the target of the repressor. A transformant of a stable *qa-1F*$^-$ recipient strain (mutant 158) has been obtained utilizing donor *qa-1F*$^+$ DNA which lacks all 5' untranscribed sequences as well as the first 104 bp of the 329 bp untranslated leader sequence (Case *et al.*, unpublished). Genetic and biochemical studies indicate that the inserted, truncated *qa-1F*$^+$ gene is not linked to the *qa* gene cluster, that the transformant produces a low level of catabolic dehydroquinase in the absence of inducer, and that this activity is subject to strong induction by quinic acid. S1 nuclease studies have shown that *qa-1F*$^+$ mRNA from the donor DNA is being produced constitutively from a new promoter and is not subject to induction (Fig. 3). By contrast, *qa-1F*$^-$ mRNA (Fig. 3), as well as mRNAs from the other *qa* genes (not shown), is subject to strong induction.

These overall results clearly indicate that the sole target of the repressor cannot be an operator site located in the sequence 5' to the initiation site of *qa-1F* mRNA. The results also exclude the possibility of a single repressor target located in the first 104 bp (ca. one-third) of the *qa-1F* leader sequence, but do not exclude the possibility of a repressor target in the remaining 225 bp (ca. two-thirds) of the leader sequence. Experiments currently underway include constructs in which the entire 5' and leader sequences of the *qa-1F*$^+$ gene will be replaced by the corresponding sequences of the constitutively expressed *am* gene (23). It is anticipated that transformation experiments with these constructs should indicate if there is a repressor target in the remaining part of the *qa-1F*$^+$ leader sequence.

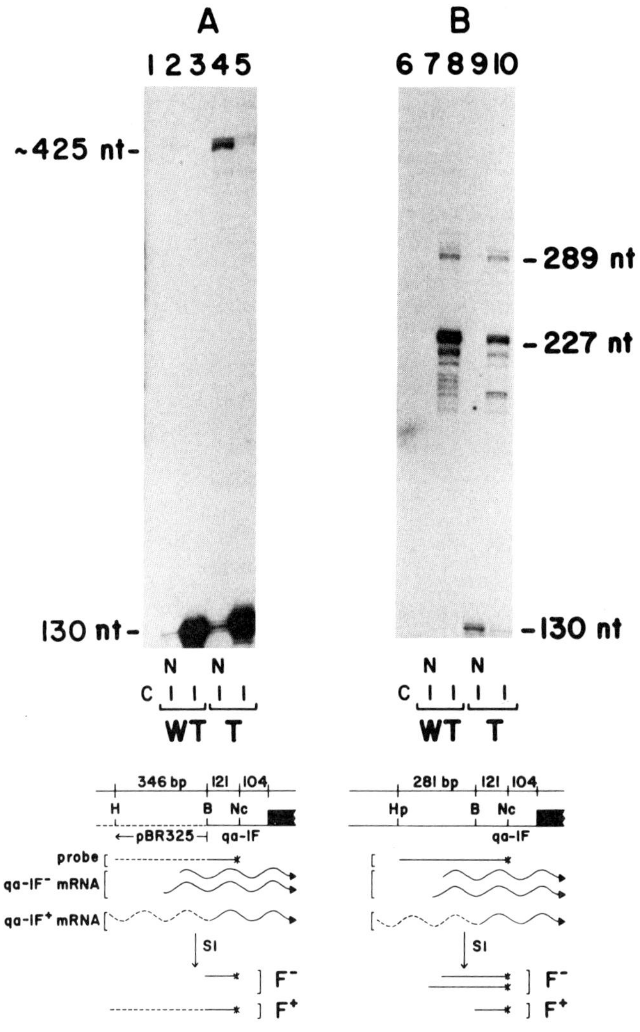

Figure 3. Nuclease S1 protection mapping of *qa-1F*$^+$ and *qa-1F*$^-$ transcripts in a recipient *qa-1F*$^-$ strain transformed with donor plasmid containing a truncated *qa-1F*$^+$ gene. Single-stranded probes 5' labeled at the *NcoI* site were isolated from (A) the *qa-1F*$^+$ donor plasmid that lacks all 5' untranscribed sequences and ca. one-third of the leader sequence, and (B) a plasmid that contains all 5' upstream *qa-1F*$^+$ sequences. Each probe was hybridized to

poly(A)$^+$RNAs prepared from wild-type (WT) and the _qa-1F_$^+$ transformant (T) strain, grown under non-inducing (NI) and inducing (I) conditions, then digested with 50 units of S1 nuclease (15). Protected fragments were resolved on a thin sequencing gel. A. autoradiogram shows fragments protected using the probe from the donor plasmid. Lanes 2 (WT,NI), 4 (T,NI), and 5 (T,I) correspond to hybridizations using 10 μg RNA, lane 3 (WT,I) 5 μg RNA. Protected fragments from the transformant are depicted schematically below. Dashed lines represent sequences complementary to pBR325, solid lines to _qa-1F_. The junction occurs at the _Bam_HI site. B. Fragments protected with a probe containing the 5' upstream sequences of _qa-1F_$^+$. The two sets of transcripts (226 and 289) protected at the _qa-1F_$^-$ locus in the recipient correspond to major (+1) and minor (-65) regions of initiation (21). Lanes 7 (WT,NI), 9 (T,NI), and 10 (T,I) 10 μg RNA, lane 8 (WT,I) 5 μg RNA. Dashed line in diagram depicts _qa-1F_$^+$ transformant RNA containing sequences complementary to pBR325. Abbreviations for restriction sites are _Bam_HI (B), _Hind_III (H), _Hpa_I (Hp), and _Nco_I (Nc). Lengths and distances are given in base pairs (bp). C (lanes 1 and 6) represents a mock hybridization with 5 μg of _E. coli_ DNA. Differences in the apparent level of protected F$^+$ transcripts between T,NI and T,I may not be significant.

Additional experiments in progress are designed to consider alternative mechanisms of regulation. One clear possibility is that the target of the repressor is the activator molecule itself. Clearly, this mechanism would involve a protein-protein interaction, and would thus resemble one aspect of regulation in the galactose system of the yeast _Saccharomyces cerevisiae_ (24). Such a mechanism in the _qa_ system would predict the possibility of obtaining constitutive mutations located within the coding region of the _qa-1F_ gene (_qa-1F_^{c} mutants). In yeast, such constitutive mutations have been detected within the coding region of the _GAL4_ gene which encodes an activator protein presumably comparable to the activator protein encoded by the _qa-1F_$^+$ gene (25). Appropriate _qa_ strains are being constructed to search for comparable _qa-1F_^{c} mutations.

Alternative interpretations concerning the target of the repressor must also be considered. For example, the possibility exists that the repressor may interact with operator sites 5' to each of the _qa_ genes, rather than directly with the activator, and thus, in the absence of inducer, block activator-facilitated transcription. Appropriate strains designed to test this possibility are also being constructed.

DNA sequences of repressor mutants. Our current hypothesis that the *qa-1S* gene encodes a repressor protein was based primarily on genetic evidence derived from the study of *qa-1S* mutants. Additional evidence that strengthens this interpretation is derived from the recent localization of three mutations within the *qa-1S* coding region. By analogy to mutations in certain prokaryotic repressors, e.g., the *lac* repressor (26), and to mutations in *gal*80 (21), the two types of mutants detected genetically in the *qa-1S* gene should define two domains, with *qa-1S*$^-$ mutations (semi-dominant, non-inducible) occurring in the inducer-binding domain, and *qa-1S*^{c} mutations (recessive, constitutive) occurring in the repressor target domain. Presumably the majority of such mutants should be the result of missense mutations. However, the occurrence of nonsense or frameshift mutations (as well as deletions) in the NH_2-terminal region should also produce constitutive mutants if the *qa* repressor has only a negative role in regulation.

The cloning and DNA sequencing of both types of *qa-1S* mutants has been initiated. To date, two *qa-1S*$^-$ mutants have been located (Huiet and Giles, unpublished) and both occur within the C-terminus of the *qa-1S* coding region. One mutant (141) results from a transition mutation changing glycine to aspartic acid at amino acid residue 627; the second mutant (M105^{ts}) results from a transversion mutation changing asparagine to tyrosine at amino acid residue 743. (The total number of amino acid residues in the repressor protein is 918.) However, *qa-1S* mutant 105^{ts} is unusual in that, while it acts as a non-inducible mutant when grown at 25°C, it behaves as a constitutive mutant when grown at 35°C. Evidence based on transcriptional studies is compatible with the hypothesis that this mutant produces a thermolabile repressor which is almost fully active at 25°C, but largely inactive at 35°C (22). The only other mutant (M105^c) for which sequence data are now available was derived as a constitutive "revertant" of 105^{ts}. As expected, this mutant behaves as a constitutive at both 25°C and 35°C (but is somewhat inducible) and produces elevated levels of *qa*- mRNAs at both temperatures (22). The DNA sequence data show that, as expected, mutant 105^c is a double mutant, containing both the original mutation present in the parental strain 105^{ts} (at residue

743) and a new transition mutation (at residue 791) that changes a proline to a leucine. Initially, the position of the second mutation in the repressor sequence of 105^c seemed somewhat surprising, since we anticipated that such a mutation might be expected to occur in the NH_2-terminal region of the repressor, i.e., in the presumptive repressor target domain. However, it appears possible that the presence of this second rather drastic amino acid change (a proline to a leucine) may produce a repressor which is partially inactive at both 25°C and 35°C, but still retains the ability to interact with the inducer.

Presumably, more critical data for determining whether the NH_2-terminal region of the repressor contains a repressor target domain will come from the sequencing of $qa\text{-}1S^c$ mutants derived directly from wild-type (27) rather than as "revertants" of various types of $qa\text{-}1S^-$, especially temperature-sensitive, non-inducible mutants. We must also consider the probability that some region (probably an amino-terminal sequence) of both the repressor and activator molecules is involved in "targeting" these proteins into the nucleus (28,29).

Mechanism of Qa Activator Function.

Evidence from $qa\text{-}2^{ai}$ (activator-independent) mutants. Evidence derived from the study of two different types of mutants, qa-2 activator-independent ($qa\text{-}2^{ai}$) and qa-1F temperature-sensitive ($qa\text{-}1F^{ts}$) mutants, has enabled a distinction to be made between two requirements for qa-1F-mediated qa gene transcription.

The $qa\text{-}2^{ai}$ mutants were selected after ultraviolet irradiation of $qa\text{-}2^+$ $qa\text{-}1F^-$ $arom\text{-}9^-$ double mutants as "revertants" now able to grow on minimal medium without the added aromatic supplements which the parental strain requires. Such "revertants" can grow on minimal medium because the partially restored catabolic dehydroquinase activity encoded by the $qa\text{-}2^+$ gene can substitute for the biosynthetic dehydroquinase activity encoded by the arom-9 gene. Several $qa\text{-}2^{ai}$ mutants have been cloned and sequenced (Table 2). The mutations all occur in a region 5' to the qa-2 gene and exhibit a variety of DNA sequence alterations including "point mutations", small duplications, and large DNA rearrangements. All the mutants exhibit constitutive expression of $qa\text{-}2^+$ (1-45% of induced wild-type) (Table 2) (17; Geever et al., unpublished).

Table 2. Location and Characteristics of qa-2^{ai} Mutants

Mutant	Mutation type	Mutation site[a]	Relative cDHQase activity[b]
158-6	Duplication (68 bp)	-175 to -108	6%
158-21	Rearrangement (reciprocal)[c]	-53	4%
158-33	Rearrangement (reciprocal)	-379	45%
158-35	Rearrangement (reciprocal)	-233	6%
158-41	Rearrangement (reciprocal)	-260	1%
158-48	Rearrangement (reciprocal)	-274	5%
158-99	Duplication (84 bp)	-189 to -106	9%
158-106	Transversion	-200	6%
158-108	Rearrangement (non-reciprocal)	-190	5%
158-110	Rearrangement (reciprocal)	-86	3%
162-4	Insertion-deletion	-97 to -89	3%

[a]Sites are given relative to the major (+1) initiation site in wild-type.

[b]Activity determined as μmol of dehydroshikimate produced per min at 37°C per mg soluble protein. Values expressed relative to wild-type levels under inducing conditions (0.3% quinic acid for 6 hrs in minimal medium).

[c]The specific type of DNA rearrangement is unknown. However, recent evidence (Geever, unpublished data) has shown that six rearrangements are consistant with a simple two point breakage and joining event (reciprocal). Mutation 158-108 is complex involving at least a three point breakage and joining (non-reciprocal).

To examine how the activator protein directs the transcription of the qa structural genes, the transcription initiation regions of these genes have been mapped in induced wild-type, in a set of qa-2^{ai} mutants, and in three qa-1F^{ts} mutants (Fig. 4). Whereas in induced wild-type, transcription of the qa structural genes is initiated from two to four distinct regions (for the purposes of this discussion, we assume that each region of transcription initiation corresponds to a specific promoter region), in qa-2^{ai} qa-1F^{-} and qa-1F^{ts} mutants only a certain subset of these promoters (designated Type II promoters) is activated by the mutations. The activation of these promoters is independent of the orientations, positions, and precise separations of the mutations with respect to the promoters,

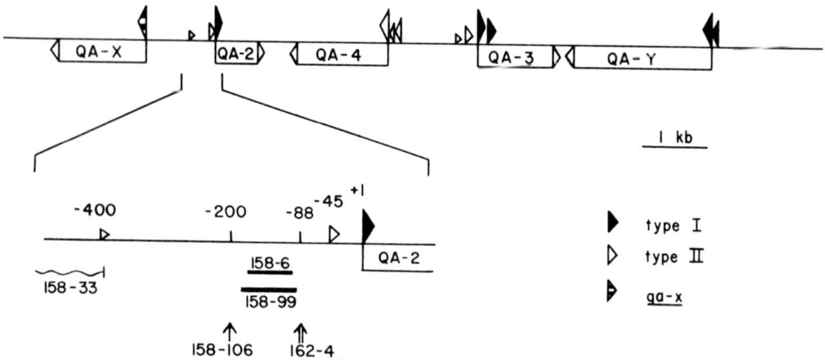

Figure 4. Structural gene promoters and qa-2^{ai} mutations in the qa gene cluster. The size of the arrowhead describes approximately the strength of a given promoter relative to the other promoters for that gene. 158-6 and 158-99 are duplications. The precise nature and locations of the qa-2^{ai} mutations are given in Table 2. The qa-x promoter is unique in that qa-1S appears to exert partial control over this promoter directly. However, it may resemble type II promoters since it is stimulated by some qa-2^{ai} mutations (15).

suggesting that the qa-2^{ai} 'point' mutations and short duplications have created enhancer-like elements in the -200 to -88 region 5' to qa-2. The subset of non-activated promoters (Type I), presumably requiring direct qa-$1F^+$ protein binding, is most notably evidenced in qa-$1F^{ts}$ mutants, which are selectively deficient in the activation of these promoters. Individual qa structural genes can have either type I promoters, type II promoters, or both. However, in wild-type, the qa-$1F^+$ activator protein is required for transcription initiation at both type I and type II promoters (15).

Recent genetic experiments have succeeded, by gene conversion (Case et al., unpublished), in replacing the qa-$1F^-$ gene with a qa-$1F^+$ gene in the original qa-2^{ai} mutant (158-33). In this new qa-2^{ai} qa-$1F^+$ strain (mutant 158-33-10), qa-2 transcripts are now initiated primarily from the -45 region (type II promoter) in the absence of inducer. However, when inducer is added, transcription initiation occurs primarily from the +1 region (type I promoter) at a level essentially equal to induced wild-type. This change in promoter usage strongly supports the prior evidence for two types of qa promoters each having different requirements for activation. The currently preferred interpretation of these overall results is that the activator protein has two distinguishable functions, one of which may involve facilitating RNA polymerase II access to type II promoters, while the other involves the enhancement of polymerase II transcription at type I promoters by direct DNA binding of a fully functional activator. On this hypothesis, the qa-1F^{ts} proteins would have selectively lost their activator function for type I but not for type II promoters.

Evidence from DNase I hypersensitive sites. Studies of DNase I hypersensitive sites in chromatin within the qa gene cluster have also provided support for the interpretations based on qa mRNA studies of qa-2^{ai} and qa-1F^{ts} mutants. The DNase I hypersensitive sites of qa-x and qa-2 chromatin have been studied in the greatest detail and both trans-acting (e.g., qa-1F^{-}) and cis-acting (qa-2^{ai}) regulatory mutations have been examined for their effect(s) on these sites (30; Baum and Giles, unpublished). Although uninduced wild-type chromatin contains hypersensitive sites in the common 5' flanking region of qa-x and qa-2, additional sites can be detected upon induction. These changes in DNase I sensitivity appear to depend on the presence of qa-1F^{+}. Significantly, induction is accompanied by increased DNase I cleavage in the -200 to -88 region of qa-2, and qa-1F^{-} mutants lack a DNase I hypersensitive site normally associated with this region (near position -160), irrespective of the genotype of qa-1S. Presumably, the qa-1F^{+} activator protein binds to a specific sequence within this region, perhaps disrupting the normal spacing of nucleosomes, and thereby rendering adjacent sequences accessible to DNase I. Experiments with the qa-2^{ai} mutants have also demonstrated that virtually all of the qa-2^{ai} mutations are associated with new or enhanced DNase I hypersensitive sites in the 5' flanking region of qa-2, suggesting that qa-2 transcription may normally depend on the existence of certain hypersensitive sites.

Interestingly, a computer search of the DNA sequences in the qa cluster has revealed an apparently conserved sequence that is present, with slight variations, one or more times 5' to each qa gene. The derived 16 nucleotide consensus sequence, GGATAANNNNTTATCC, exhibits dyad symmetry extending six nucleotides. Furthermore, this sequence is closely associated with certain inducible hypersensitive sites. For example, one sequence is located adjacent to the region 5' to qa-2 (near position -160) which becomes hypersensitive to DNase I digestion in the presence of qa-1F$^+$. Transformation experiments designed to restore the qa-1F$^+$ gene into various qa-1^{ai} qa-1F$^-$ rearrangement mutants should establish the significance of this region for transcriptional activation. Of particular interest are mutants 158-108 and 158-110, which have rearrangement breakpoints at positions -190 and -86, respectively (see Table 2). If these mutants flank the region required for activator recognition, then 158-108 (qa-2^{ai} qa-1F$^+$) should be inducible while 158-110 (qa-2^{ai} qa-1F$^+$) should not.

Only one such sequence, also associated with hypersensitive sites, exists in the 380 bp common 5' flanking region of qa-1F and qa-1S, raising the possibility that activation of these genes proceeds bidirectionally from a single controlling sequence. A dual control over qa-1F and qa-1S transcription would be consistent with the observation that these two genes exhibit comparable levels of induction (ca. 50-fold). The ability of this sequence to confer inducibility upon qa-1F$^+$ is currently under investigation. Of course, it is possible that this sequence represents a target for the qa-1S repressor and that some other sequence(s) is responsible for activator recognition.

N. crassa Homologous in vitro Transcription Systems.

A complete dissection at the molecular level of gene regulation in the qa cluster will eventually require a homologous in vitro transcription system in which polymerase II faithfully transcribes cloned qa genes. Such a system must be capable of reproducing the in vivo transcription of these genes as has been accomplished so successfully in prokaryotic in vitro transcription systems. Consequently, studies have been undertaken to develop such a system for N. crassa. Initial experiments demonstrated that it is possible to develop homologous in vitro transcription systems for both polymerase III (31) and polymerase I (32). Of most interest for an understanding of qa gene regulation is the recent development of a homologous in vitro transcription system for polymerase II (33). This system accurately initiates transcription of three constitutively expressed N. crassa protein-coding genes -- the am gene, which encodes glutamate dehydrogenase (22) and the two genes which encode histones H3 and H4 (34). This transcription is sensitive to 1 mg/ml α-amanitin. The 5' heterogeneity of the in vitro am and histone H3 transcripts closely matches that of the corresponding in vivo mRNAs, indicating that the in vitro initiation reaction is highly specific.

When this polymerase II in vitro transcription system was tested with each of the seven qa genes, only one of the seventeen previously identified inducible mRNA initiation sites in the qa gene cluster was utilized by the extracts. This was a minor (Type II) qa-3 initiation site (at -390) (34). Synthesis of the transcripts was inhibited by α-amanitin but not by rifampicin and transcription proceeded through a T5 RNA polymerase III termination signal in a ØX174 fragment inserted into the template. These results indicate that transcription was due to RNA polymerase II. The 5' heterogeneity of the in vitro qa-3 transcripts was similar to that of the corresponding in vivo transcripts, although the major initiation nucleotide in vivo is poorly utilized in vitro. It is not yet clear why transcripts were initiated in vitro at only a single site within the qa-gene cluster. Further refinement of this crude in vitro system should provide a basis for determining in more detail the proteins and DNA sequences required for transcription from the different qa gene initiation sites, as well as for transcription of strongly expressed constitutive genes like am and the histone genes.

It is hoped that experiments currently underway in which the two qa regulatory genes are being inserted into expression vectors will ultimately permit the isolation and characterization of the repressor and activator molecules. The availability of significant amounts of these two proteins for use in the RNA polymerase II in vitro transcription system will clearly be essential for a complete characterization of gene regulation at the molecular level in the qa system.

References

1. Rines HW (1968). Genetics 60:215A.
2. Rines HW (1969). Ph.D. Thesis, Yale Univeristy.
3. Rines HW, Case ME, Giles NH (1968). Genetics 61:789-800.
4. Case ME, Giles NH (1975). Proc Natl Acad Sci USA 72:553-557.
5. Case ME, Giles NH (1976). Molec gen Genet 147:83-89.

6. Giles NH, Alton NK, Case ME, Hautala JA, Jacobson JW, Kushner SR, Patel VB, Reinert WR, Strøman P., Vapnek D (1978). In Redei GP (ed): "Stadler Genetics Symp Vol 10," Columbia: Univ of Missouri, pp 49-63.
7. Strøman P, Reinert W, Case ME, Giles NH (1979). Genetics 92:67-74.
8. Hawkins AR, Reinert WR, Giles NH (1982). Biochem J 203:769-773.
9. Vapnek D, Hautala JA, Jacobson JW, Giles NH, Kushner SR (1977). Proc Natl Acad Sci USA 74:3508-3512.
10. Schweizer M, Case ME, Dykstra CC, Giles NH, Kushner SR (1981). Proc Natl Acad Sci USA 78:5086-5090.
11. Case ME, Schweizer M, Kushner SR, Giles NH (1979). Proc Natl Acad Sci USA 76:5259-5263.
12. Huiet L (1984). Proc Natl Acad Sci USA 81:1174-1178.
13. Patel VB, Schweizer M, Dykstra CC, Kushner SR, Giles NH (1981). Proc Natl Acad Sci USA 78:5783-5787.
14. Alton NK, Buxton F, Patel V, Giles NH, Vapnek D (1982). Proc Natl Acad Sci USA 79:1955-1959.
15. Tyler BM, Geever RF, Case ME, Giles NH (1984). Cell 36:493-502.
16. Alton NK, Geever RF, Giles NH, Vapnek D (1985). Gene (in press).
17. Geever RF, Case ME, Tyler BM, Buxton F, Giles NH (1983). Proc Natl Acad Sci USA 80:7298-7302.
18. Rutledge BJ (1984). Gene 32:275-287.
19. Huiet L (1984). Ph.D. Thesis, Univeriesty of Georgia.
20. Laughon A, Gesteland RF (1984). Mol Cell Biol 4:260-267.
21. Nogi Y, Fukasawa T (1984). Nuc Acids Res 12:9287-9298.
22. Patel VB, Giles NH (1985). Mol Cell Biol (submitted).
23. Kinnaird JH, Fincham JRS (1983). Gene 26:253-260.
24. Johnston SA, Hopper JE (1982). Proc Natl Acad Sci USA 79:6971-6975.
25. Matsumoto K, Adachi Y, Toh-e A, Oshima Y (1980). J Bact 141:508-527.
26. Miller JH, Reznikoff WS (eds) (1978). "The Operon." New York: Cold Spr Harb Lab.
27. Partridge CWH, Case ME, Giles NH (1972). Genetics 72:411-417.
28. Hall MN, Hereford L, Herskowitz I (1984). Cell 36:1057-1065.
29. Silver PA, Heegan LP, Ptashne M (1984). Proc Natl Acad Sci USA 81:5951-5955.
30. Baum J, Giles NH (1985). Jour Mol Biol (in press).

31. Tyler BM, Giles, NH (1984). Nuc Acids Res 12:5737-5755.
32. Tyler BM, Giles, NH (1985). Nuc Acids Res (submitted).
33. Tyler BM, Giles, NH (1985). Proc Natl Acad Sci USA (in press).
34. Woudt LP, Pastink A, Kempers-Veenstra AE, Jansen AEM, Mager WH, Planta RJ (1983). Nuc Acids Res 11:5347-5360.

Molecular Genetics of Filamentous Fungi, pages 117–125

THE *AM* (NADP-SPECIFIC GLUTAMATE DEHYDROGENASE) GENE OF *NEUROSPORA CRASSA*

J.R.S. Fincham[1], Jane H. Kinnaird and P.A. Burns

Department of Genetics, University of Edinburgh, Edinburgh EH9 3JN

ABSTRACT 1. The special advantages of the Neurospora *am* gene as a system for studying effects of specific amino acid replacements on enzyme properties are described, and some of the results outline. 2. The internal consensus sequence characteristic of the introns found in *am* and other Neurospora genes is described and its significance discussed. 3. The strongly selective codon usage seen in *am* and other constitutively expressed Neurospora genes is pointed out. A double-frameshift mutant in which three consecutive "rare" codons are introduced into the coding sequence shows an approximately 35 per cent level of gene expression. The significance of this is discussed.

INTRODUCTION

The *am* gene, mutations in which cause a generalized requirement for α-amino nitrogen[1], has recently been cloned and sequenced[3]. An efficient system for positive selection of *am* mutants has been worked out[4]. A fine-structure map of mutational sites, based both on recombination frequencies combined with conversion polarity[5] and on over-lapping deletions[6] has been related to the restriction site map of the cloned DNA. It is thus possible to locate any point mutation within a defined segment of the physical DNA map and to determine its molecular nature fairly rapidly. The amino acid sequence of the polypeptide chain of wild type Neurospora NADP-

[1] Present address: Department of Genetics, University of Cambridge, Downing Street, Cambridge CB2 3EH

specific glutamate dehydrogenase (GDH), which consists of six identical chains each 453 residues in length, was determined a number of years ago[7], and the amino acid replacements in a number of mutants have also been determined[8,9,10].

This paper reviews three aspects of the structure of the gene in relation to its function: the spectrum of effects of amino acid replacements on enzyme activity, the nucleotide sequences of the introns, and the functional significance of the biassed codon usage.

EFFECTS ON ENZYME PROPERTIES OF AMINO ACID REPLACEMENTS

Table 1 summarizes the variety of effects of single amino acid replacements on the properties of Neurospora GDH. One can distinguish three broad categories of effect in the primary mutants. First, there are a few changes that eliminate enzyme activity unconditionally. Of these, the change $Ser^{336} \rightarrow Phe^{336}$ (am[1]) abolishes the binding of NADPH. Secondly, a number of changes, in widely separated (at least in terms of primary structure) parts of the polypeptide chain, shift the allosteric equilibrium constant of the enzyme in favour of the inactive conformational state[11]. Thirdly, one mutant (am[14]) can be interpreted as producing GDH with critically destabilized quaternary structure. This mutant is unique in the collection in being partially reparable by high osmotic pressure of the growth medium[9].

To a first approximation it is true to say that each of these three types of mutant can complement either the of other two in heterokaryons to produce functional enzyme. Evidence has been obtained that such allelic complementation is due to mutual compensation of differently defective polypeptide chains in mixed hexamers[12,13].

Now that the amino acid sequences of two closely comparable GDHs, from Escherichia coli[14] and Saccharomyces cerevisiae[15] respectively, are available, it is possible to see that the amino acid changes that have pronounced effects on enzyme function tend to affect residues that have been highly conserved in evolution (see Table 1).

In addition to the relatively drastic changes found in the primary series of am auxotrophs, a number of effects falling short of critical functional deficiency have been found associated with secondary changes in revertants. A series of double frame-shift mutants with alterations in

TABLE 1
EFFECTS OF AMINO ACID REPLACEMENTS ON NEUROSPORA NADP-SPECIFIC GDH

Mutant No.	Residue No.	Alteration	Effect	Residue in corresponding position in E.coli[14]	Yeast[15]
14	25	Leu → His ↓ Tyr	Destablization of quaternary structure. Osmotically reparable. Restored stability[9].	Glu	Leu
130	75	Pro → Ser	Stabilization of inactive conformation[10].	Pro	Pro
19	141	Lys → Met	Strong stabilization of inactive conformation[8].	Arg	Arg
2	142	His → Gln	Stabilization of inactive conformation[8].	His	His
17*	313	Gln →"Amber"→Leu, →Tyr	Increased Km's for NH_4^+ and glutamate[17].	Gln	Gln
1	336	Ser → Phe	Complete sctivity loss - failure to bind NADPH[8].	Ala	Ser
7	372	Gly → Ser	Complete activity loss[8].	Gly	Gly
19*	391	Gln → Arg	Partial compensation for Lys[141] → Met[8].	?‡	Ser
3	393	Glu → Gly	Stabilization of inactive conformation[8].	Glu	Arg

* revertants from these mutants ‡ <u>E.coli</u> has a 1-residue deltion in this position.

several consecutive residues near the N-terminus of the chain[16] have nearly normal GDH except for different degrees of enhanced thermolability. This portion of the polypeptide chain is hardly conserved at all betwcen E.coli and Neurospora, and is largely absent from the yeast enzyme, which is relatively truncated at the N-terminus. To take another example, two different classes of "mis-sense" revertant, selected from the "amber" chain-termination mutation am[17], have, respectively, leucine and tyrosine substituted for the highly-conserved glutamine at residue 313. The GDH varieties produced by these revertants have K_m values with respect to ammonium ion increased by a factor of 20 to 30 in comparison with the wild type[17].

All of this information will be much more readily interpreted when we know more about the three-dimensional structure of the protein. In this connection the recent report[18] of a preliminary crystallographic analysis of the substantially similar glutamate dehydrogenase from Clostridium symbosium is encouraging.

INTRONS - THE INTERNAL CONSENSUS SEQUENCE

The Neurospora am gene gave the first hint of an internal consensus sequence in a filamentous fungus. Its two introns have in common the sequence GCTGACT ending, respectively, 14 and 10 nucleotides from the 3' splice site. Comparison with nine other Neurospora nuclear introns, which became available shortly afterwards, showed a general consensus no more than five nucleotides in length, CT$^{A}_{G}$AC, ending between 9 and 17 residues upstream of the splice site[3]. The first, second, fourth and fifth positions are invariant in the Neurospora introns sequenced so far.

Langford et al[19] found that the sequence TACTAACA was invariant in Saccharomyces nuclear introns and were able to show that it was necessary for splicing. More recently, a much less tightly-conserved sequence, approximating to CTGAC but with only the second and fourth positions apparently invariant, has been implicated in the formation of a "lariat" splicing intermediate in animal cell nuclei[20,21]. The fourth-position A is the branch point of the lariat, forming a 2'-5' phosphodiester bond with the 5' end of the intron.

The introns of filamentous fungi may occupy an intermediate position between those of Saccharomyces on the one hand and animals on the other, their internal sequence less tightly conserved than the former but more so than the latter. If the degree of conservation of the internal sequence is an indication of the stringency of the requirement for splicing, it may be that filamentous fungal systems will not be efficient at splicing out the introns of Saccharomyces but will be able to cope better with those of higher eukaryotes.

FRAMESHIFTS AND CODON PREFERENCES

Differences in codon usage between different fungal species and between different genes within a species have been previously noted[22,23]. In Saccharomyces[23] the rule appears to be that genes that are expressed more or less constitutively at a high level ("housekeeping" genes) almost confine their codon usage to a preferred set of little more than 30, while genes subject to strong regulation (involved, for example, in amino acid biosynthesis or special sugar utilization) show very much less bias. Much the same seems to be true of Neurospora. The am gene and several others that can be placed in the "housekeeping" category[3] use hardly more than half of the possible codons. The bias in the trp-1 gene[24] is, though in the same direction, very much less pronounced and the strongly regulated qa-4 comes still closer to random codon usage[25].

The Neurospora pattern of preferred codon usage, which is substantially different from that of Saccharomyces, can be summarized in terms of three rules. Firstly, the AGN arginine and serine codons and the UUPu leucine codons are rarely used. Secondly, pyrimidines are almost always used in preference to purines in third positions of codons where that choice exists. Thirdly, guanine is almost always used in the third position in preference to adenine where a purine is obligatory. In fact, among the 453 codons of the am gene there is only one with third-position A. It is of some interest to ask whether the introduction of one or more "rare" codons into such a highly biassed gene will have any effect on its level of expression. Some recent work on frameshift mutagenesis in our laboratory gave us the opportunity of addressing this question.

The mutant am^{15} has long been known (L.E. Kelly and A. Radford, unpublished) to be highly revertible with the frameshifting mutagen ICR170. We have now cloned and

TABLE 2
LEVELS OF EXPRESSION OF DOUBLE-FRAMESHIFT SEQUENCES

Strain	Sequence ("rare" codons underlined)	Yield GDH protein (% wild)	Specific activity (% wild)	Stability % remaining after 5 min at 60C
Wild type	53 60 GluAspAspAsnGlyAsnValGln --GAGGACGACAACGGCAACGTCCAG-- ↓ -1	100	100	80
am[15]	GAGGACGACAAGGCAACGTCCAG-			
am[15]R12	GluAspAspLysGlyAsnValGln GAGGACGACAAGGGCAACGTCCAG ↑ +1	103	100	75
am[15]R15	GluAspAspLysAlaThrSerGln --GAGGACGACAAGGCAACGTCCCAG- ↑ +1	84	68	10
am[15]R11	--GluGlyArgGlnGlyAsnValGln -GAGGGACGACAAGGCAACGTCCAG-- ↑ +1	36	78	4

sequenced the gene from several revertants as well as from the original mutant, and shown that the primary mutation is a single base-pair deletion and the revertants nearby single base-pair insertions generating, between the frameshifts, one to three sequential amino acid replacements. The GDH formed in each case is fully functional, though with unusual electrophoretic mobility due to loss of acidic and/or gain of basic residues and, in the cases of the more extensive sequence changes, marked thermolability. The interesting point in the present context is that the longest reading-frame alteration generates three successive codons with third-position A. The sequences are summarized in Table 2, together with data on the levels both of GDH activity and of GDH protein as measured by "rocket" electrophoresis. The introduction of three successive rare codons does appear to reduce the level of translation to about 35 per cent of normal. There has to be some reservation about this conclusion, since it is not ruled out that the loss of protein may be post-translational. It seems unlikely that the GDH was lost by thermal denaturation during extraction since, although labile at 50°, it is quite stable under the conditions used; but enhanced susceptibility to in vivo proteolysis is not ruled out.

The results suggest, at all events, that restricted availability of rarer tRNA species does not impose a very stringent limitation on the messenger RNAs that can be translated. That it may exert some modulating effect (as recently suggested for E.coli [26]) remains a likely possibility

REFERENCES

1. Fincham JRS (1950). Mutant strains of Neurospora deficient in aminating ability. J Biol Chem 182:61-73.
2. Kinnaird JH, Keighren MA, Eaton M, Fincham JRS (1982). Cloning of the am (glutamate dehydrogenase) gene of Neurospora crassa through the use of a synthetic DNA probe. Gene 20:387-396.
3. Kinnaird JH, Fincham JRS (1983). The complete nucleotide sequence of the Neurospora crassa am (NADP-specific glutamate dehydrogenase) gene. Gene 26:253-260.
4. Kinsey JA (1977) Direct selective procedure for isolating Neurospora crassa mutants defective in NADP-specific glutamate dehydrogenase. J Bact 132:751-756.

5. Fincham JRS (1967). Recombination within the am gene of Neurospora crassa. Genet Res Camb 9:49-62.
6. Rambosek JA, Kinsey JA (1983). Fine structure mapping of the am (GDH) locus of Neurospora. Genetics 105:293-307.
7. Holder AA, Wootton JC, Baron AJ, Chambers GK, Fincham JRS (1975). The amino acid sequence of Neurospora NADP-specific glutamate dehydrogenase. Peptic and chymotryptic peptides and the complete sequence. Biochem J 149:757-773.
8. Brett M, Chambers GK, Holder AS, Fincham JRS, Wootton JC (1976). Mutational amino acid replacements in Neurospora crassa NADP-specific glutamate dehydrogenase J Mol Biol 106:1-22.
9. Fincham JRS, Baron AJ (1977). The molecular basis of an osmotically reparable mutant of Neurospora crassa producing unstable glutamate dehydrogenase. J Mol Biol 110:627-642.
10. Kinsey JH, Fincham JRS, Siddiq MAM, Keighren MA (1980). New mutational variants of Neurospora crassa NADP-specific glutamate dehydrogenase. Genetics 95:305-316.
11. Ashby B, Nixon JS, Wootton JC (1981). Mutational variants of the Nuerospora crassa NADP-specific glutamate dehydrogenase altered in a conformational equilibrium. J Mol Biol 149:521-570.
12. Coddington A, Fincham JRS (1965). Proof of hybrid protein formation in a case of inter-allelic complementation in Neurospora crassa. J Mol Biol 12:152-161.
13. Coddington A, Sundaram TK, Fincham JRS (1966). Multiple active varieties of Neurospora glutamate dehydrogenase formed by hybridization between two inactive mutant proteins in vivo and in vitro. J Mol Biol 17:503-512.
14. Mattaj IW, McPherson MJ, Wootton JC (1982). Localization of strongly conserved section of coding sequence in glutamate dehydrogenase. FEBS Let 147:21-25 and McPherson MJ (1983) Ph D Thesis, Leeds.
15. Nagasu T, Hall BD (1985). DNA sequence of the GDH gene coding for the NADP-specific glutamate dehydrogenase of Saccharomyces cerevisiae. Gene (in press).
16. Siddig MAM, Kinsey JA, Fincham JRS, Keighren MA (1980). Frameshift mutations affecting the N-terminal sequence of Neurospora NADP-specific glutamate dehydrogenase. J Mol Biol 137:125-135.
17. Seale TW, Brett M, Baron AJ, Fincham JRS (1977) Amino acid replacements resulting from suppression and missense reversion of a chain terminator mutation in Neurospora. Genetics 86:261-274.

18. Rice DW, Hornby DP, Engel P (1985). Crystallization of an NADP-dependent glutamate dehydrogenase from Clostridium symbiosium. J Mol Biol 181:147-149.
19. Langford CJ, Klinz FJ, Donath C, Gallwitz D (1984). Point mutations identify the conserved intron-contained TACTAAC box as an essential splicing signal sequence in yeast. Cell 36:645-653.
20. Keller EB, Noon WA (1984). Intron splicing: a conserved internal signal in introns of animal pre-mRNAs. Proc Nat Acad Sci USA 81:7417-7420.
21. Keller W (1984). The RNA lariat: a new ring to the splicing of mRNA precursors. Cell 39:423-425.
22. Bennetzen JL, Hall BD (1982). Codon selection in yeast. J Biol Chem 257:3026-3031.
23. Kammerer B, Guyonvarch A, Hubert JC (1984). Yeast regulatory gene PPR1 nucleotide sequence, restriction map and codon usage. J Mol Biol 180:239-250.
24. Schechtman MG, Yanofsky C (1983). Structure of the trifunctional trp-1 gene from Neurospora crassa and its aberrant expression in Escherichia coli. J Mol Appl Genet 2:83-99.
25. Rutledge BJ (1984). Molecular characterisation of the qa-4 gene of Neurospora crassa. Gene 32:275-287.
26. Pederson A (1984). Escherichia coli ribosomes translate in vivo with variable rate. EMBO J 3:2895-2898.

Molecular Genetics of Filamentous Fungi, pages 127–143

REGULATION OF PYRIMIDINE METABOLISM IN NEUROSPORA[1]

A. Radford, F.P. Buxton[2], S.F. Newbury
and J.A. Glazebrook

Department of Genetics, The University of Leeds,
Leeds LS2 9JT, U.K..

ABSTRACT The paper reviews pyrimidine biosynthesis, uptake and salvage pathways in *Neurospora crassa* and their genetic basis. Regulation of these systems by mechanisms apparently specific to pyrimidine metabolism and by general nitrogen metabolite repression is discussed. Recent work on cloning of the *pyr-4* gene, the structural gene for orotidine 5'-P decarboxylase, is presented, with localisation of the gene within the plasmids pFB4 and pFB6. The functional gene is identified by insertional inactivation with the transposon Tn 1000. On the basis of DNA sequencing through the region, more precise localisation is made on the basis of base composition and by comparison of homology at the amino acid level with the equivalent *Saccharomyces cerevisiae ura3* gene. Finally, the base sequence in the promoter region is discussed in comparison with equivalent sequences from other fungal genes, and conserved sequences identified.

INTRODUCTION

The pyrimidine biosynthetic pathway is composed of six enzyme-catalysed steps, the first of which is common to the arginine biosynthetic pathway(1). By these pathways glutamine carbon dioxide and a phosphate group from ATP are combined into carbamoyl phosphate, and this intermediate is converted

[1]This work was supported by the Science and Engineering Research Council of the U.K.
[2]Present address: Allelix Inc, 6850 Goreway Drive, Mississauga, Ontario L4V 1P1, Canada.

through a further five steps to uridine-5'-P, or through three steps to arginine. Saprophytes and rhizosphere organisms such as Neurospora have an alternative source of pyrimidines, namely nucleic acids in their environment, which may be broken down to their constituent nucleotides, nucleosides and bases by extracellular nucleases(2), the breakdown products being then taken up and incorporated into mainstream metabolism by uptake and salvage pathways(3). In this way, both nucleosides and free bases are recycled. Selection and analysis of mutants blocked in biosynthesis, uptake and salvage, and the regulation of these systems, has given us a fairly comprehensive understanding of the

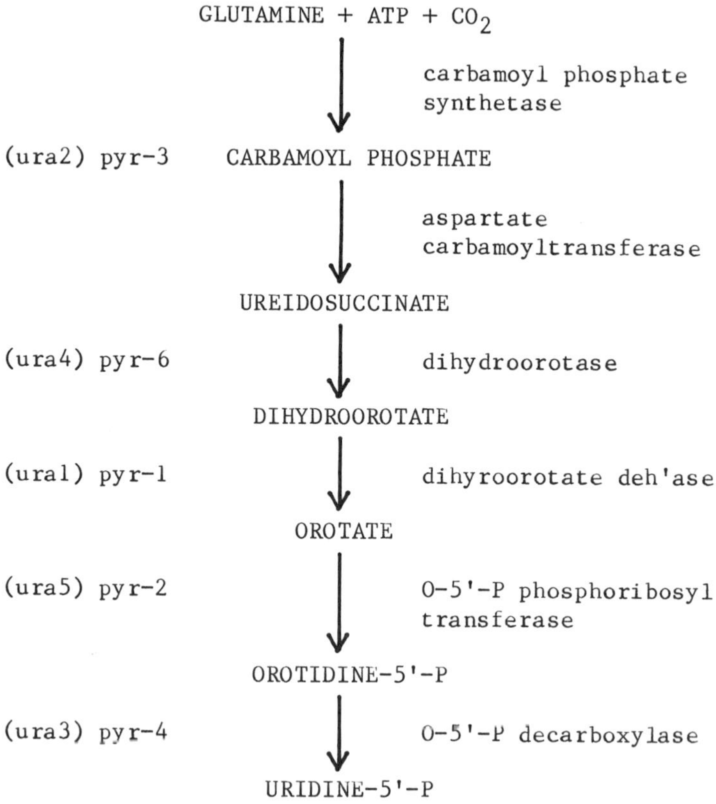

FIGURE 1. The pyrimidine biosynthetic pathway, with the genes of Neurospora, the genes of Saccharomyces in parentheses, and enzymes.

pathways and their regulation.

PYRIMIDINE BIOSYNTHESIS

The pyrimidine biosynthetic pathway is shown in figure 1, together with the enzymes of Neurospora, the genes that encode them(4), and their yeast equivalents(5). This shows that the first two enzyme functions of the pathway in Neurospora are encoded by the same gene, *pyr-3*. This bifunctional gene is not universally found, as many bacteria including *Escherichia coli*, and all higher plants have separate genes for CPSase and ACTase, with a single CPSase generating a single carbamoyl phosphate pool which supplies both pyrimidine and arginine pathways(6,1). In fungi and animals there are two CPSase genes, one regulated by arginine and the other by a pyrimidine, and the pyrimidine-regulated enzyme is bifunctional, specifying also the next enzyme in pyrimidine biosynthesis, ACTase(1,7). Yeast has but one pool of carbamoyl phosphate although duplicate genes and enzymes, with effective control of pyrimidine metabolism via feedback on the ACTase(8). Neurospora, other filamentous fungi, and animals, normally have two pools of carbamoyl phosphate, one for each pathway, with effective feedback control only on the CPS enzymes(1).

There is a continuing tendency to gene fusion in the pyrimidine pathway, as shown in figure 2, with Drosophila and higher animals having the gene for the enzyme for the third step in the pathway, DHOase, fused to CPSase and ACTase in the *rudimentary* locus(9), and also the last two enzymes fused in the *rudimentary-like* gene(10). These fusions appear to result in not merely polycistronic genes, but probably multifunctional single cistrons.

	CPS	ACT	DHO	DHD	O5PP	O5PD
E. coli	carA,B	pyrB,I	pyrC	pyrD	pyrE	pyrF
Neurospora	pyr-3		pyr-6	pyr-1	pyr-2	pyr-4
Drosophila	r			dhod	r-1	

FIGURE 2. Gene equivalents in pyrimidine biosynthesis

CHANNELLING

Channelling in the pyrimidine and arginine pathways is not a feature of E. coli or S. cerevisiae. Only in the filamentous fungi does channelling first appear. This is to some extent compartmental, as Davis and associates have shown(11), with the arginine-specific pool of carbamoyl phosphate in the mitochondria and the pyrimidine-specific pool in the nucleus. The fusion of pyrimidine-specific CPSase with ACTase must also aid channelling, as the production of carbamoyl phosphate adjacent to the ACTase substrate-binding site together with the high substrate-affinity of ACTase lead to the generation of very little free carbamoyl phosphate in this pathway(12).

Channelling in Neurospora can be broken down, and the accumulation of carbamoyl phosphate in either pool as the result of a post-pool genetic block can lead to cross-feeding in either direction(13,14). This cross-feeding potential has been extensively exploited in selection of post-pool mutants as suppressors of pre-pool mutants in the other pathway. The situation is illustrated in figure 3.

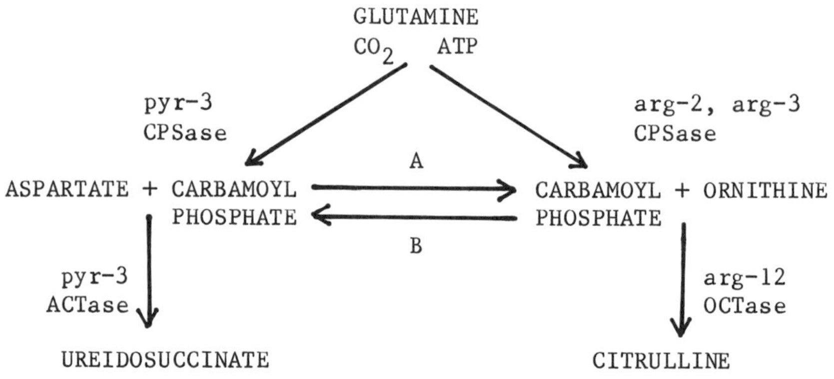

FIGURE 3. Channelling and cross-feeding. In the presence of a pyr-3 (CPS) mutant, the pyrimidine requirement is suppressed by a leaky mutant at arg-12 via cross-feeding "B". Likewise, a pyr-3 (ACT) mutant will suppress the arginine requirement of an arg-2 or arg-3 mutant via shunt "A".

ENZYMES OF BIOSYNTHESIS

The most intensively studied gene in the biosynthetic pathway has been until recently pyr-3, the equivalent to the ura2 gene of yeast. The former has a molecular weight of circa 650,000 d as the native enzyme, but proteolysis and other active fragments of various sizes and characteristics have been identified. These are a half-size protein with both activities(15), a 100,000 d proteolysis fragment with ACTase activity, and a frame-shift fragment of 180,000 d with CPSase(16). The yeast native enzyme is circa 800,000 d, the extra size probably being accounted for by the feedback-binding region for ACTase, a property absent from the Neurospora enzyme. Other enzymes of the pathway have been isolated and described(4), but have not been subjected to the intensive studies performed on the complex pyr-3 gene.

PYRIMIDINE UPTAKE

Nucleotides are not effectively taken up as such, but nucleosides are taken up by a pathway which appears not to discriminate between purine and pyrimidine nucleosides, nor between the various pyrimidine nucleosides(3). Mutants at the Neurospora ud-1 gene are defective in such uptake, being resistant to inhibition by fluoropyrimidine nucleosides, and unable to utilise pyrimidine nucleosides to supplement a requirement caused by a mutant block in the pyrimidine biosynthetic pathway(4,17). Similarly, these mutants are unable to use adenosine to compensate for a purine biosynthetic block. Those allelic mutants selected by analogue-resistance were originally called fdu-1(18).

Pyrimidine bases are taken up by a separate system, specified by the uc-5 gene(4,17). This block confers resistance to fluoropyrimidine bases, and an inability to use normal pyrimidine bases to supplement growth of a pyrimidine auxotroph.

SALVAGE PATHWAYS

Although the first step in extracellular salvage of pyrimidines is uptake, other blocks in salvage beyond this point have also been selected(4,17). The first of these is uc-2, which blocks conversion of deoxyuridine to uridine and thymidine to thymine ribonucleoside, the mutant defect being in deoxypyrimidine-2'-hydroxylase. The uc-3 mutant specifies a defective thymine-7-hydroxylase, and accumulates and excretes thymine due to its inability to convert it via 5-hydroxymethyl uracil and 5-formyl uracil to uracil 5-carboxylic acid. The uc-4 mutant is deficient in phosphoribosyltransferase, and udk lacks uridine kinase. Figure 4 shows uptake and salvage pathways.

REGULATORY SYSTEMS

The regulation of enzymes in the pyrimidine biosynthetic pathway is effected in various ways. The first gene, pyr-3 specifying CPSase/ACTase, shows a 4-5x difference in repressed and derepressed levels, and is derepressed by end-product depletion(17,18). Fine control of flux through the pathway is by UTP-feedback onto the CPSase activity, and unlike in yeast, the feedback binding site has so far proved inseparable from the CPSase catalytic site(12,19). The only other unequivocally regulated gene in the pathway is pyr-1, the gene for dihydroorotate dehydrogenase, which shows substrate-induction(18).

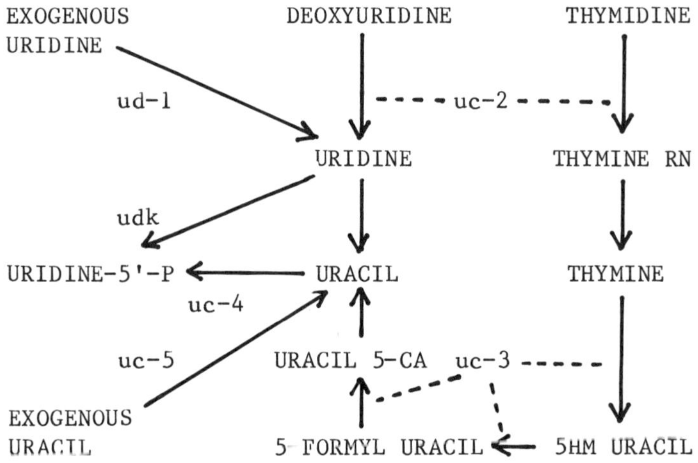

FIGURE 4. Pyrimidine uptake and salvage pathways

Regulation specific to the salvage pathway is controlled by uc-1, a gene defined by a mutant with elevated enzyme levels for the oxidative demethylation of thymine specified by uc-3(3,20).

A regulatory gene specific as far as is known to pyrimidine metabolism is fdu-2(17,18,21). This affects levels of activity of CPSase/ACTase in pyrimidine biosynthesis by rendering the complex pyr-3 gene insensitive to repression. It also has the effect of reducing by circa 20% the nucleoside uptake activity specified by ud-1, and abolishing the activity of the udk gene(17).

The final known regulatory effect on pyrimidine metabolism is the general nitrogen metabolite regulation system. Nitrogen source and concentration had a dramatic effect on pyrimidine base uptake. The effect of nitrogen starvation, after a lag of 30 minutes, was a rapid 50x increase in rate of uracil uptake(22). The uptake of uridine responded to nitrogen-starvation in a qualitatively very similar manner, but the increase was an order of magnitude less. This would seem to explain the early results on supplementation of auxotrophs, when on normal salts solutions (high in nitrogen), bases such as uracil were found to be so much poorer supplements than equimolar amounts of nucleosides like uridine.

Further evidence for the nitrogen metabolite regulation of various aspects of pyrimidine metabolism come from studies with nit-2, a mutant in the presumed nitrogen metabolite regulator gene, and with gln-1a, in which mutant depleted glutamine levels alter the properties of the nit-2 gene product(23). Uracil uptake in a nit-2 strain remains at low level on nitrogen-starvation, and in a gln-1a background shows only slight increase in response to a similar induction. Uridine uptake shows reduced levels of induction by nitrogen starvation of either nit-2 or gln-1a compared to wild type(22).

CLONING OF THE PYR-4 GENE

Using a Neurospora genomic library of a Sau3A partial digest in the plasmid pRK9, obtained from Dr M Schechtman, an attempt was made to clone the pyrimidine biosynthetic genes of Neurospora by complementation of the equivalent mutants of *Escherichia coli*. The plasmid pRK9 is identical to pBR322 except that the 380 bp EcoR1 - BamH1 fragment has been replaced by a 96 bp *Serratia marcescens trp* promoter in order to maximise transcription of DNA inserted into the BamH1 site. After amplification in *E. coli* HB101, purified library DNA was used to transform competent cells of *E. coli* mutant at one of *pyrC*, *pyrD*, *pyrE*, or *pyrF*. In parallel experiments, selection was in one case direct for pyrimidine prototrophy, and in the other initially for the ampicillin resistance of the vector, with subsequent testing for pyrimidine-independence(24).

Only for *pyrF* were prototrophs obtained, and these at a frequency of 0.04% of that of ampicillin-resistance. These transformants were putative clones of the equivalent gene of Neurospora, *pyr-4*$^+$. Southern analysis of the plasmids from these transformants confirmed that they now contained inserts of Neurospora DNA.

Restriction mapping of two of the putative *pyr-4*$^+$ plasmids, pFB4 and pFB6 was carried out, and the maps

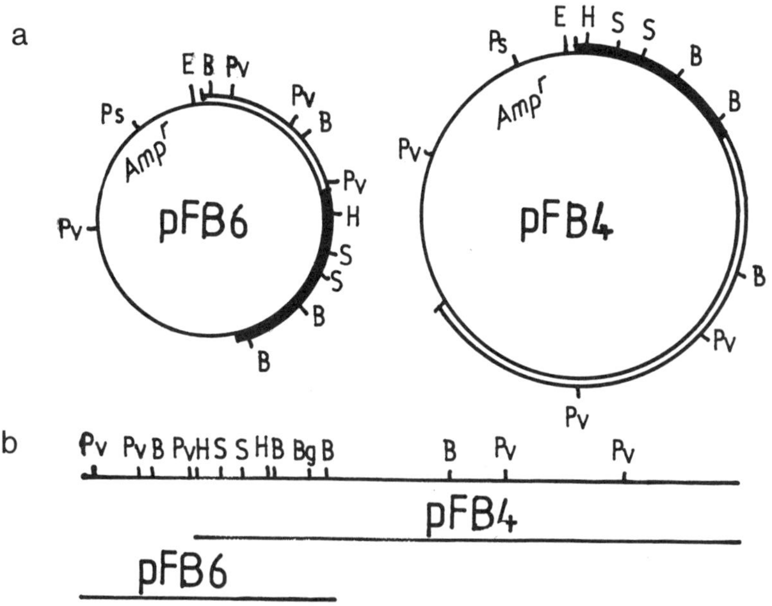

FIGURE 5. (a) Maps of the plasmids pFB4 and pFB6. (—) plasmid vehicle, (═) Neurospora DNA, (▬) Neurospora DNA in common.

(b) Restriction map of the Neurospora sequences. B: BamHI, Bg : BglII, E: EcoRI, H: HindIII, Ps: PstI, Pv: PvuII, S: SalI.

derived are shown in figure 5a. A comparison of the two Neurospora inserts is shown in figure 5b. This shows that the two clones have a region in common of approximately 2.5 kb, in which the functional pyr-4^{+} gene must be located. The insert in pFB6 was subsequently inverted, but expression was no different than in pFB6 itself, although of course the two were in opposite orientation with respect to the Serratia trp promoter. It would seem likely that expression of the Neurospora gene was autonomous, not under control of a vector promoter. The verification of the fact that the clone was in fact the pyr-4^{+} gene of Neurospora was obtained by the retransformation of Neurospora and complementation of the pyrimidine auxotrophy of a pyr-4 mutant. However this was not simple, due to the leakiness of the only available mutant, and had to be accomplished by construction of a plasmid containing both pyr-4^{+} and qa-2^{+}, selecting for transformants of qa-2, and subsequent testing for pyrimidine prototrophy.

Not only was the pyr-4^{+} clone able to complement Neurospora pyr-4 and *E. coli* pyrF, but it was subsequently found to be capable of complementing a pyrG mutant of *Aspergillus nidulans* and a ura3 mutant of *Saccharomyces cerevisiae*(25). The ability to complement its yeast equivalent is most interesting, as there is no complementation in the reciprocal transfer(26,27).

TRANSPOSON-INACTIVATION OF PYR-4^{+}

The pyr-4^{+} gene was localised within the pFB4 and pFB6 cloned saquences by obtaining and mapping Tn1000 inserts into the gene. Plasmids which do not encode transfer functions for *Escherichia coli* and are not mobilised can be transferred at low frequency by the F plasmid. This process is associated with transposition of the F plasmid transposon Tn1000, otherwise known as "gamma-delta", into the plasmid so transferred(28). It is thought that the co-integrate formed during transposition of Tn1000 is actually transferred.

Tn1000 inserts into pFB4 and pFB6 were obtained by crossing an *E. coli* strain carrying the F plasmid and also one of the pyr-4^{+} plasmids with strain PC1773 (pyrF), and selecting for transfer of the ampicillin-resistance gene encoded by the vector sequence into PC1773. Single colonies so selected were screened for the ability to grow without pyrimidine-supplementation. Strains unable to grow without this uracil supplementation were presumed to have Tn1000 inserted into the pyr-4 gene, thus inactivating it.

Inserts which inactivated the pyr-4 gene of pFB4 were obtained at low frequency, at approximately one in 500 exconjugants screened. Only one Tn1000 insert into the gene in pFB6 was obtained from approximately 4,000 screened.

Plasmid DNA was prepared from the strains which were found to have undergone inactivation of pyr-4$^+$, and the positions of the inserts were mapped by restriction analysis. The results are shown in figure 6.

It will be seen that two insertions have gone into the HindIII-SalI region, one into the SalI-SalI region, four into the SalI-HindIII region, four into the SalI-HindIII, a further seven into one or other of the previous two regions, and four into the HindIII-BamHI region. Subsequent results described below indicate that the first of these regions spans the promoter, and the remaining regions cover all except 75bp of the coding region.

The target sequence for Tn1000 insertion, as determined by the flanking duplicated sequence, is a 5bp sequence rich in AT base pairs(29). As is shown by sequencing studies below, the promoter region is rich in such short runs, having one sequence of five AT base pairs and several with four out of five. The coding region is low in AT content, and low in runs of AT base pairs, having no runs of five and few runs of four consecutive or four out of five. The region downstream from the stop codon is again rich in target sequences, but insertions here would be unlikely to lead to inactivation.

SEQUENCING OF THE PYR-4$^+$ CLONE

The DNA sequence in the region common to both pFB4 and pFB6 was determined from the EcoRI site of pFB4 to midway through the BamHI-BglII region. Restriction enzyme fragments from pFB4 and pFB6 were subcloned into the M13 vectors mp8 and mp11 for sequencing. Approximately 90% of the above region was sequenced on both strands, and where sequencing was in one direction only, consistency between at least three separate sequencing runs was used. All restriction sites used in the sub-cloning were sequenced through. The complete DNA sequence included an open reading frame of 1,191 bp from a 5' ATG to a 3' TAG.

Support for the hypothesis that this open reading frame was the actual pyr-4$^+$ coding region came from several

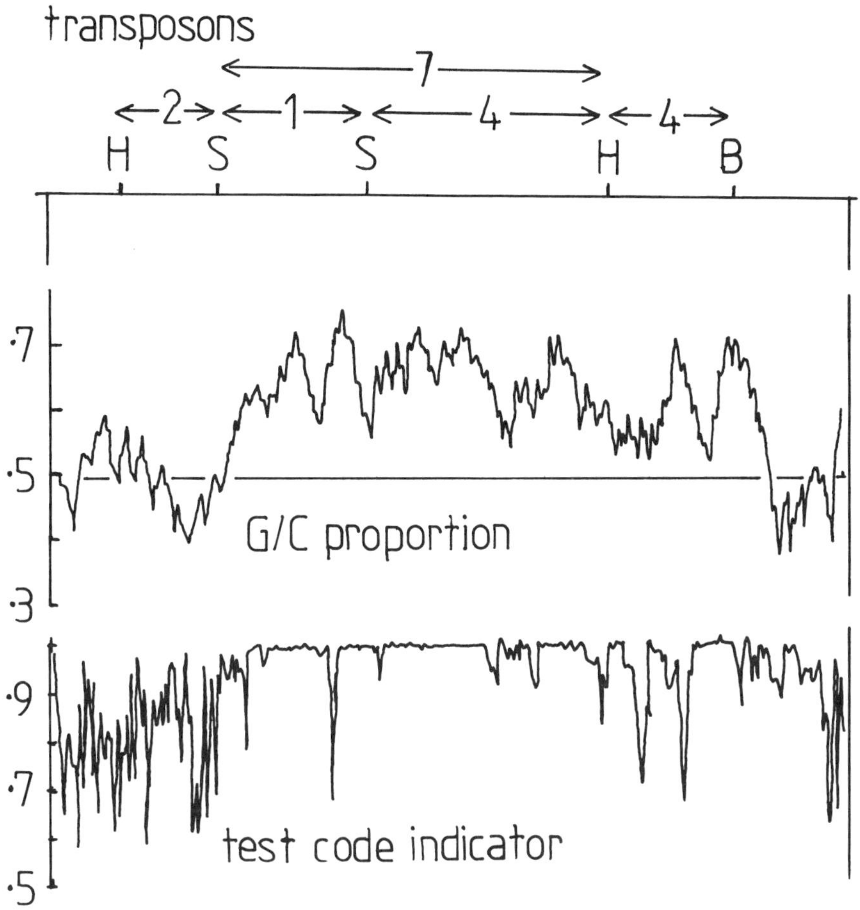

FIGURE 6. Restriction map of the region common to pFB4 and pFB6, showing positions of Tn1000 inserts, GC content, and a plot of the test code indicator value.

different analytical methods. The first of these was the comparison of the open reading frame with the site of the Tn1000 inactivational insertions described above. Two insertions were in the HindIII-SalI region immediately upstream from the start codon, in the promoter region. The remainder were within the open reading frame. Also included in figure 6 are plots of the GC content and of the test code indicator value across the sequence, a metric of the likelihood that a given sequence is in fact a coding sequence.

CODON USAGE

Analysis of codon usage within the open reading frame was undertaken, and the results are given in Figure 7. Also included are codon usage data from two other Neurospora biosynthetic genes, $\underline{am}^{+}$ and $\underline{trp\text{-}1}^{+}$(30,31). It will be seen that codon usage in the $\underline{pyr\text{-}4}^{+}$ gene is generally similar to that of the other Neurospora genes, with a marked bias against A in the third position of the codon, and some bias

phe				ser				tyr				cys			
TTT	2	7	1	TCT	0	14	3	TAT	1	6	0	TGT	1	0	1
TCT	7	15	16	TCC	4	20	21	TAC	11	10	16	TGC	3	10	5
				TCA	0	1	0								
leu				TCG	9	8	1	his				trp			
TTA	0	0	0					CAT	1	5	0	TGG	6	5	8
TTG	3	10	1	pro				CAC	9	13	10				
CTT	3	21	13	CCT	3	11	4					arg			
CTC	10	26	25	CCC	5	20	15	gln				CGT	3	12	16
CTA	0	2	1	CCA	5	5	0	CAA	2	7	0	CGC	10	12	14
CTG	11	18	1	CCG	6	4	0	CAG	17	23	19	CGA	0	0	0
												CGG	4	3	0
ile				thr				asn							
ATT	5	11	6	ACT	0	5	8	AAT	0	4	1	ser			
ATC	15	33	11	ACC	16	22	8	AAC	9	20	24	AGT	2	4	0
ATA	0	0	0	ACA	7	2	0					AGC	9	12	4
				ACG	10	5	0	lys							
met								AAA	1	1	1	arg			
ATG	9	15	10	ala				AAG	23	43	27	AGA	0	4	0
				GCT	3	19	16					AGG	4	6	0
val				GCC	23	37	32	asp							
GTT	4	11	14	GCA	2	5	0	GAT	8	16	3	gly			
GTC	17	39	16	GCG	13	7	0	GAC	17	24	11	GGT	1	18	22
GTA	1	3	0									GGC	19	34	17
GTG	4	13	0					glu				GGA	3	7	0
								GAA	9	3	0	GGG	3	3	0
								GAG	24	48	32				

FIGURE 7. Codon usage in, from left to right, $\underline{pyr\text{-}4}^{+}$, $\underline{trp\text{-}1}^{+}$, and $\underline{am}^{+}$.

against either purine in that position. Codon usage appears somewhat less stringent than at am+, and of a level of relaxation comparable to trp-1+.

COMPARISON OF PYR-4+ WITH URA-3+

The determined DNA sequence of pyr-4+ predicts a polypeptide gene product of molecular weight circa 44,000 d, but no estimate of the size of the enzyme is available from the literature. This predicted amino acid sequence was compared with that of the S. cerevisiae equivalent, ura3+. The ura3+ DNA sequence predicts a polypeptide of 267 amino acid residues, of molecular weight circa 29,200 d, somewhat smaller than that of Neurospora.

Comparison of the DNA sequences of the two genes was not particularly informative, as the DNA base composition of the two organisms differs significantly, with Neurospora with 54% GC and yeast 41% GC. Comparison of the amino acid sequences was much more revealing. Both organisms start the gene with methionine followed by serine. The next 18 residues in Neurospora and 9 in yeast show no obvious homology, but there then follows a region of circa 111 residues with 58 identities and a further 20 conservative alternatives. There then follows a sequence of 102 residues in pyr-4+, and only 3 in ura3+ with no homology, before the next homologous region. In this, in a sequence of 52 residues, there are 27 identities and 7 conservative changes. Another non-homologous sequence then follows, of 30 residues in Neurospora but 11 in yeast, after which comes a short region of 10 residues which has 6 identities and one conservative change. This is followed by another short non-homologous region of 8 residues in Neurospora and 2 in yeast. The final sequence, of 57 - 59 residues to the carboxy-terminal end, is again homologous, with 31 identities and 9 conservative changes. Figure 8 shows a plot of the pyr-3+ / ura3+ comparison.

UPSTREAM CONSENSUS SEQUENCES

Analysis of the region 5'- to the translation initiation codon has revealed several typical conserved lower eukaryote nucleotide sequences in this promoter region of the pyr-4+ gene.

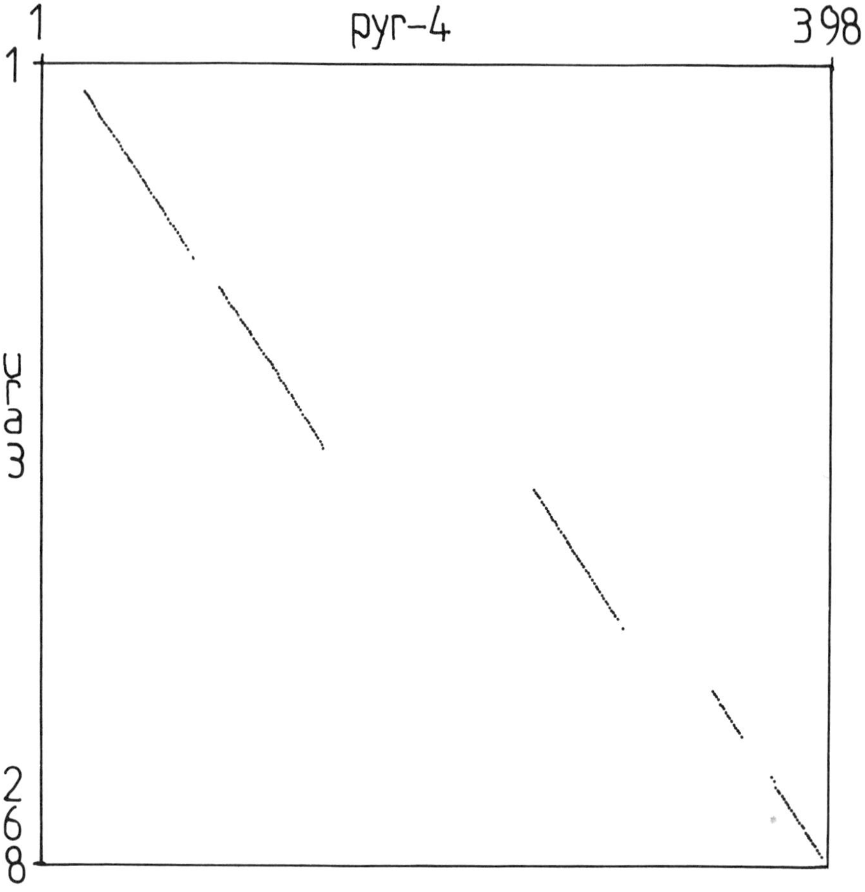

FIGURE 8. A "diagon" plot of the comparison of pyr-3$^+$ and ura3$^+$.

The first of these is the lower eukaryote "CAAT" box. In pyr-4$^+$ the sequence is GGCAAT, centred on position -138. Equivalent GGXAAPy sequences are found at -96 in trp-1$^+$, -155 and -165 in qa-2$^+$, -175 in qa-3$^+$, -150 in yeast ura3$^+$, and at -195 in the cell-1$^+$ gene of Trichoderma reesii. The related sequence CGXAAPy is found at -270 in am$^+$ and at -270 and -140 in histone H4(31,32,33,34,35).

Two possible "TATA" boxes are apparent in pyr-4$^+$, one of TACAATA centred on -90, the other of CAATAT centred on

-60. Nuclease S1 mapping to locate the origin of transcription is in progress.

REFERENCES

1. Makoff AJ, Radford A (1978). Genetics and biochemistry carbamoyl phosphate biosynthesis and its utilization in the pyrimidine biosynthetic pathway. Micro Revs 42:307.
2. Hasunuma K, Ishikawa T (1972). Properties of two nuclease genes of Neurospora crassa. Genetics 70:371.
3. Williams LG, Mitchell HK (1969). Mutants affecting thymidine metabolism in Neurospora crassa. J Bacteriol 100:383.
4. Caroline DF (1969). Pyrimidine synthesis in Neurospora crassa: gene-enzyme relationships. J Bacteriol 100:1371.
5. Mortimer RK, Hawthorne DC (1966). Genetic mapping in Saccharomyces. Genetics 56:165.
6. Pierard A, Glansdorf N, Mergeay M, Wiame JM (1965). Control of biosynthesis of carbamyl phosphate in Escherichia coli. J Mol Biol 14:23.
7. Jones ME (1971). Regulation of pyrimidine and arginine biosynthesis in mammals. Adv Enzyme Regul 9:19.
8. Lacroute F, Pierard A, Grenson M, Wiame JM (1965). The biosynthesis of carbamyl phosphate in yeast. J Gen Microbiol 40:127.
9. Rawls JM, Fristrom JW (1975) A complex genetic locus that controls the first three steps of pyrimidine biosynthesis in Drosophila. Nature (London) 255:738.
10. Rawls JM (1980). Identification of a small genetic region that encodes orotate phosphoribosyltransferase and orotidylate decarboxylase in Drosophila melanogaster. Molec Gen Genet 178:43.
11. Davis RH (1975) Compartmentation and regulation of fungal metabolism: genetic approaches. Ann Rev Genet 9:39.
12. Rigby DJ, Radford A (1982). Investigation of binding sites in the complex pyrimidine-specific carbamoyl phosphate synthetase / aspartate carbamoyltransferase enzyme of Neurospora crassa. Biochim Biophys Acta 709:154.
13. Davis RH, Woodward VW (1962) The relationship between gene suppression and aspartate transcarbamylase activity in pyr-3 mutants of Neurospora. Genetics 47:1075.

14. McDougall KJ, Ostman R, Woodward VW (1969). The isolation of one functional type of pyr-3 mutant of Neurospora. Genetica 40:527.
15. Williams LG, Bernhardt SA, Davis RH (1970). Copurification of pyrimidine-specific carbamyl phosphate synthetase and aspartate carbambyltransferase in Neurospora. Biochemistry 9:4329.
16. Makoff AJ, Buxton FP, Radford A (1978). A possible model for the structure of the Neurospora carbamoyl phosphate synthetase / aspartate carbamoyltransferase complex enzyme. Molec Gen Genet 161:297.
17. Buxton FP, Radford A (1982). Partial characterisation of 5-fluoropyrimidine-resistant mutants of Neurospora crassa. Molec Gen Genet 185:132.
18. Hoffman GR, Malling HV, Mitchell TJ (1973). Genetics of 5-fluorodeoxyuridine-resistant mutants of Neurospora crassa. Can J Genet Cytol 15:831.
19. Makoff AJ, Radford A (1976). The location of the feedback-specific region within the pyrimidine-3 locus of Neurospora crassa. Molec Gen Genet 146:247.
20. Liu CK, Hsu CA, Abbott MT (1973). Catalysis of three sequential dioxygenase reactions by thymine-7-hydroxylase. Arch Biochem Biophys 159:180.
21. Buxton FP, Radford A (1982). Isolation and mapping of fluoropyrimidine-resistant mutants of Neurospora crassa. Molec Gen Genet 185:129.
22. Buxton FP, Radford A (1982). Nitrogen metabolite repression of fluoropyrimidine-resistance and pyrimidine uptake in Neurospora crassa. Molec Gen Genet 186:257.
23. Grove G, Marzluf GA (1981). Identification of the product of the major regulatory gene of the nitrogen control circuit of Neurospora crassa as a nuclear DNA-binding protein. J Biol Chem 256:463.
24. Buxton FP, Radford A (1983). Cloning of the structural gene for orotidine 5'-phosphate carboxylyase of Neurospora crassa by expression in Escherichia coli. Molec Gen Genet 190:403.
25. Ballance DJ, Buxton FP, Turner G (1983). Transformation of Aspergillus nidulans by the orotidine-5'-phosphate decarboxylase gene of Neurospora crassa. Biochem Biophys Res Comm 112:284.
26. Bach ML, Lacroute F, Botstein D (1979). Evidence for transcriptional regulation of orotidine 5'-monophosphate decarboxylase in yeast by hybridisation of mRNA to the structural gene cloned in Escherichia coli. Proc Natl Acad Sci USA 76:386.

27. Sazci A (1983). Transformation of the pyr-3 gene of Neurospora crassa. PhD thesis, University of Leeds.
28. Guyer MS (1978). The "gamma-delta" sequence of F is an insertion sequence. J Mol Biol 126:347.
29. Calos MP, Miller JH (1980). Transposable elements. Cell 20:595.
30. Rose M, Grisafi P, Botstein D (1984). Structure and function of the yeast URA3 gene: expression in Escherichia coli. Gene 29:113.
31. Kinnaird J, Fincham JRS (1983). The complete nucleotide sequence of the Neurospora crassa am (NADP-specific glutamate dehydrogenase) gene. Gene 26:253.
32. Schechtman M, Yanofsky C (1983). Structure of the trifunctional trp-1 gene from Neurospora crassa and its aberrant expression in Escherichia coli. J Mol Appl Genet 2:83.
33. Geever RF, Case ME, Tyler BM, Buxton F, Giles NH (1983). Point mutations and DNA rearrangements 5' to the inducible qa-2 gene of Neurospora allow activator protein-independent transcription. Proc Natl Acad Sci USA 80:7298.
34. Woudt LP, Pastink A, Kempers-Veenstra AE, Jansen AEM, Mager WH, Planta RJ (1983). The genes coding for histone H3 and H4 in Neurospora crassa are unique and contain intervening sequences. Nucl Acids Res 11:5347.
35. Shoemaker S, Schweickart V, Ladner M, Gelfand D, Kwok S, Myambo K, Innis M (1983). Molecular cloning of exo-cellobiohydrolase 1 derived from Trichoderma reesii strain L27. Biotechnology 1:691.

Molecular Genetics of Filamentous Fungi, pages 145–156

REGULATION OF POLYAMINE SYNTHESIS IN *NEUROSPORA CRASSA*[1]

Rowland H. Davis

Department of Molecular Biology and Biochemistry
University of California. Irvine
Irvine. California 92717

ABSTRACT. The polyamines, putrescine, spermidine and spermine, and the initial enzyme of their synthesis, ornithine decarboxylase (ODC) are found in all organisms. The roles of the polyamines are varied and as yet obscure, but they and ODC increase during periods of rapid growth and differentiation in organisms. We have studied the pathway in *Neurospora crassa* in order to answer questions not easily approached in other organisms. We have found that the high degree of regulation of ODC is controlled by spermidine, which negatively governs the increase in the amount of enzyme activity, and by putrescine, which promotes inactivation of the enzyme. The two polyamines are highly compartmented in mycelial cells of *N. crassa*, and the pools active in regulation are a small portion of the total polyamine content. We find that the inactivation of ODC when putrescine accumulates is accompanied by loss of enzyme protein, but there is some evidence that modification may precede destruction of the enzyme. The enzyme is encoded in a unique genetic locus, *spe*-1.

INTRODUCTION

The naturally occurring polyamines, putrescine, spermidine and spermine, are indispensable to growth of eucaryotes (1-4). In most animals and fungi, putrescine (1,4-diami-

[1]This work was supported by grant BC-366 from the American Cancer Society.

nobutane) forms from ornithine in the ornithine decarboxylase (ODC) reaction. Spermidine and spermine form with the addition of aminopropyl moieties to putrescine. The latter come from S-adenosylmethionine (SAM), after its decarboxylation by another polyamine-specific enzyme, SAM decarboxylase (Fig. 1). Putrescine activates SAM decarboxylase (3, 5), thereby coordinating the substrates of spermidine synthesis. ODC is present in small quantities in eucaryotic cells. It requires ornithine and pyridoxal-P and is not inhibited by spermidine or spermine, and only weakly by putrescine, the product of the reaction.

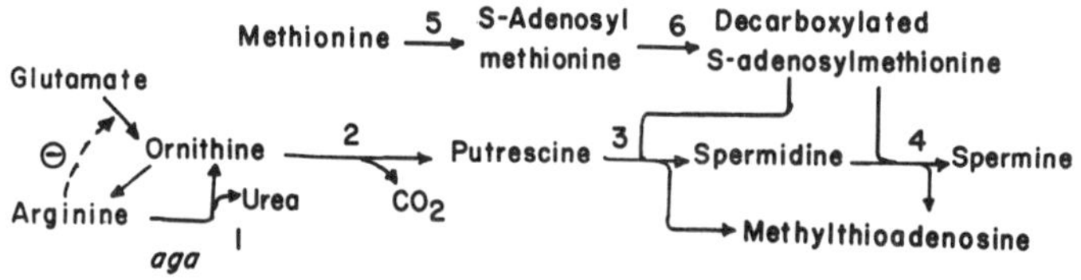

Figure 1. Polyamine synthesis in N. crassa, showing enzymes and intermediates, and the relationship of arginine, glutamate and ornithine, the last initiating the polyamine pathway. In N. crassa, de novo synthesis of ornithine is feedback-inhibited by arginine, as indicated by a circled minus sign. The aga mutation blocks arginase (Enzyme 1), thus contributing to ornithine starvation of strains carrying this mutation when they are grown in the presence of arginine. Enzymes: 1, arginase; 2, ODC; 3, spermidine synthase; 4, spermine synthase; 5, S-adenosylmethionine synthase; 6, S-adenosylmethionine decarboxylase. Methylglyoxal-bis-guanylhydrazone (MGBG) inhibits enzyme 6 and dicyclohexylamine (DCHA) inhibits enzyme 3 in most organisms tested.

Polyamines, being strong cations at cellular pH, bind ribosomes, DNA and membranes. Ionic binding makes it difficult to relate the polyamines to control of their synthesis or to specific functions (6). Spermidine and spermine promote fidelity in protein synthesis (7); they participate in DNA gyrase and other nucleic acid enzyme reactions (8); and may be promoters of membrane fusion (9). None of the many postulated biological roles of polyamines have been definitively proved: polyamines serve a number of roles, some of which can be performed by other cations such as

Mg^{2+} (6). Because polyamine-deficient mutants and cells treated with polyamine inhibitors do not grow, however, polyamines are clearly indispensable metabolites (2).

ODC usually displays large and rapid variation in activity. Many stimuli, such as polyamine starvation, hormone treatment, factors stimulating the onset of growth, differentiation and transformation or conditioning cells to the neoplastic state, lead to increased ODC activity in mammalian cells (2). In some cases, this has been related to *de novo* enzyme synthesis (10), but a portion of the increase appears to happen without transcription or translation (11). The mechanisms of augmentation and inactivation of ODC appear to vary from one system to the next, and in most cases even the metabolic signal is not known. Two issues complicate this matter. First, the change in polyamine pools is very poorly correlated with ODC variation. This has cast doubt upon the roles of polyamines in ODC control. Second, although ODC variation has been related to administration of polyamines, more fundamental changes in growth, macromolecular synthesis or differentiation are also correlated with ODC variation. The effects of growth stimuli and cessation of growth upon ODC may or may not be mediated by polyamines.

Work with *N. crassa*, *Saccharomyces cerevisiae* and *Aspergillus nidulans* does not yield a unified picture of fungal polyamine synthesis. However, *N. crassa* displays much of the phenomenology of mammalian systems. Accordingly, we have used this tractable organism to investigate four problems which had remained unsolved in higher systems:

a) Why is the regulation of the ODC reaction not correlated with the polyamine content of cells?

b) What are the metabolic signals governing changes in ODC activity?

c) Is there a single genetic locus encoding ODC in the *N. crassa* genome?

d) What is the mechanism of inactivation of ODC?

THE POOL PROBLEM

In Neurospora, the ODC reaction is the sole route of putrescine synthesis (12-14). Putrescine is transformed to spermidine, the major polyamine, and then to spermine (Fig. 1). At steady-state, the pool sizes of the polyamines in exponentially growing cultures are (per mg, dry weight) 0.5 - 1.0 nmoles putrescine, 15-21 nmoles spermidine, and 0.3 -

0.5 nmoles spermine (15). No turnover of polyamines (except as intermediates in the pathway) has been detected by isotopic studies (16, 17).

Arginase-less (aga) mutants of N. crassa grow normally on minimal medium, but become polyamine-starved in the presence of arginine (12). This reflects feedback-inhibition of de novo ornithine synthesis, combined with the genetic block in arginase, the alternate source of ornithine (Fig. 1). The cultures grow indefinitely in the presence of arginine at about half the normal rate, and ODC is augmented about 70-fold under these conditions (18). The weak ability of ODC to decarboxylate lysine is expressed sufficiently to produce polyamine analogs which support the slow growth observed. We have taken advantage of the aga mutant in all of our work. In this work, we measure ODC activity in permeabilized cells and polyamines in extracts, both taken from the same exponential culture. Polyamines and ODC accumulation can therefore be expressed on the basis of culture volume, in addition to a dry-weight or protein basis.

The aga (or wild-type) strain, grown in minimal medium, accumulates polyamines at a low rate (0.06 - 0.09 nmoles per min per mg dry weight); the activity of ODC, a cytosolic enzyme (19), is also in this range. Upon addition of arginine, which creates a state of ornithine deprivation, ODC becomes 70- to 100-fold elevated in activity before polyamine pools change significantly (20). After prolonged ornithine deprivation, addition of ornithine causes rapid polyamine synthesis and rapid loss of ODC activity. The onset of inactivation takes place before the polyamine content is restored to normal (20). The data suggested that, if polyamines control the behavior of ODC, spermidine and/or putrescine must be highly compartmented in the cells, and only a small "free" fraction of the content could be active in regulation. According to this view, free molecules of polyamine would be those in a transient state between their appearance in biosynthesis and their sequestration. The free pool would therefore quickly disappear upon ornithine deprivation, and it would quickly reappear upon restoration of ornithine.

We sought evidence for "free" and "bound" spermidine pools by isotopic studies in vivo (16). A trace of [^{14}C] ornithine was added to living cells so that it would enter the polyamine pathway and become incorporated into spermidine. It was reasoned that if the large, resident spermidine pool freely diluted the [^{14}C]spermidine molecules as they were used for the synthesis of the terminal polyamine, spermine, label would not be detected in spermine for over two hours. The result was that the label appeared in spermine within 7 minutes. This indicated that the resident pool of spermidine was unable to trap newly synthesized spermidine molecules before they were used for spermine synthesis. The specific radioactivity of new molecules of spermine was over 8-fold that of the extractable spermidine early in the experiment. Thus 80-90% of the spermidine in N. crassa is not on the main line of polyamine biosynthesis (16). These data show that newly synthesized spermidine is a distinct metabolic fraction whose magnitude depends upon the rate of synthesis and the rate of sequestration. In fact, we found that at least one-third of the sequestered spermidine could be accounted for by the vacuolar pool, where it is membrane-confined, non-exchangeable even in vitro, and ionically bound to polyphosphate. Our findings have a considerable implication for other organisms, even those lacking vacuoles, because the non-vacuolar, bound fraction may have a similar disposition in all organisms. This fraction may be the one found associated with ribosomes and nuclei in cell extracts, a finding which has been impossible so far to distinguish from artifacts of the extraction process. Sequestration of putrescine in N. crassa had been proved earlier (21).

It is logical to suggest that the small, free polyamine pools defined as a metabolic intermediates might be the postulated controlling signals for ODC (20). At least the isotope experiments demonstrate a plausible mechanism by which polyamines can act as a metabolic signals despite the small changes in their cellular content during regulatory excursions of ODC.

METABOLIC SIGNALS GOVERNING ODC

We have seen that ODC activity is augmented up to 100-fold in response to ornithine deprivation. This does not prove that polyamines exert regulatory effects. We reasoned that other ways of inhibiting polyamine synthesis would allow more definite and specific assignments of roles of the polyamines. Methylglyoxal bis-guanylhydrazone (MGBG), an inhibitor of S-adenosylmethionine decarboxylase, and dicyclohexylamine (DCHA), an inhibitor of spermidine synthetase (see Fig. 1), were used (20). Addition of either caused only 10- fold increase in ODC activity, and neither interfered with the 100-fold effect of ornithine deprivation. Common to the effect of both of these compounds, and in contrast to the effect of ornithine deprivation, was the accumulation of putrescine. This suggested that while spermidine starvation caused augmentation of activity, putrescine interfered in some way with maximal augmentation.

When untreated cultures or those treated with arginine (ornithine-deprived), MGBG or DCHA were treated with cycloheximide, they differed markedly. ODC was relatively stable in ornithine-deprived cultures, having very high ODC activity. In contrast, ODC was rapidly inactivated in the other cultures, with much lower activity. The implication was that putrescine actually opposed the augmentation of activity by inducing its constant inactivation (20). The results were consistent with the observation that addition of ornithine to an ornithine-starved culture caused both a huge intracellular pool of putrescine to develop (some is excreted) and a rapid loss of activity. Experiments in which putrescine was added directly to cycloheximide-treated, ornithine-starved cultures, confirmed our interpretation (20).

The upshot of these experiments was that spermidine exerts its control by blocking formation of active enzyme (possibly by effects upon transcription), while putrescine has its effect by promoting enzyme inactivation. It is the free pools of these compounds which appear to exert these effects, because they are felt rapidly upon any interference with synthesis of these metabolites. Our findings complement those of workers in mammalian systems because until recently there were no ways of interfering with polyamine synthesis without interfering with polyamine enzymes. Our ability to starve cells specifically for ornithine allowed us to discern a distinct role for putrescine, and to integrate this knowledge with our discovery of a small, "free" pool of the metabolites controlling ODC.

GENETIC CONTROL OF ODC

In higher organisms, it has been difficult to identify the structural gene for ODC. Mutations affecting ODC have been isolated, but none yields evidence that it affects the structure, as opposed to the regulation, of the enzyme. This problem has been encountered in fungi: until recently, there were mutants lacking ODC entirely in *N. crassa* (18, 22) and *A. nidulans* (23), none of which defined itself as a structural variant. In *S. cerevisiae*, mutants lacking ODC were also found (24, 25), but through studies of revertant derivatives, it was concluded arguably that they were regulatory mutants (25). Because antibodies to ODC have become available only recently in certain species, the ODC-less mutants could not be investigated immunologically with respect to the presence and structure of mutant proteins. The problem is compounded by the extreme difficulty of obtaining mutants in the first place. This reflects a long phenotypic lag, in many cases, during which a polyamine requirement is gradually established.

Improved methods of filtration-enrichment in our laboratory yielded five ODC-deficient mutants, most of them temperature sensitive (18, 26). One has an almost complete dependence on spermidine or putrescine supplementation and no detectable ODC (18); the other four grow slowly without supplementation at 25°C, and little or not at all at 35°C. All of the latter class have detectable ODC activity when grown without polyamines at 25°C, but this activity is not temperature-sensitive (26). (The temperature-sensitive phenotype is presumably owing to the severe ODC deficiency, whose effects are exacerbated at higher temperature.) No mutant displays activity after growth in the presence of polyamines, indicating that the activity seen in the starved cultures is, as expected, regulated upward at least 10-fold. The residual ODC of one mutant, PE85, has a higher K_m for ornithine and an increased thermostability, which in mixed extracts with wild type are shown to be intrinsic properties of the mutant ODC. This mutant grows much better if ornithine is added to the medium, and in fact, ornithine permits growth at half the normal rate at 35°C. All mutants, when starved, display ODC cross-reacting material on Western blots, using antiserum to the pure enzyme (see below). The level of antigen is normally regulated by polyamines (26). These properties are evidence that the mutation lies in the structural gene for ODC. At 35°, there is no complementation among any of the mutants, including two earlier ODC-less mutants isolated by MacDougall (22). This suggests that the 7 existing mutations are alleles of a single genetic locus, unique in specifying ODC. The locus maps Linkage Group V, in a 1.5 map-unit interval between *am* (glutamate dehydrogenase) and *his-1* (18, 22, 26).

Since we found these mutations, others have cloned the gene from mammals (27) and from *S. cerevisiae* (28). The cloned yeast gene complements one of the ODC-less mutants of this organism mentioned above, and specifies an ODC with the properties of the yeast enzyme in an E. coli host. Thus the locus at which ODC-less mutants of yeast lie is very likely to be the structural gene for ODC in this organism. Given knowledge of the structural gene and having mutations lacking the enzyme, it will be relatively straightforward to recognize clones of the gene by complementation of ODC-less recipients of N. crassa.

INACTIVATION OF ODC

We have purified ODC to homogeneity, starting with the ODC-augmented aga mutant. The enzyme is 550-700 fold the activity of the original extract, or over 35,000 fold the activity of cells grown on minimal medium. Gel filtration suggests a native molecular weight in the 100 kd range; the preparation shows a major 52 kd band and a minor 44 kd band (proteolytic artifact) on 1D SDS gels (20, 26). The native enzyme is a dimer. Its physical properties are similar to those of the mammalian enzyme, and somewhat different from those of the yeast enzyme, which appears to be a monomer of 86 kd (29).

A rabbit antiserum was raised to the purified ODC of N. crassa. Western blots of crude extracts run on SDS gels reveal the 52 kd band which varies with the polyamine status (20). The 52 kd band is highly augmented when ODC is augmented. The 52 kd band is lost upon restoration of polyamine synthesis in ornithine-deprived aga cells. There is only a small amount of the 52 kd in wild-type or ODC-less mutant cells grown in the presence of spermidine (20).

Counting the radioactivity associated with the 52 kd band of Western blots shows that loss of ODC protien is slower than loss of activity (20). It may be that ODC is first inactivated, then proteolyzed. We presume that increased ODC synthesis or activation contributes to augmentation of activity, but that lower rates of inactivation account for some of the increase. Our material gives us an effective entree into the mechanisms of ODC inactivation, and this is the focus of our current efforts.

OPEN QUESTIONS

N. crassa has contributed to the solution of several problems in polyamine biochemistry and regulation. The organism ofers one of the most coherent accounts of this pathway, in which biochemical, genetic and molecular techniques can be coordinately applied. The picture of polyamine biochemistry we are developing is not a complete one, nor can it be universalized. Variations among organisms in the details of regulation of the path can be expected; even some fundamental phenomena may differ. Nevertheless, N. crassa has allowed us to define the issues of compartmentation, metabolic control and genetics clearly, and the experimental logic is applicable to more complex systems if the complexities are controlled.

What we are left with are some of the oldest questions in polyamine biochemistry. First, what are the roles of these compounds in the cell? Can any of them be definitively demonstrated? Or are polyamines destined to be accepted only slowly as significant compounds having varied roles, such as potassium, magnesium and other inorganic ions?

Second, how is growth tied to the polyamine pathway? The circumstantial connection is great, but the mechanistic one is almost wholly unknown. Polyamines may have roles as stoichiometric entities, cofactors and regulators. One possibility is that polyamine-dependent (or -stimulated) protein kinases, recently demonstrated, may have roles in regulating rates of macromolecular synthesis. An even tighter connection was suggested earlier by Kuehn, working with the slime mold, Physarum polycephalum. He has presented evidence that a spermidine-dependent protein kinase phosphorylates a nucleolar form of ODC (30, 31). This inactivates the enzyme with respect to putrescine synthesis, but activates it as a cofactor in rRNA synthesis. This is one of the earliest suggestions of how the polyamine pathway participates in the onset of growth, and others may be discovered soon.

Third, how does ODC respond to stimuli leading to the onset of growth? Are these stimli mediated by polyamines, either by synthesis, depletion or sequestration of existing pools? Or does the onset of growth include its own signals, parallel to the polyamines, in stimulating ODC?

We hope that continued study of the polyamine pathway in filamentous fungi will make further contributions to the problems common to all groups of organisms, including plants and mammals.

ACKNOWLEDGMENTS

The work reviewed here was done with Tom Paulus, Carole Cramer, Gary Krasner, Joe DiGangi, Pam Eversole and Janet Ristow. I thank them warmly for their contributions.

REFERENCES

1. Cohen SS (1971). "Introduction to the Polyamines" Englewood Cliffs, NJ: Prentice-Hall.
2. Tabor CW and Tabor H (1984). Polyamines. Ann Rev Biochem 53:749-790.
3. Williams-Ashman HG and Canellakis ZN (1979). Polyamines in mammalian biology and medicine. Perspectives Biol Med 22:421-453.
4. Stevens L and Winther MD (1979). Spermine, spermidine and putrescine in fungal development. Adv Microbial Physiol 19:63-148.
5. Hart D, Winther M and Stevens L (1978). Polyamine distribution and S-adenosylmethionine decarboxylase activity in filamentous fungi. FEMS Microbiol Lett 3:173-175.
6. Canellakis ES, Viceps-Madore D, Kyriakidis DA and Heller JS (1978). The regulation and function of ornithine decarboxylase and of the polyamines. Curr Topics Cell Reg 15:155-202.
7. Tabor H, Tabor CW, Cohn MS and Hafner EW (1981). Streptomycin resistance (rpsl) produces an absolute requirement for polyamines for growth of an Escherichia coli strain unable to synthesize putrescine and spermidine (△speA-speB) △speC. J Bacteriol 147:702-704.
8. Krasnow MA and MR Cozzarelli (1982). Catenation of DNA rings by topoisomerases. Mechanism of control by spermidine. J. Biol Chem 257:2687-2693.
9. Schuber F, Hong K, Duzgunes N and Papahadjopoulos D (1984). Polyamines as modulators of membrane fusion: aggregation and fusion of liposomes. Biochemistry 22: 6134-6140.
10. Persson L, Seeley JE and Pegg AE (1984). Investigation of structure and rate of synthesis of ornithine decarboxylase protein in mouse kidney. Biochemistry 23: 3777-3783.
11. Tabor CW and Tabor H (1976). 1,4-Diaminobutane (putrescine), spermidine and spermine. Ann Rev Biochem 45: 285-306.

12. Davis RH, Lawless MB and Port LA (1970). Arginaseless Neurospora: genetics, physiology and polyamine synthesis. J Bacteriol 102:199-305.
13. Karlin JN, Bowman BJ and Davis RH (1976). Compartmental behavior of ornithine in Neurospora crassa. J Biol Chem 251:3948-3955.
14. Bowman BJ and Davis RH (1977). Cellular distribution of ornithine in Neurospora: anabolic and catabolic steady states. J Bacteriol 130:274-284.
15. Davis RH and Paulus TJ (1983). Uses of arginaseless cells in the study of polyamine metabolism (Neurospora crassa). Meth Enzymol 94:112-116.
16. Paulus TJ, Cramer CL and Davis RH (1983). Compartmentation of spermidine in Neurospora crassa. J Biol Chem 258:8608-8612.
17. Paulus TJ and Davis RH (1981). Regulation of polyamine synthesis in relation to putrescine and spermidine pools in Neurospora crassa. J Bacteriol 145:14-20.
18. Paulus TJ, Kiyono P and Davis RH (1982). Polyamine-deficient Neurospora crassa mutants and synthesis of cadaverine. J Bacteriol 152:291-297.
19. Weiss RL and Davis RH (1973). Intracellular localization of enzymes of arginine metabolism in Neurospora. J Biol Chem 248:5403-5408.
20. Davis RH, Krasner GN, DiGangi JJ and Ristow JL (1985). Distinct roles of putrescine and spermidine in the regulation of ornithine decarboxylase in Neurospora crassa. Proc Natl Acad Sci US (in press).
21. Paulus, TJ and Davis RH (1982). Metabolic sequestration of putrescine in Neurospora crassa. Biochem Biophys Res Commun 104:228-233.
22. McDougall KJ, Deters J and Miskimen (1977). Isolation of putrescine-requiring mutants of Neurospora crassa. Antonie van Leeuwenhoek J Microbiol Serol 43:143-151.
23. Sneath PHA (1955). Putrescine as an essential growth factor for a mutant of Aspergillus. Nature 175:818-819.
24. Whitney PA and Morris DR (1978). Polyamine auxotrophs in Saccharomyces cerevisiae. J Bacteriol 134:214-220.
25. Cohn MS, Tabor CW and Tabor H (1980). Regulatory mutations affecting ornithine decarboxylase in Saccharomyces cerevisiae. J Bacteriol 142:791-799.
26. Eversole P, DiGangi JJ, Menees T and Davis RH (1985). Structural gene for ornithine decarboxylase in Neurospora crassa. Mol Cell Biol (in press).

27. McConlogue L, Gupta M, Wu L and Coffino P (1984). Molecular cloning and expression of the mouse ornithine decarboxylase gene. Proc Natl Acad Sci US 81:540-544.

28. Fonzi WA and Sypherd PS (1985). Expression of the gene for ornithine decarboxylase of *Saccharomyces cerevisiae* in *Escherichia coli*. Mol Cell Biol 5:161-166.

29. Tyagi AK, Tabor CW and Tabor H (1981). Ornithine decarboxylase from *Saccharomyces cerevisiae*: purification, properties and regulation. J Biol Chem 256:12156-12163.

30. Atmar VJ and Kuehn GD (1981). Phosphorylation of ornithine decarboxylase by a polyamine-dependent protein kinase. Proc Natl Acad Sci US 78:5512-5518.

31. Kuehn GD (1984). Regulation of ornithine decarboxylase by reversible phosphorylation: an example of an end-product mediated protein kinase reaction. *In Enzyme Regulaton by Reversible Phosphorylation. Further Advances*. (Cohen, ed.) Elsevier, NY pp 185-207.

Molecular Genetics of Filamentous Fungi, pages 157–169

STRUCTURE AND EXPRESSION OF THE ASPERGILLUS[1] *amd*S GENE

Michael J. Hynes, Joan M. Kelly,[2]
Catherine M. Corrick, and Timothy J. Littlejohn

Department of Genetics, University of Melbourne,
Parkville, Victoria 3052, Victoria.

ABSTRACT The *amd*S gene codes for an amidase enzyme allowing utilization of acetamide. *Cis* and *trans*-acting mutations affecting regulation have been isolated and used to show that the gene is subject to control by multiple independently acting regulatory genes. Cloning and sequencing of some mutant *amd*S genes has shown that alterations at the 5' end of the gene, but within the coding region, can result in reduced levels of the *amd*S transcript. A single base pair substitution 171 base pairs upstream of the transcription start point occurs in the *amd*I9 mutation which results in increased *fac*B dependent acetate inducibility. Evidence is presented for this mutation resulting in increased affinity for a regulatory gene product-presumably the *fac*B gene product. The *amd*I93 mutation abolishes ω-amino acid induction mediated by the *amd*R (or *int*A) regulatory gene. The *amd*I93 mutation contains an approximately 33 base pair deletion of a sequence which includes a direct 9 base pair repeated sequence. Studies using transformation indicate that this region of the *amd*S gene is involved in binding of the *amd*R gene product.

[1]This work was supported by the Australian Research Grant Scheme.

[2]Present address: Department of Genetics, University of Adelaide, South Australia.

INTRODUCTION

The structure and regulation of many genes in lower eukaryotes have been characterized by genetic methods. The advantage of such systems is that defined regulatory genes may be identified by mutation. In the majority of systems in higher eukaryotes such genes are not readily identified. A very wide range of regulation systems involved in catabolism have been studied in *Aspergillus nidulans*. One of the most complex but well studied of these systems is discussed here.

The *amd*S gene codes for an acetamidase enzyme which enables *A. nidulans* to use acetamide as a sole nitrogen or a sole carbon source. Regulation of this gene by multiple control circuits has been shown by studies of enzyme induction and regulatory mutations (1, 2). These circuits are summarized in Table 1. A particular feature of this system has been the success in isolating specific *cis*-acting mutations which affect one regulatory circuit but not the others. The isolation of *amd*S DNA clones (2) has allowed the study of the sequences affected by these mutations. Furthermore, the development of *A. nidulans* transformation using the *amd*S gene (3), has made possible studies in which multiple copies of *amd*S DNA result in apparent titration of regulatory gene products.

TABLE 1
*amd*S REGULATORY CIRCUITS

Regulatory gene	Effector	Co-regulated genes
*amd*R (*int*A)	ω-Amino acids	*gat*A, *gab*A, *lam*A
*fac*B	Acetate	*fac*A, *acu*D, *acu*E
*amd*A	Acetate	*aci*A
*are*A	Ammonium (Glutamine)[a]	Many nitrogen catabolic systems
Not known	Benzoate	Unknown genes
*cre*A	Catabolite repression	Many carbon catabolic systems

[a]Recent evidence suggests that repression by glutamine may be separable from ammonium repression but this requires confirmation.

RESULTS

Mutations in the Coding Region Leading to Lowered Levels of *amd*S RNA.

Previous genetic analysis of *amd*S revealed a group of mutations at the 5' end of the gene which completely abolished *amd*S expression (1). Several of these mutations have now been sequenced (Figure 1). The *amd*-407 mutation was found to be a deletion encompassing the start-point of transcription as well as the proposed TATA and CAAT boxes. No detectable *amd*S transcript was observed in Northern analysis. The *amd*-709, 710 and 513 mutant sequences contained alterations downstream from the start point of transcription, and each of these mutations resulted in reduced levels of the *amd*S transcript. The 80 base pair deletion present in *amd*-513 was reflected in a shorter transcript. These mutations were all up-stream from the putative first intron sequence (Corrick and Hynes, unpublished results) indicating that their effects are unlikely to be on RNA splicing.

Characterization of the *amd*I9 and *amd*I93 *cis*-acting Mutations.

The *amd*I9 mutation results in increased inducibility of *amd*S expression by acetate, without affecting other inducer responses (5). Recessive mutations in the *fac*B gene are epistatic to the *amd*I9 mutation (5), and the semi-dominant *fac*B88 mutation results in constitutive *amd*S expression with respect to acetate induction (Hynes, unpublished results). These results indicate that the *amd*I9 mutation causes an alteration in the response of *amd*S to *fac*B mediated activation. Acetate induction was shown to cause increased *amd*S RNA levels in an *amd*I9 strain compared with wildtype (Figure 2). A single base pair substitution at position -171 relative to the transcription start point was found in *amd*I9 (Figure 2).

The *amd*I93 mutation specifically abolishes induction by ω-amino acids without affecting induction by other inducers. As predicted, *amd*I93 is epistatic to the semi-dominant *amd*R^c mutations which are partially constitutive for ω-amino acid induction (6). Northern analysis showed that *amd*S RNA was not inducible by β-alanine in an *amd*I93 mutant (Figure 2). Sequencing of *amd*I93

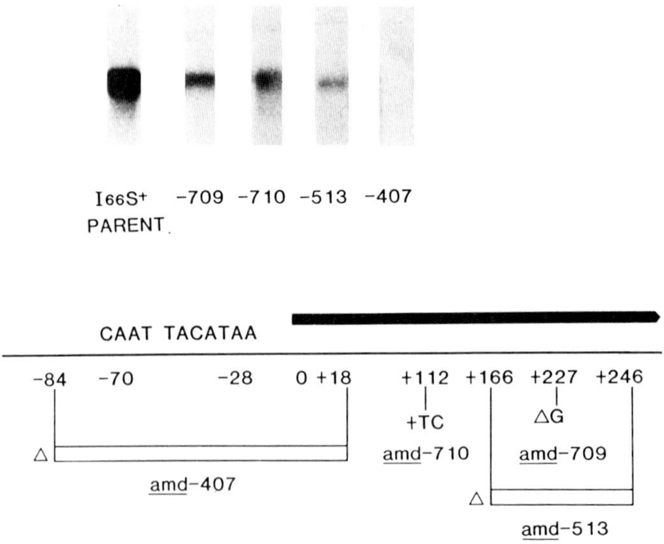

FIGURE 1. Characterization of *amd*S mutations. RNA was isolated from the parent *amd*I66 *amd*S$^+$ and the *amd*-407, 709, 710 and 513 mutant strains grown in the presence of the inducers acetate and β-alanine for 4 hours and subject to Northern blot analysis using ^{32}P-labelled plasmid p3SR2 as probe (1). Mutant DNA was isolated from lambda gene banks generated by EcoR1 digestion. Sequencing was by the M13-dideoxy method (4).

revealed an approximately 33 base pair deletion up-stream from the transcription start point (Figure 2). A 9 base pair directly repeated sequence was found within the deletion and this may be significant for the *amd*R mediated induction response.

The use of Transformation to Study *amd*S Gene Regulation.

Acetamide is only a very poor sole source of nitrogen for wildtype strains in the presence of sucrose. In fact this has proved extremely useful in the characterization of regulatory mutants with increased acetamidase levels. This led to the finding that transformation of wildtype strains using the procedure of Tilburn et al. (3) with the *amd*S$^+$ containing plasmid p3SR2 was possible. Transformants were readily detected on sucrose-acetamide

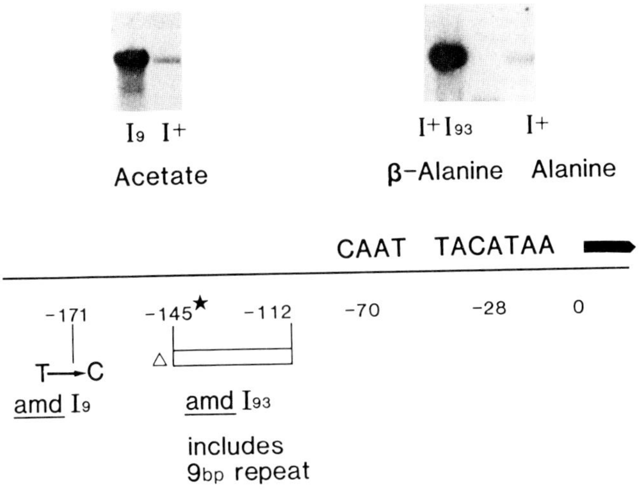

FIGURE 2. Characterization of *amd*I9 and *amd*I93 mutations. Northern blot and sequencing analyses were performed as described in the legend to Figure 1. Induction by acetate of *amd*I9 and wildtype strains was for 4 hours in the presence of ammonium chloride in the absence of glucose. Wildtype and *amd*I93 strains were grown in glucose containing medium with β-alanine or L-alanine as the sole nitrogen source for 20 hours. The asterix marks a region of uncertainty in sequencing the end point of the *amd*I93 deletion.

medium containing 15 mM CsCl. Southern and dot blot analysis showed that these transformants contained multiple tandemly integrated plasmid copies. The transformant T8/11 was found to have a plasmid copy number in excess of 100.

When tested for growth on various nitrogen and carbon sources it was found that transformants grew as well as the wildtype strain. The exception was transformant T8/11 which was impaired in growth on the lactam, 2-pyrrolidinone, and the ω-amino acids, β-alanine and γ-aminobutyric acid as either nitrogen or carbon sources. This is illustrated in Figure 3. Utilization of these compounds is dependent on the functions of the *lam*A, *gat*A and *gab*A genes which are under the control of the *amd*R (or *int*A) gene (7, 8). The results therefore suggested that

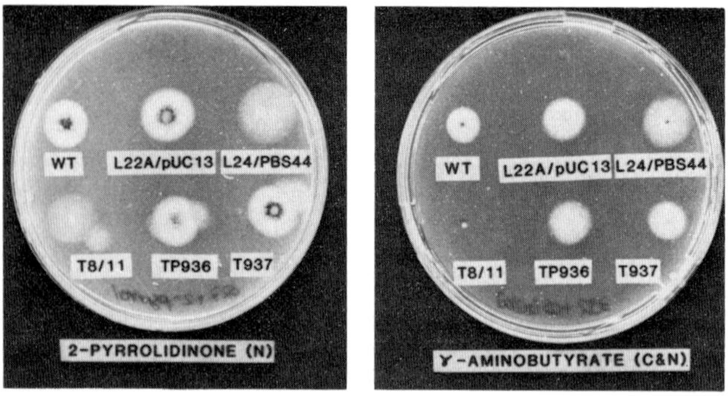

FIGURE 3. Growth of transformants on 2-pyrrolidinone and γ-aminobutyric acid media. 2-pyrrolidinone (10 mM) was present as the sole nitrogen source in the presence of 1% glucose. γ-aminobutyric acid (50 mM) was present as the sole carbon and nitrogen source. WT = wildtype strain; L22A/pUC13 = multicopy transformant containing plasmid SmR1-pUC9-9; L24/PBS44 = multicopy transformant containing plasmid p3BS1; T8/11 = multicopy transformant containing plasmid p3SR2; TP936 and T937 = multicopy transformants containing plasmid *amd*I93-322 (see Table 2). Plates were incubated for 2 days at 37°C.

multiple copies of the p3SR2 plasmid resulted in titration of the positively acting *amd*R gene product leading to decreased activation of the genes responsible for ω-amino acid utilization. Growth of T8/11 was not affected on other nitrogen or carbon sources.

When tested for growth on various nitrogen and carbon sources it was found that transformants grew as well as the wildtype strain. The exception was transformant T8/11 which was impaired in growth on the lactam, 2-pyrrolidinone and the ω-amino acids, β-alanine and γ-aminobutyric acid as either nitrogen or carbon sources. This is illustrated in Figure 3. Utilization of these compounds is dependent on the functions of the *lam*A, *gat*A and *gab*A genes which are under the control of the *amd*R (or *int*A) gene (7, 8). The results therefore suggested that multiple copies of the p3SR2 plasmid resulted in titration of the positively acting *amd*R gene product leading to decreased activation of the genes responsible for ω-amino acid utilization. Growth of T8/11 was not affected on other nitrogen or carbon sources.

The *cis*-acting *amd*I9 mutation was cloned along with a wildtype *amd*S gene on a 5kb EcoR1-SalI fragment inserted into pBR322. This plasmid was used to transform a wildtype strain and transformants selected, as before, for strong growth on sucrose-acetamide media. Again multiple integrated plasmid copies were detected by blot analysis. Two transformants had copy numbers in excess of 20 - T5/18 (20-25 copies) and T5/6 (greater than 70). These transformants grew more poorly than wildtype on lactam and ω-amino acid media with the effects being much greater for T5/6 than T5/18. In addition both of these transformants grew poorly on media in which acetate or acetamide was the sole carbon source. This contrasted with the T8/11 transformant containing multiple copies of the wildtype *amd*S sequence. This is illustrated in Figure 4. These results suggested that multiple copies of the *amd*I9 but not the wildtype plasmid resulted in titration of the product of the *fac*B regulatory gene leading to a reduced ability to express the genes required for acetate utilization.

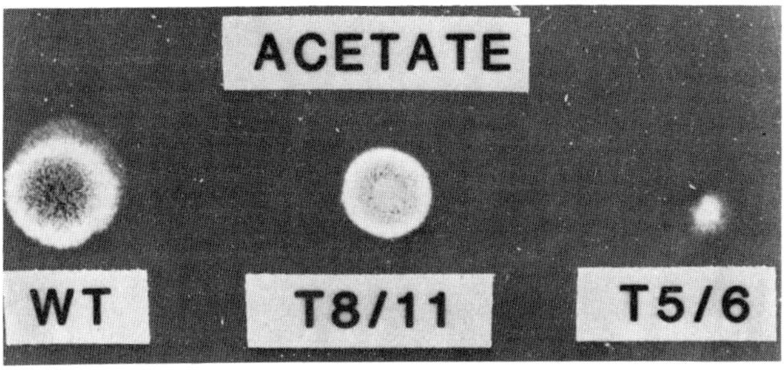

FIGURE 4. Growth of transformants on acetate medium. The plate contained 50 mM sodium acetate as the sole carbon source with 10 mM NH4Cl present as the sole nitrogen source. Transformants T8/11 and T5/6 contain multiple copies of the p3SR2 (*amd*I$^+$) and *amd*I9-322 plasmids respectively.

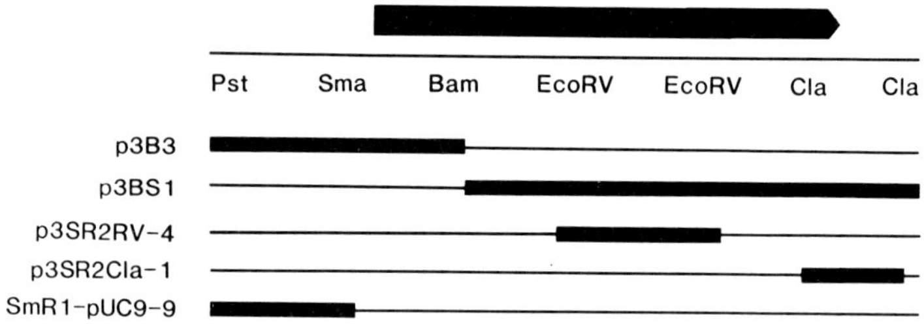

FIGURE 5. Structure of *amd*S plasmid derivatives. The upper arrow shows the direction and extent of *amd*S transcription relative to the restriction sites. The open boxes show the sequences deleted in the plasmids. The SmR1-pUC9-9 plasmid was obtained by cloning the 3.6kb SmaI-EcoRI fragment of p3SR2 (2) into pUC9. Transformants containing this plasmid were found to express *amd*S.

It was predicted that the sequence responsible for titration of the *amd*R product would be localized at the 5' end of the *amd*S gene. Various *in vitro* manipulated plasmids were constructed by subcloning the p3SR2 plasmid and the generation of deletions within the *amd*S gene and at the 3' end of the gene (Figure 5). Since these plasmids lacked functional *amd*S sequences direct transformation studies were not possible. However it was found that co-transformation frequencies in excess of 50% were possible when a proline utilization defective strain (*prn*$^-$) was transformed with a mixture of an *amd*S plasmid with a plasmid containing *prn*$^+$ genes (kindly provided by Peter Green and Claudio Scazzocchio). Therefore *prn*$^-$ protoplasts were transformed with a *prn* plasmid and the *amd*S derived plasmids. Transformants were growth tested on ω-amino acid media. The results (Table 2 and Figure 3) showed that only plasmids containing the 5' *amd*S sequence resulted in transformants with reduced growth on ω-amino acids.

Furthermore the results for the SmR1-pUC9-9 plasmid indicated that the sequence responsible for *amd*R product titration was upstream of the SmaI site, consistent with the localization of the *amd*I93 mutation (see above). Finally, in approximately 40 transformants containing the *amd*I93 mutant sequence none were found to have reduced ω-amino acid utilization. These results provided support for the proposal that the sequence deleted in the *amd*I93 mutant is necessary for binding of the *amd*R product.

TABLE 2
PROPERTIES OF TRANSFORMANTS CONTAINING *amd*S DERIVED PLASMIDS

Plasmid	No. of Transformants	No. with Reduced Growth on ω-amino acids
p3B3	59	0
p3BS1	39	2
p3SR2-RV-4	40	10
p3SR2Cla-1	20	4
SmR1-pUC9-9	60[b]	0
*amd*I93-322[a]	40[b]	0

[a]This plasmid was a 5kb EcoR1-SalI subclone of a lambda clone obtained from an *amd*I93 strain. Transformants containing this plasmid were obtained by Meryl Davis.

[b]Some of these transformants were obtained by direct selection on acetamide medium. The remainder were derived by co-transformation using a *prn*$^+$ containing plasmid and the particular *amd*S derived plasmid to transform a strain of relevant genotype *amd*S368; *prn*-309.

DISCUSSION

The complexities of the regulation of the *amd*S gene of *A. nidulans* are beginning to be understood at a molecular level. The results summarized here represent two lines of approach - DNA sequencing of *in vivo* generated mutations with defined phenotypes and the use of transformation to generate multiple copies of sequences apparently involved in regulatory gene product binding.

The sequencing of the *amd*-709, *amd*-710 and *amd*-513 mutations has revealed that these result from alterations downstream from the start-point of transcription, within the *amd*S coding region. The reduced amounts of *amd*S transcripts observed in these mutants are not expected. More detailed studies on the regulation of transcription in these mutants is required to see if the level of transcription is affected under all induction conditions. The possibility is raised by these results that there is some mechanism coupling translation and transcription.

The *amd*I9 mutation results from a single base pair substitution upstream from the transcription start point. Multiple copies of this mutant sequence, but not the wildtype sequence, in transformants cause reduced growth on acetate. It is proposed that this results from titration of the *fac*B gene product such that there is now insufficient levels of this activator to allow expression of genes required for acetate utilization. The *amd*I9 mutation results in greatly increased affinity for *fac*B product binding. This is reflected in the increased *fac*B dependent inducibility of *amd*S in *amd*I9 containing strains. A prediction of this model is that multiple copies of a small DNA sequence containing the *amd*I9 mutation should result in reduced growth on acetate. Determination of the size of the fragment required for the effect will allow assessment of the sequences necessary for regulatory product binding. In addition it should be possible to address the question of whether regulatory product binding is dependent on RNA polymerase binding. If this were the case then a large sequence including the promoter region would be the minimum sized DNA fragment required.

Induction of the acetamidase by ω-aminoacids via the *amd*R gene product is relatively strong compared to acetate induction in wildtype strains (1). Multiple copies of the 5' end of the *amd*S gene have been found to cause reduced growth on lactam and ω-aminoacids as carbon or nitrogen sources. Therefore it is proposed that titration of the *amd*R gene product by binding to the copies of this sequence occurs, leading to reduced activation of genes under *amd*R control. This effect has not been observed in transformants containing multiple copies of a plasmid lacking sequences more than 70 base pairs upstream from the transcription start point; nor in transformants containing the *amd*I93 mutant sequence. This *amd*I93 mutation has been found to contain a deletion of about 33 base pairs between positions -112 and -145. The evidence strongly suggests that this sequence is required for binding of the *amd*R gene product. However, it has not yet been shown that this sequence is sufficient for binding.

C

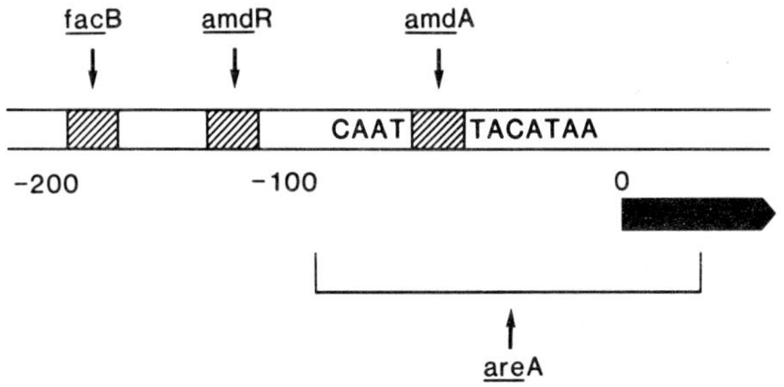

FIGURE 6. Model for *amd*S regulation. The sites of action of the products of regulatory genes are indicated. The evidence for the binding sites of the *fac*B and *amd*R products are presented in this paper. The proposed site of action of *amd*A is based on sequence comparisons between the 5' regions of the *amd*S and *aci*A genes (P.W. Atkinson and M.J. Hynes, unpublished data). Sequences downstream from the SmaI site located at position -75 have been shown to be sufficient for ammonium repression and for *are*A control (T.J. Littlejohn and M.J. Hynes, unpublished data).

There are at least two other proposed activators of *amd*S expression - the *are*A and *amd*A gene products (Table 1). Studies using a cloned gene, *aci*A, which is under *amd*A control indicate that the *amd*A product can be titrated by multiple copies of binding sequences (P.W. Atkinson and M.J. Hynes, unpublished). There is no equivalent evidence for the *are*A gene product as shown by no detectable effects on nitrogen catabolism in transformants containing more than 100 copies of the *amd*S gene. This may reflect a high level of *are*A product normally present, or possibly autogenous repressive control of *are*A expression such that titration of *are*A product results in increased *are*A gene expression.

The current model for regulation of the *amd*S gene is presented in Figure 6. Regulatory gene products are proposed to bind at sequences at the 5' end of the gene and result in increased transcription. These binding sequences are both upstream and within the proposed CAAT-TATA region of the gene. It remains to be determined whether these sequences resemble enhancer or upstream activation sites. The opportunities for *in vitro* manipulations followed by transformation and expression studies now makes it possible to address this problem.

ACKNOWLEDGMENTS

The assistance of Andrea Twomey and Karen Linton in this work is gratefully acknowledged.

REFERENCES

1. Hynes MJ (1978). Multiple independent control mechanisms affecting the acetamidase of *Aspergillus nidulans*. Mol Gen Genet 161:59.
2. Hynes MJ, Corrick CM, King JA (1983). Isolation of genomic clones containing the *amd*S gene of *Aspergillus nidulans* and their use in the analysis of structural and regulatory mutations. Mol Cell Biology 3:1430.
3. Tilburn J, Scazzocchio C, Taylor GG, Zabicky-Zissman JH, Lockington RA, Davies RW (1983). Transformation by integration in *Aspergillus nidulans*. Gene 26:205.
4. Sanger F, Coulson AR, Barrell BG, Smith AJ, Roe BA (1980). Cloning in single stranded bacteriophage as an aid to rapid DNA sequencing. J Mol Biol 143:161.

5. Hynes MJ (1977). Induction of the acetamidase of *Aspergillus nidulans* by acetate metabolism. J Bacteriol 131:770.

6. Hynes MJ (1980). A mutation, adjacent to gene *amd*S, defining the site of action of positive control gene *amd*R in *Aspergillus nidulans*. J Bacteriol 142:400.

7. Arst HN (1976). Integrator gene in *Aspergillus nidulans*. Nature 262:231.

8. Arst HN, Penfold HA, Bailey CR (1978). Lactam utilization in *Aspergillus nidulans* : evidence for a fourth gene under the control of the integrator gene *int*A. Mol Gen Genet 166:321.

Molecular Genetics of Filamentous Fungi, pages 171–184

MOLECULAR ANALYSIS OF ALCOHOL METABOLISM IN *ASPERGILLUS*

John A. Pateman, Colin H. Doy, Jane E. Olsen,
Heather J. Kane and Ernest H. Creaser

Department of Genetics, and Centre for Recombinant
DNA Research, Research School of Biological Sciences,
Australian National University,
PO Box 475, Canberra, A.C.T., 2601, Australia.

ABSTRACT

Genetic mapping has identified two closely linked genes, *alc*A, the structural gene for alcohol dehydrogenase (ADH) and *alc*R, a positive *trans*-acting regulatory gene for ethanol metabolism. The expression of *alc*A is repressed at the level of transcription by carbon catabolites. A genomic restriction fragment from the *alc*A-*alc*R region was cloned in pBR322 and used to select from a genomic bank in λ EMBL3A three overlapping clones covering 24 kb DNA. Southern genomic analysis of wild-type, *alc*A and *alc*R mutants showed that the mutants contained extra DNA at sites near the center of the cloned DNA. Transcription from the cloned DNA and hybridization with a clone carrying the *Saccharomyces cerevisiae* gene for ADHI (*ADC*1) are both confined to the *alc*A-*alc*R region. One of several species of mature mRNA is large enough to code for ADH. For all species, carbon catabolite repression over-rides control by induction. There is significant DNA sequence homology between part of the *S. cerevisae* and *Shizosaccharomyces pombe* ADH structural genes and part of the *alc*A gene.

INTRODUCTION

In *Aspergillus nidulans* there is a single major alcohol dehydrogenase (ADH) and a single major aldehyde dehydrogenase (AldDH) (1). Both enzymes are induced by ethanol or threonine, both are subject to carbon catabolite repression and both are necessary for the utilization of ethanol and threonine. This indicates that both compounds are utilized *via* acetaldehyde, which may be the endogenous inducer. The subunit molecular weight is about 38K for ADH and about 54K for AldDH (2).

ADH is specified by a structural gene *alc*A on chromosome VII and AldDH by a structural gene *ald*A on chromosome VIII (1,2). A regulatory gene, *alc*R, is adjacent to *alc*A and acts in *trans* to control the expression of *alc*A and *ald*A (and possibly other genes) (1,3). The product of *alc*R is necessary for the expression of *alc*A and *ald*A, and the regulation is at the level of transcription.

Apart from the intrinsic interest of ADH, the cloning of *alc*A is important because one would expect to be able to clone with it the adjacent gene *alc*R, a eukaryotic trans-acting positive regulatory gene. Sequencing and molecular studies should also be of interest in understanding carbon catabolite repression.

MATERIALS AND METHODS

Strains

Aspergillus nidulans was cultured as described in (1). The wild type was biA1 (Glasgow, 051). Mutants were obtained using either diepoxyoctane (DEO) or γ radiation (γ) and characterised (1). Mutants *alc*A51 (DEO) and *alc*A83 (γ) behave as overlapping deletions within *alc*A. They lack ADH activity and do not make the ADH protein(s). The mutant *alc*A55 *alc*R3 (DEO) is a deletion of all of *alc*A and *alc*R. Mutant *alc*R54 (DEO) has the typical *alc*R phenotype.

DNA Preparation

The procedures for the preparation of *Aspergillus* DNA are described in (4). Gel analysis of such DNA showed a single band of DNA of greater molecular weight than the top band of *Hind* III digested λ DNA and had a A_{260}/A_{280} ratio of 1.75 to 2.0.

Plasmid DNA was prepared as described in (5) or in (6). Phage DNA was prepared by the plate lysate method (7). For DNA prepared by the Quigley (5) or plate lysis (7) methods it was sometimes necessary to purify the DNA by use of an Elutip-d-Column-Set (NA 010/0 Schleicher and Schull) before successful probes could be made. Restriction enzymes were used according to the procedures given in (8).

Genomic Banks

A λ EMBL3A bank of *A. nidulans* wild type DNA was made by the procedures described in (4). This bank gave small uniform plaques and has a 99% chance of containing any one 16 kb fragment of the *A. nidulans* genome.

RNA Preparation, Probes, Blotting and a pBR322 Bank

The procedures for all of these preparations are described in (4).

RESULTS

The levels of functional mRNA for ADH and AldDH in cells grown under induced or repressed conditions suggested a plus/minus approach to the cloning of *alc*A. A positive cDNA probe (+) was derived from threonine induced and carbon derepressed wild-type cultures. Negative cDNA probes (-) were derived from uninduced and carbon repressed cultures of the wild-type and also from threonine induced and carbon derepressed cultures of the deletion mutant *alc*A55 *alc*R3. Attempts to apply this approach to *A. nidulans* genomic banks in Charon 28 or Charon 30 failed, although we did obtain a clone, pAN3-7-1, which could represent *ald*A or another gene regulated by *alc*R. Subsequently, cloning of the *alc*A-*alc*R region was achieved by a method which relied on finding a restriction fragment changed by a mutation in the *alc*A-*alc*R region, cloning this from a partial genomic bank in pBR322 by screening with the +/- probes and then using this cloned DNA to probe a genomic bank in λ EMBL3A.

Cloning of a 3.85 kb *Bam* HI Fragment from the *alc*A-*alc*R Region

Genomic Southern blots of wild type, *alc*A51, *alc*A83 and *alc*A55 *alc*R3 were analysed with the +/- probes, described above. The clearest example of the desired change was a 3.85 kb *Bam* HI fragment which occurred in wild type, *alc*A51 and *alc*A83 DNAs but not in *alc*A55 *alc*R3 DNA with the positive probe. This fragment was missing from all with either of the negative probes. The 3.85 kb *Bam* HI fragment represented a transcribed (but carbon catabolite repressed) region of DNA which was missing from *alc*A55 *alc*R3 but was present in both *alc*A mutants tested. Therefore this fragment was from the *alc*A-*alc*R region but was not necessarily part of either *alc*A or *alc*R.

A partial genomic bank was made from the 3.85 kb region of DNA isolated from a complete *Bam* HI digest run on a gel and then ligated into the *Bam* HI site of pBR322. The bank was screened with the positive and negative cDNA probes. Several clones were obtained and one, pAN4-8-1, was used to make a cDNA probe for screening a λ EMBL3A *A. nidulans* genomic bank.

About 5 x 10^4 plaques of a λ EMBL3A bank were probed with cDNA made from pAN4-8-1. Five positives were obtained. Southern blot analysis showed that there were 3 different clones 4K1, 4K3 and 4K4. In all, about 24 kb of DNA was cloned and restriction mapping showed that the 3.85 kb *Bam* HI fragment was in the middle.

Recognition and Location of *alc*A and *alc*R on the Cloned DNA

It was not known if the cloned DNA contained *alc*A or *alc*R, since the 3.85 kb *Bam* HI probe (Fig. 1) was covered in *alc*A55 *alc*R3 by a deletion of unknown size. Genomic

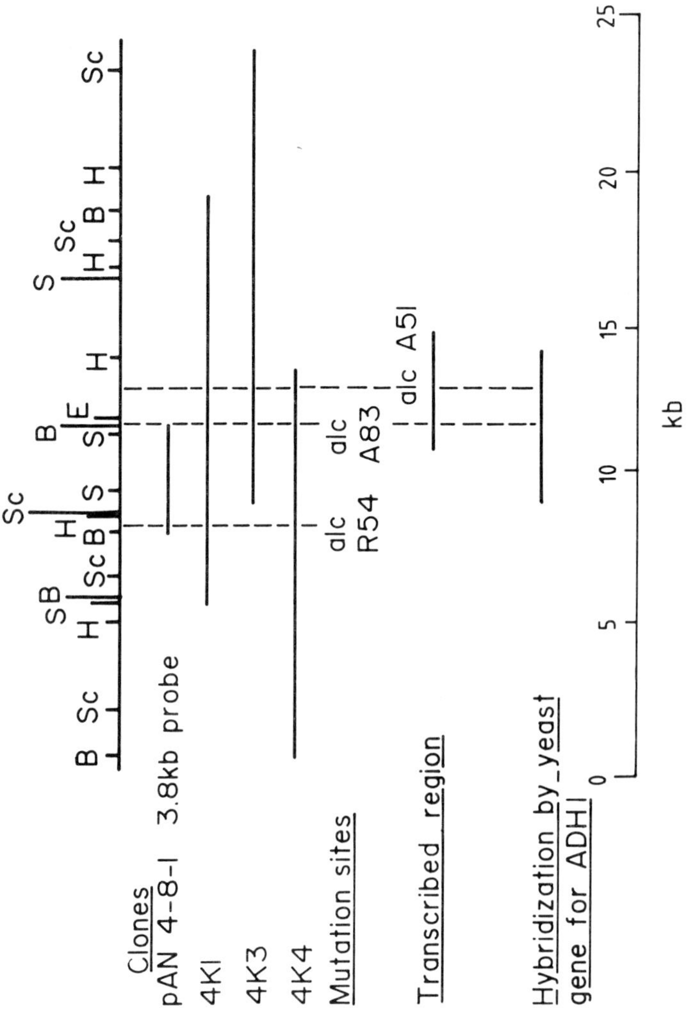

Fig. 1: The restriction site map and characteristics of the *alc*R-*alc*A region and flanking sequences.

blots of *alc*A55 *alc*R3 failed to label with probes from all three clones, 4K1, 4K3 and 4K4, showing that the mutation was a deletion which covered more than the 24 kb DNA cloned. This result also shows that none of the sequences in the cloned DNA are strongly represented elsewhere in the genome. To determine if the clones contained *alc*A and/or *alc*R, a number of *alc*A and *alc*R mutants were examined for a different restriction enzyme pattern from wild type when probed with the clones.

The 3.85 kb *Bam* HI fragment cloned in pAN4-8-1 is unaltered in *alc*A51 and *alc*A83 but in both mutants there is extra DNA outside the 3.85 kb *Bam* HI fragment and inserted into the 7.5 kb *Bam* HI fragment.

Results with individual probes (clones 4K1, 4K3 and 4K4) cover the entire 24 kb DNA cloned. They confirm the restriction map (Fig. 1) and by comparison allow the orientation of the two *Eco* RI fragments. This allows the mutations *alc*A51 and *alc*A83 to be ordered and *alc*A83 to be positioned between the *Eco* RI site and the adjacent *Bam* HI site (Fig. 1). Both *alc*A mutations occur in the same 5.5 kb *Hind* III fragment (Fig. 1).

Doubly digested DNA was probed on Southern blots with the 3.85 kb *Bam* HI fragment (contained in pAN4-8-1). The restriction fragment changes show that *alc*R54 also contains inserted DNA. The physical map for the locations of *alc*A51, *alc*A83 and *alc*R54 is consistent with the genetic studies that showed that *alc*A and *alc*R were closely linked, possibly adjacent genes (1).

Transcription from the Cloned DNA

A Southern blot of cleaved 4K1 DNA was probed with oligo (dT) primed cDNA from total mRNA from a culture grown on threonine (ADH inducer) and limiting glucose (ADH derepression conditions). There was little labelling of the 7.5 kb *Bam* HI fragment and the 5.5 kb *Sal* I fragment which contain the *alc*A51 and *alc*A83 mutation sites. The 3.8 kb *Bam* HI and the 1.9 kb *Sal* I fragments were relatively strongly labelled. The common fragment strongly labelled was the 1.9 kb *Sal* I. When random primers were used instead of oligo (dT) primers the results with the *Bam* HI and *Sal* I fragments were reversed. Synthesis of cDNA from oligo(dT) primers is known to be biased in favour of the 3' end of the transcript whereas the random primer method should be without bias (7). The unbiased results therefore establish that transcription occurs mainly in the region of both *alc*A mutation sites. The results with the 3'-biased oligo(dT) primers suggest that the 3' end of *alc*A (and perhaps of other genes) is in the 1.9 kb *Sal* I fragment (Fig. 1). Similar experiments with probes 4K3 and 4K4 show that the fragments mentioned above are the only portions of the 24 kb DNA cloned which are significantly transcribed under the conditions of growth. Irrespective of the priming method, no labelling occurs when the mRNA is derived from cultures grown on threonine plus excess glucose, showing that carbon catabolite repression over-rides induction and operates at the level of transcription. There is little evidence for the transcription of *alc*R. Maybe the 3' end of this gene also is in the 1.9 kb *Sal* I fragment and *alc*R is transcribed at a much lower level than *alc*A (as might be expected for a regulatory gene). Taking these data into account and weighting the intensity of labelling to the size of the fragments, the approximate transcribed region is given in Fig. 1.

Hybridization of a Yeast *ADC1* Clone to the *alc*A Region

A cDNA probe made from pJD14 (9), which carries the entire *Saccharomyces cerevisiae ADC1* (the gene for ADH1) labels *A. nidulans* DNA restriction enzyme fragments. Comparison with photographs of ethidium bromide stained gels shows that labelling is selective for bands and not proportional to the size of the fragment. The labelling is therefore specific but weak compared to the *A. nidulans* clone on itself. Inspection of the data leads to the approximate labelling limits shown in Fig. 1. Note that the 1.9 kb *Sal* I fragment is labelled.

mRNA Transcribed from the 24 kb Cloned DNA

Northern blot analysis of total wild-type mRNA probed with clones 4K3 and 4K4 shows that cultures induced (threonine) and carbon catabolite derepressed (limiting glucose) for ADH contain multiple species of mRNA. The major component is large enough to code for the ADH subunit. Two minor, smaller species label more heavily with probe 4K4 than with 4K3. All forms of mRNA are eliminated when cultures are grown with threonine in the presence of excess glucose. This demonstrates that carbon catabolite repression operates at the level of transcription and over-rides induction. When threonine is omitted from cultures grown with limiting glucose, a small amount of the largest mRNA is detected, especially by probe 4K4. The mutant *alc*A55 *alc*R3 is unable to produce any of these species of RNA when grown under conditions optimal for wild-type ADH induction/derepression.

*alc*A55 *alc*R3 is Deleted for DNA Duplicated and Transcribed Elsewhere in the Genome

A genomic Southern blot of *alc*A55 *alc*R3 showed no hybridization with 4K1, 4K3 and 4K4 and therefore this deletion extends over all the DNA cloned and beyond each end. These results also show that the cloned DNA is not strongly duplicated elsewhere in the genome. Southern blots of *Hin*d III cut DNA from wild type and mutants *alc*A83, *alc*A51 and *alc*A55 *alc*R3 were probed with cDNA from cultures of wild type and *alc*A55 *alc*R3 grown under threonine induction and carbon derepression. A striking difference was that *alc*A55 *alc*R3 DNA lacked (with either probe) a 4.1 kb *Hin*d III fragment although *alc*A55 *alc*R3 still produced the mRNA required to hybridize to that fragment in the wild type. Thus the deleted DNA region contains a sequence which hybridizes with cDNA derived from oligo(dT) primed mRNA transcribed elsewhere in the genome. Therefore this sequence is repeated at least once in the genome outside the 24 kb DNA cloned.

DNA sequencing

A start has been made on sequencing the *alc*A-*alc*R region and the flanking DNA. All the sequence analysis is by primed DNA synthesis on single-stranded recombinant M13 mp18 and M13 mp19 DNA templates in the presence of dideoxynucleotide triphosphates as described in (10,11). The sequence data has been analysed using the Lipman and Wilbur algorithm and a data base search similar to that of Dayhoff on a VAX 750 computer.

There is significant homology between the 445 bp sequence of the *Eco* RI-*Sal* I fragment (Fig. 1) covering the *alc*A83 mutation site and the region 265-710 bp of the *S. pombe* ADH structural gene (12). There is also homology with the similar region of the *S. cerevisiae* ADH1 structural gene (9). Translation of the *Aspergillus* sequence gives a reading frame in which 61 amino acids match the *S. pombe* and *S. cerevisiae* amino acid sequences. There is a similar degree of homology for this region between *Aspergillus* and each of the yeasts as there is between the two yeasts. The homologous regions in the *S. pombe* and *S. cerevisiae* genes cover most of the sequence which codes for the NAD binding site so this might well be one of the most conserved regions in the ADH genes. These homologies unequivocally locate part of the coding region of the ADH gene *alc*A in the *Eco* RI-*Bam* HI fragment spanning the *alc*A83 mutation. In addition the homology establishes the orientation of the *alc*A gene with the 3' end next to the *alc*R gene. This finding is in agreement with the supposition based on strong hybridization to an oligo (dT) primed cDNA probe that the 3' flanking sequence of *alc*A is in the 1.9 kb *Sal* I fragment (Fig. 1) between the *alc*A and *alc*R genes.

There are two other regions of possible homology that have been observed so far. A 270 bp sequence in the *Bam* HI-*Hin*d III fragment which contains the *alc*R54 mutation aligns with part of the 3' end of the coding region and shows about 39% homology with the *S. cerevisiae* and *S. pombe* ADH genes. If this homology is real it orientates the *alc*R gene with the 3' end towards the *alc*A gene. This opposite orientation of the *alc*R and *alc*A genes is consistent with the 3' ends of both genes being in the strongly hybridizing 1.9 kb *Sal* I fragment. The possible homology between *alc*R and an ADH structural gene may indicate that the regulatory gene and the ADH structural gene share a common ancestor.

The other region of possible homology is a sequence of 178 bp starting at a *Bgl* II site about 870 bp to the right of the *Eco* RI site (Fig. 1). This shows 38% homology to the 5' end of a *S. cerevisiae* ADH gene and aligns with the juncture region of the 5' flanking and coding regions of both the *S. cerevisiae* and *S. pombe* genes. It also has a possible TATA box sequence GTATACCTAA in an analogous position to the TATA boxes in the two yeast genes. If this apparent homology is real it implies the existence of an intron in the 5' end of the *alc*A gene.

Genomic clones comprising the *alc*A-*alc*R of *A. nidulans* have also been obtained by R.A. Lockington, H.M. Sealy-Lewis, C. Scazzocchio and R. Wayne Davies (Gene, in press).

DISCUSSION

It is clear that the ADH and AldDH system in *A. nidulans* is a favourable one for the study of the molecular basis of the regulation of gene expression. The *alc*A and *alc*R genes have been cloned. The restriction maps of the cloned region and preliminary sequence data indicate that the two genes are close together with at most 1-2 kb separating the coding regions. The relationship between genetic recombination distance and physical size in kilobases is not straightforward, but the position of the two genes on the DNA is in good agreement with the genetic data (1).

It is probable that the 3' end of the *alc*A gene is in a 1.9 kb *Sal* I fragment in between the two genes. The evidence for this is twofold. There is a marked difference in the labelling pattern of Southerns probed with mRNA derived cDNAs made by the oligo (dT) primer method and the random primer method. Also the sequence homologies between the *alc*A gene and the ADH genes of *S. cerevisiae* and *S. pombe* indicate that the *alc*A gene is orientated with the 3' end towards *alc*R. The heavy differential labelling by oligo (dT) primed probes of only the 1.9 kb *Sal* I fragment may indicate that the 3' end of *alc*R also lies in this region. If that is so and *alc*A and *alc*R are in opposite orientations, then there must be separate transcripts for the two genes.

There are several mRNA species derived from the *alc*A-*alc*R region. The major species is probably the *alc*A transcript. The minor mRNA species are smaller molecules and both hybridize more strongly to the *alc*R region of the cloned DNA than elsewhere. It is possible that one of these mRNA species is the *alc*R transcript. This is consistent with the genetic analysis which indicates that the *alc*R gene is smaller than *alc*A. However *alc*R is a regulatory gene and might be transcribed at a low level such that its transcript has not been detected.

Hybridization with a cDNA probe made from a *S. cerevisiae* ADH gene containing no introns but some 5' and 3' flanking sequences was spread over about 4 kb of the *alc*A-*alc*R region. This suggests that the *alc*A gene may contain some introns. Another possibility is that the *alc*A-*alc*R region may contain a number of partially homologous sequences or genes related to the *alc*A gene. Some preliminary sequence data is consistent with both of these possibilities.

The *Sal* I-*Eco* RI fragment (Fig. 1) which covers the *alc*A83 mutation site has been sequenced. This fragment shows homology with a region covering the NAD binding site in the coding regions of both the *S. pombe* and the *S. cerevisiae* ADH genes. This locates and orientates 445 bases of the *alc*A coding region on the cloned DNA. A second fragment of 178 bases adjoins a *Bgl* II site some 870 bases to the right of the *Eco* RI site. This fragment aligns with the end of the 5' region and the start of the coding region of both the *S. cerevisiae* and *S. pombe* ADH genes. Also there is a putative TATA box in an analogous position in the *Aspergillus* fragment to the TATA boxes in the two yeast genes. If this alignment represents real homology it implies the presence of a 200-300 bp intron in the *Eco* RI-*Hind* III fragment which contains the *alc*A51 mutation site (Fig. 1). A 270 bp portion of the *Bam* HI-*Hind* III fragment containing the *alc*R54 site has been sequenced. It shows about 39% homology with the 3' end of the coding region of both the yeast genes. This apparent partial homology of part of the *alc*R region with the *S. cerevisiae* coding region is consistent with the hybridization of the yeast gene over some 3-4 kb of the *Aspergillus* DNA. A simple hypothesis to account for this homology is that the *alc*A structural gene for the ADH protein and the *alc*R gene for a positive acting regulatory protein are derived by duplication from a common ancestor.

During the course of our work on the ADH system we have made a number of interesting observations which require and merit further investigation. The most important of these are:

1) Genetic analysis has shown that recombination values in the range 0.7 - 2.2cM are found in the great majority of crosses between *alc*A and *alc*R mutant alleles. This and other genetic data indicate that the *alc*A-*alc*R region is about 2cM in size (1).

However, some crosses specifically involving certain *alc*A and *alc*R alleles give much larger recombination frequencies in the range 5-35cM. A possible explanation is that recombination may occur between common insertion elements present only in certain mutant *alc*A and *alc*R alleles.

2) SDS polyacrylamide gels show two or three polypeptides of similar molecular weight associated with ADH and two peptides associated with AldDH. The *in vitro* translation of mRNA from induced, carbon derepressed cells also gives two or three ADH associated polypeptides. This suggests the possibility of post-transcriptional modification and/or post-translational modification steps in the expression of the *alc*A and *ald*A genes (1).

3) The demonstration of several species of mRNA derived from the *alc*A-*alc*R region is of interest in connection with possible multiple transcripts from the *alc*A gene and the *alc*R transcript.

4) The mutants *alc*A51, *alc*A83 and *alc*R54 appear from the genetic data to be deletions but restriction enzyme analysis indicates that they contain inserted DNA. It may be that the additional DNA has transposon properties and/or has effects on the recombination frequency in the neighbourhood of the insertion.

We conclude by summarising the more important lines of work for the future.

a) The topography of the *alc*A-*alc*R region and flanking regions will be determined by sequencing and the amino-acid sequence of the ADH protein will provide valuable complementary information.

b) A search for regulatory sequences shared by *alc*A and *ald*A and carbon regulated genes elsewhere in the genome.

c) The expression of the regulatory gene *alc*R in *E. coli* or *S. cerevisiae*. This will provide material for studying DNA - regulatory protein interactions.

d) The search for and the analysis of further carbon regulated genes. There is evidence for the presence of such genes both within the region covered by the large *alc*A55*alc*R3 deletion and elsewhere in the genome.

e) The regulation of the AldDH structural gene, *ald*A, is closely co-ordinated with that of *alc*A. The cloning and sequencing of the *ald*A region for comparison with *alc*A is clearly important for understanding this regulation.

f) The nature of the inserted DNA in mutants *alc*A51, *alc*A83 and *alc*R54 and the molecular basis of the mutations in the *alc*A and *alc*R alleles which show high frequence recombination.

ACKNOWLEDGEMENTS

We thank many of our colleagues, particularly J. Shine, M. Cronk, N. Deacon, K. Reed and B. Evans for helpful advice and/or materials. E. Wimmer is thanked for excellent technical assistance in the maintenance and growth of stock cultures. B. Hall is thanked for pJD14.

REFERENCES

1. Pateman JA, Doy CH, Olsen JE, Norris U, Creaser EH, Hynes M (1983a). Regulation of alcohol dehydrogenase (ADH) and aldehyde dehydrogenase (AldDH) in *Aspergillus nidulans*. Proc R Soc Lond B217:243-264.
2. Creaser EH, Porter RL, Britt KA, Pateman JA, Doy CH (1985). Purification and preliminary characterization of alcohol dehydrogenase from *Aspergillus nidulans*. Biochem J 225:449-454.
3. Pateman JA, Doy CH, Olsen JE, Kane HJ (1983b). Genes for alcohol utilization in the lower eukaryote *Aspergillus nidulans*. In Nagley P, Linnane AW, Peacock WJ, Pateman JA (eds): "Manipulation and Expression of Genes in Eukaryotes"; Sydney: Academic Press, pp 171-178.
4. Doy CH, Pateman, JA, Olsen JE, Kane HJ, Creaser EH (1985). Genomic clones of *Aspergillus nidulans* containing *alc*A, the structural gene for alcohol dehydrogenase and *alc*R, a regulatory gene for ethanol metabolism. DNA (in press).
5. Holmes DS, Quigley M (1981). A rapid boiling method for the preparation of bacterial plasmids. Anal Biochem 114:193-197.
6. Scott KF, Rolfe BG, Shine J (1983). Biological nitrogen fixation: Primary structure of the *Rhizobium trifolii* protein gene. DNA 2:149-155.

7. Maniatis T, Fritsch EF, Sambrook J (1982). "Molecular cloning: A laboratory manual", New York: Cold Spring Harbor Laboratory.
8. Davis RW, Botstein D, Roth JR (1980). "A manual for genetic engineering: Advanced Bacterial Genetics", New York: Cold Spring Harbor Laboratory.
9. Bennetzen JL, Hall BD (1982). The primary structure of the *Saccharomyces cerevisiae* gene for alcohol dehydrogenase I. J Biol Chem 257:3018-3025.
10. Sanger F, Nicklen S, Coulson AR (1980). DNA sequencing with chain terminating inhibitors. Proc Natl Acad Sci USA 74:5463-5467.
11. Sanger F, Coulson AR, Barrell BG, Smith AJH, Roe BA (1980). Cloning in single-stranded bacteriophage as an aid to rapid DNA sequencing. J Mol Biol 143:161-178.
12. Russell PR, Hall, BD (1983). The primary structure of the alcohol dehydrogenase gene from the fission yeast *Schizosaccharomyces pombe*. J Biol Chem 258:143-149.

Molecular Genetics of Filamentous Fungi, page 185

SEQUENCE ANALYSIS OF THE 5' ENDS OF THE ALC A AND ALD A GENES FROM ASPERGILLUS NIDULANS

David I. Gwynne, Mark Pickett and R. Wayne Davies

Allelix Inc., Mississauga, Ontario, Canada
Robin Lockington[1], Heather Sealy-Lewis[2] and Claudio Scazzocchio[1]
University of Essex, Colchester, U.K.

Alc A and ald A encode alcohol dehydrogenase I and and aldehyde dehydrogenase in Aspergillus nidulans. A third gene alc R is closely linked to alc A on linkage group VII. Mutations within alc R result in no detectable alc A or ald A messenger RNA, thus alc R seems to elicit the expression of both alc A and ald A. Both genes are also controlled by the carbon catabolite repression system mediated by cre A. Alc A, alc R and ald A have been isolated from A. nidulans gene libraries, and the transcripts have been mapped. Since both alc A and ald A are under at least two common control systems it is to be expected that the regulatory elements within their promoters would have similar functional elements. In order to investigate this possibility, we have sequenced the 5' coding and non-coding regions of both genes and have indicated some areas of homology. Future work will involve deletions of various putative promotor elements in order to identify the functional regulatory sites within these genes.

[1]Present address: Institut de Microbiologie Batiment 409, Universite Paris Sud 91405 Orsay France
[2]Present address: Biochemistry Dept. University of Hull, Hull, U.K.

Molecular Genetics of Filamentous Fungi, pages 187–192

FUNGAL DIFFERENTIATION AND DEVELOPMENT: PROBLEMS AND PROSPECTS

James S. Lovett

Department of Biological Sciences, Purdue University, West Lafayette, Indiana 47907

ABSTRACT The introduction to the subject summarizes past and present approaches to the study of fungal differentiation and development, including metabolic studies, biosynthesis of cell structures, electron microscopy, transmission genetics, two-dimensional gel analysis of stage-specific proteins, and RNA/DNA hybridization. Although progress has been made, many important questions remain to be answered and the speakers in this session will describe a variety of experimental systems being used to study developmental problems by the use of genetic, biochemical, cell biological and molecular techniques.

In the previous sessions of this meeting we have heard about the exciting progress being made on the structure and regulation of fungal genes at the molecular level. The important advances being made on cloning and sequencing of fungal genes, and on transformation, might lead us to an optimistic view that these kinds of methods will lead to a rapid solution of a number of important and complex problems in fungal biology, such as development, which is the subject of today's session. Work has already begun in this direction and it seems certain that molecular genetics will make very important contributions to the study of fungal development. Perhaps this will be most beneficial in the identification and characterization of regulatory genes and the control of their transcription. However, as an introduction to the session today, I would like to take a few minutes to provide some background on the subject that suggests a somewhat cautious forcast concerning the ease with which fungal development will quickly succumb to analysis by the use of molecular genetic techniques alone.

One can look at the study of fungal development as an evolutionary series of overlapping and increasingly narrow approaches dating back over one hundred years to when the emphasis of mycology was primarily descriptive: identifying new fungi, and establishing the details of their life cycles. This was followed during the late 1800's to mid-1900's by many studies of fungal physiology, and to the enumeration of conditions that can lead to the development of asexual and sexual reproductive structures in culture. In the same period descriptive morphology of the fungi continued, and continues even today, to be an active field of mycology. In the past 30 years or so, the emphasis has been increasingly placed on a more and more detailed description of the metabolism of a few species during sporulation, on the details of fungal cell organization and composition, and on the biosynthesis of important structural macromolecules such as the chitin, glucans and glycoproteins that make up the cell wall. Much of this has been done by biochemists and microbiologists with relatively little interest in fungal development, or in mycology as a field.

This work has steadily increased in its sophistication, and the level of detail it can reveal, in parallel with the development of high resolution electron microscopy and modern biochemical methods. As a result, a wealth of detailed information is now available; but, though enormously interesting, it is still largely descriptive of the processes of developmental events, rather than an explanation of how such events are regulated at the cellular level. We know how to induce many fungi to sporulate or produce sexual fruits in culture, but we know very little about how these organisms detect and transduce a change in their environment into a delopmental response such as conidiation or fruiting.

During the same period of time, an enormous amount of genetic information was being accumulated for *Neurospora crassa* and *Saccharomyces cerevisiae*, but relatively little attention was placed on development. However, the production and selection of mutants specifically blocked in various stages of development was undertaken (1, 2) and the results supported the idea that differential gene expression was important in the regulation or coordination of developmental events. The results of such studies also provided the basis for estimates of the number of genes that might uniquely be involved in the asexual sporulation of *Aspergillus nidulans*, 45 -150 loci (1), and the sexual sporulation of yeast, 48 ±27 loci (2). It has not led to the identification of the genes or their products. More recent work by Dr. Champe's group has identified a number of interesting mutants that influence the "competence" of *A. nidulans* to begin conidiation when shifted to an inductive environment. In his talk Dr. Champe will describe some work which is leading to the characterization of products that may be related to the functions of these early-acting regulatory genes for conidiation in *A. nidulans*.

Another recent approach to the role of differential gene action has involved the pulse-labeling, and two dimensional separation of proteins by gel electrophoresis, to detect new proteins, or losses of those produced by growing cells, as a reflection of differential gene expression. The results have been somewhat disappointing because the method can only detect proteins produced from abundant cytoplasmic messenger RNA's which are generally the least interesting because they probably represent "housekeeping" functions that are present throughout; nevertheless, decreases in some proteins and the stage-specific appearance of a small number of new proteins have been reported for yeast sporulation, in *Neurospora* conidiation, and in the early stages of fruiting by *Schizophyllum commune*. None of the proteins detected by these experiments have been identified, and only a handful of putative stage-specific gene products have been characterized by biochemical methods (e.g., a-glucosidase, and B-glucosidase in yeast; storage proteins in *Botryodiplodia* and *Sclerotinia*; gamma particle proteins in *Blastocladiella*)(see 5 for review).

An alternative strategy has been to hybridize poly(A)RNA or total RNA from developing cultures in RNA excess/tracer DNA reactions to obtain estimates of the fraction of the total genome that is represented in the form of RNA transcripts at any given stage. The results have shown that fungi not only differ from higher organisms by having small genomes (only a few times larger than *E. coli*), with a relatively small fraction of highly repetitive dispersed DNA sequences, but, that they also "express" a large fraction of their genomes at all stages. When the base sequence complexity numbers that result from such experiments are converted into estimates of the numbers of average-sized messages that are present only during a developmental stage, e.g., 1300 in *A. nidulans* sporulation (6), ca 6000 in *Blastocladiella* (4), and ca 7000 in *Achlya* (3), the predictions are ten-fold larger than those that were suggested by the frequencies with which developmental mutants were isolated.

If the estimates from the hybridization results are a reflection of reality, it would appear that many low abundance genes are specifically expressed during sporulation, but that their individual functions may not be essential (because so few mutants have been isolated). At present one can only speculate about what their functions might be. To take conidiation as an example, it is reasonable to propose that most of what occurs during the production of conidiophores and conidia is a reorientation of the biosynthetic patterns and the assembly of cell walls to yield derivative cells of different shapes from the growing hyphae, but which are not very different in their overall composition (e.g., the vertical conidiophore stalk and vesicle, sterigmata and phialides of *Aspergillus*).

Thus, a relatively small number of gene products may be involved in reorientating the growth and nuclear division patterns involved (1). In addition, only a very few completely new products may be required in any quantity, such as for production of special spore wall layers, pigments, etc. (as for example the green conidial pigment of A. nidulans for which the laccase I is essential). One obvious way to go after such abundant gene products is to use the prevalence of the mRNAs as a basis for cloning the genes involved. Dr. Wessels will discuss his group's work with one such gene, and the use of fruiting-stage cDNAs to probe for other stage-specific mRNAs, in the mushroom Schizophyllum commune.

Most asexual and sexual spores are metabolically and biosynthetically quiescent, if not actually dormant. It therefore seems possible that a number of the putative, developmentally-specific, low copy number genes detected via hybridization of sporulation phase RNA, could code for factors that are involved in modulation, or even arrest of the activities of biosynthetic and metabolic pathways while retaining these "systems" in a functional state for reactivation when the spore germinates. Dr. Robert Brambl will discuss some interesting work that illustrates how the synthesis and activity of mitochondria may be regulated in the spore stage.

The production of the cell wall is an essential biosynthetic process that underlies most fungal development. The complexities involved in the coordinated transport of various enzymes, precursors, and preformed polymers to the surface of the growing cell and their integration to form the functional cell wall, are impressive. And, it seems probable that it must be at this level that new structural and directional patterns are established in development. Dr. Bartnicki-Garcia in his talk will discuss this problem of basic structural determination.

This leads to a final point. It is evident that the ability to isolate and characterize "developmental genes" and study their transcription adds a potentially powerful new tool to our armamentarium for developmental analysis, and a start in this direction has already been made. What I hope will result from today's session is an appreciation not only of the interesting problems and complexities of development in fungi, but, even more important, a realization that the use of transmission genetics and molecular genetics, as well as biochemical and cell biological techniques, will all be required if we are to understand how fungal development is accomplished.

REFERENCES

1. Clutterbuck AJ (1978) Genetics of vegetative growth and asexual reproduction. In Smith JE, Berry DR, (eds):"The Filamentous Fungi. Vol. III. Developmental Mycology." New York: Wiley, p 240.
2. Esposito RE, Klapholz S (1981) Meiosis and ascospore development. In Strathern JN, Jones EW , and J.R. Broach JR (eds.): "The Molecular Biology of the yeast Saccharomyces: Life Cycle and Inheritance." New York: Cold Spr. Harb. Lab., p 27.
3. Gwynne DI, Brandhorst BP (1982) Changes in gene expression during sporangium formation in Achlya ambisexualis. Dev Biol 91:263.
4. Johnson SA, Lovett JS (1984) Gene expression during development of Blastocladiella emersonii. Exp Mycol 8:132.
5. Lovett JS (1985) The molecular biology of fungal development. In Kosuge T, and Nester GW (eds):"Plant-Microbe Interactions," Vol 2. New York: MacMillan Pub Co., In Press.
6. Timberlake WE (1980) Developmental gene regulation in Aspergillus nidulans. Devel Biol 78:497.

Molecular Genetics of Filamentous Fungi, pages 193–206

GENE EXPRESSION DURING BASIDIOCARP FORMATION IN SCHIZOPHYLLUM COMMUNE

J.G.H. Wessels

Department of Plant Physiology, Biological Center, University of Groningen, Haren, The Netherlands

ABSTRACT Monokaryons and the derived dikaryon of *Schizophyllum commune*, genetically different only in the incompatibility genes they possess, were grown vegetatively in submerged cultures or in surface cultures permitting basidiocarp formation in the dikaryon. The mycelia were then compared with respect to the proteins they synthesize and the RNA populations they contain. Few differences in proteins were detected and regulation of RNA sequences appeared confined to the period of basidiocarp formation in the dikaryon. This regulation involved a limited number of abundant mRNAs for a number of which cloned cDNA sequences were obtained and analyzed. Regulation in the class of rare mRNAs, if occurring, was too limited to be detected.

INTRODUCTION

Basidiocarp formation is of interest to the mushroom grower (1) but also provides an amenable system for studying the genetic and molecular basis of multicellular development. Although the basidiocarps or fruit bodies of basidiomycetes are the most complex structures produced among the fungi, an important advantage is that the basic genetic system that governs the formation of basidiocarps is well known. This is particularly true for *Schizophyllum commune* and *Coprinus cinereus* (2-6), unfortunately two species of no economic importance.

Sexual spores (basidiospores) of e.g. *S. commune* normally germinate into monokaryotic mycelia with septate hyphal compartments each with a single haploid nucleus. Such monokaryons each carry a particular *A*- and *B* incompatibility factor. Although each factor is composed of two equivalent loci, α and β, for simplicity these factors will be called genes. The presence of certain alleles outside the incompatibility genes has been shown to elicit the formation of (aberrant) fruit bodies (7,8) but the monokaryons normally do not fruit and grow indefinitely as vegetative mycelia.

In nature, many different *A* and *B* genes (carrying different alleles at one or both of the α and β loci) exist. When two monokaryons with different *A* and *B* genes are confronted they merge into a new type of mycelium, the dikaryon. In this dikaryon each hyphal compartment contains two different nuclei carrying the different *A* and *B* genes. Mitosis of these paired nuclei in the apical compartments is synchronous while the partitioning of daughter nuclei is mediated by a clamp connection which is formed at each septum. This dikaryon can also be grown as a vegetative mycelium but, unlike the monokaryons from which it was derived, it readily forms basidiocarps in surface culture under appropriate conditions (9). In the basidial cells of these basidiocarps the paired nuclei fuse to form a diploid nucleus which then undergoes meiosis to form four haploid basidiospores in which the incompatibility genes are independently reassorted.

This brief account of sexual morphogenesis in *S. commune* serves to indicate that the continued vegetative growth of the monokaryons and the ability to switch to basidiocarp formation in the dikaryon is basically controlled by the incompatibility genes. The presence of only one wild-type allele for the *A* and the *B* gene leads to continued vegetative growth while the presence of two different alleles for both genes releases the ability to produce basidiocarps under appropriate conditions. All other genes in the two types of mycelium can be the same. In practice this is accomplished by using two co-isogenic monokaryons and the derived dikaryon.

The notion that the incompatibility genes act as master switches which control sexual development is strengthened by the observed effects of certain mutations in these genes (5,6). The presence of such mutations switches on dikaryotic hyphal morphology and basidiocarp formation in a homokaryon, that is a mycelium with a single nuclear type. Since these mutated alleles are dominant over their progenitor alleles they presumably activate genes involved in sexual morphogenesis. In a wild-type dikaryon such activation would result from the interaction of the products of the two different alleles of the *A* and *B* genes. In summary:

AxBx → homokaryotic monokaryon → no basidiocarps
AxBx/AyBy → heterokaryotic dikaryon → basidiocarps
AmutBmut → homokaryotic dikaryon → basidiocarps

The monokaryon and the dikaryon grow at the same rate (10). Their protein/RNA ratios are the same but the volume of the hyphal compartments of the dikaryon and the amount of RNA they contain is 1.3 times larger than of monokaryotic compartments, although the former contain twice the amount of DNA (11).

In the work reviewed here use was made of the possibility to compare mycelia of *S. commune* differing only in the incompatibility genes, to see how these genes influence the expression of other genes in relation to hyphal morphology and basidiocarp formation. All analyses were made with actively growing mycelia and it was anticipated that the procedure of comparing monokaryons and dikaryon at the same culture age would cancel out any change in gene expression not related to morphogenesis but e.g. due to culture age or changing conditions during culture. In general, the proteins and RNAs from two compatible co-isogenic monokaryons were mixed and compared with the proteins and RNAs from the dikaryon freshly synthesized from these monokaryons. Occasionally a comparison was made between a monokaryon and a co-isogenic homokaryotic dikaryon (*AmutBmut*). However, the former procedure was preferred because the expression of morphogenesis was more regular and because it lessened the chance of differences being due to incomplete isogeny of the mycelia.

GENE EXPRESSION IN MONOKARYON AND DIKARYON IN THE ABSENCE OF BASIDIOCARP FORMATION

Basidiocarps develop on the dikaryotic mycelia of *S. commune* but not on those of the parent monokaryotic mycelia. It was thus important to examine to what extend the incompatibility genes effect differential expression of genes before the initiation of basidiocarps in the dikaryon, possibly related to the difference in hyphal morphology between monokaryon and dikaryon.

Because differences in proteins and RNAs were expected to become maximally expressed in dividing apical compartments, where the difference in morphology originates, the monokaryons and the synthesized dikaryon were grown in liquid shaken culture in a step-up regime such that the final 16-20 h mycelia contained 20-30 % apical hyphal compartments. An extensive analysis of [^{35}S]-methionine-labelled proteins of these mycelia revealed few differences (12). Among 700 polypeptides analyzed on two-dimensional gels, only about 20 polypeptides appeared specifically synthesized in the monokaryons while 23 different polypeptides appeared specifically synthesized in the dikaryon. That these differences in protein synthesis were regulated by the incompatibility genes was strongly indicated by the behaviour of the *AmutBmut* strain. This double mutant synthesized at least 13 of the 23 polypeptides typical for the wild-type dikaryon. On the other hand, the *AmutBmut* strain still synthesized many of the polypeptides specific for the monokaryon. This may be related to the fact that the expression of the dikaryotic phenotype was only partial and many monokaryotic cells were still produced. Mutants carrying mutations in either the *A* or the *B* gene produced overlapping subsets of the polypeptides synthesized in the double mutant.

Assuming that the magnitude of differences observed on the protein gels is representative for the whole range of proteins synthesized, this would set the magnitude of regulation in the mRNA populations at about 6 %. Surprisingly, nucleic acid hybridizations and *in vitro* translations failed to detect any differences in the RNA populations from vegetatively growing monokaryon and dikaryon (13,14,15).

With a single copy DNA complexity in *S. commune* amounting to 31.5×10^6 base pairs (16), RNA:scDNA hybridizations revealed that about 33 % of this complexity (10.4×10^6 nucleotides) is contained in complex RNA. With a number-average length of the RNA molecules of 1100 nucleotides, as determined for poly(A)-containing RNAs, this would suggest the presence of about 10 000 different RNAs in the complex RNA. Only 25 % of this complex RNA appeared polyadenylated but RNA:scDNA hybridizations did not reveal a difference in the complexities of total RNA, poly(A)-containing RNA, and polysomal RNA. This is a common situation encountered in fungi (ref in 15). Such hybridizations apparently are not sensitive enough to reveal e.g. the presence of small introns in some of the primary transcripts (17 and references therein). As expected these hybridizations also did not reveal any difference between the RNAs from monokaryons and dikaryon because differences of less than 10 % go undetected by this method.

A higher sensitivity was obtained by using RNA:cDNA (complementary DNA) hybridizations, particularly by using cDNA enriched for putative cell-type specific sequences (14). No differences were found neither in the class of abundant sequences (calculated as 600 sequences at 100 copies per cell on average) nor in the class of rare sequences (calculated as 13 000 sequences at 5 copies per cell on average). These hybridizations should have revealed differences exceeding 1 % in each of the classes. In the same study (14) a difference of 300-400 RNA sequences between monokaryons and dikaryon was found when using poly(A)-containing RNA instead of total RNA to drive cDNA into hybrids. Further studies should indicate wether this points to an effect of the incompatibility genes on the polyadenylation of individual RNA sequences or results from a failure to reproducibly isolate the whole complexity of the poly(A)-containing RNA fraction due to the short lenghts of the poly(A) tracks (33 nucleotides on average). In agreement with the apparent absence of differences in the class of abundant RNA sequences it was found that among 400 polypeptides synthesized *in vitro* under direction of total RNA (14) or polysomal RNA (15) from monokaryons and dikaryon no differences were found.

These experiments seem to rule out that the small differences in proteins synthesized by vegetative monokaryons and dikaryon are brought about by differences in mRNA populations. In general, qualitative differences in the RNAs produced by these mycelia must be small, if they exist. As outlined above in these analyses a difference in 100-200 rare sequences would go undetected and even a difference in a few abundant RNA sequences could have been missed if their *in vitro* protein products were not visualized on the two-dimensional gels. In fact, probing with cloned sequences isolated as specific for basidiocarp formation has shown that these sequences are present in the vegetative dikaryon but are not detected in the monokaryons (18 and see below).

A fundamental difficulty in these analyses is that proteins and RNAs specifically regulated by the incompatibility genes and related to hyphal morphogenesis, e.g. the formation of clamps, may be expressed only in apical hyphal compartments and during a limited time of the duplication cycle. The percentage of apical compartments was 20 to 30 but the subapical compartments were as active in RNA synthesis as the apical compartments (11). The concentration of relevant protein and RNA species in the extracts from whole mycelia might thus have dropped under the detection level.

GENE EXPRESSION IN RELATION TO BASIDIOCARP FORMATION

To detect proteins and RNAs regulated by the incompatibility genes and specifically expressed during basidiocarp formation, cultures of monokaryons and dikaryon were grown in surface cultures in the light from mycelial fragments. Under these conditions the dikaryon forms numerous basidiocarps (40-50 cm^{-2}) between the third and fourth day of cultivation whilst the monokaryons only produce a mycelial mat.

General Analysis of Protein and RNA Patterns

Prolonged exposure (4-5) h) to [^{35}S]-sulphate gave adequate labelling of proteins in mycelium and basidiocarps. Analysis of about 400 polypeptides on two-dimensional gels, labelled on the fourth day of culturing, indicated that eight polypeptides were exclusively synthesized in the monokaryons but the fruiting dikaryon now synthesized 37 polypeptides not seen in the monokaryons (19). Many of these polypeptides were very prominent and only a few of these were seen in a parallel experiment in which the mycelia were grown in shaken culture.

Figure 1 shows that 15 of the novel polypeptides of the dikaryon were exclusively synthesized in the basidiocarps while three polypeptides of the monokaryons were synthesized only in the supporting vegetative mycelium of the dikaryon but not in the basidiocarps. The remaining 22 novel polypeptides were synthesized in both the supporting mycelium and the basidiocarps. Nine of these polypeptides, with low molecular weights (between 10 000 and 26 000), were abundantly excreted into the medium and two of the excreted polypeptides were also found firmly associated with the hyphal walls of the dikaryon. None of these excreted polypeptides was produced in shaking cultures of the dikaryon.

Although RNA:scDNA hybridizations did not reveal any differences in RNA populations between monokaryons and fruiting dikaryon, RNA:cDNA hybridizations showed that about 5 % of the complex RNA mass of the fruiting dikaryon was missing from the monokaryons (21). On the other hand,

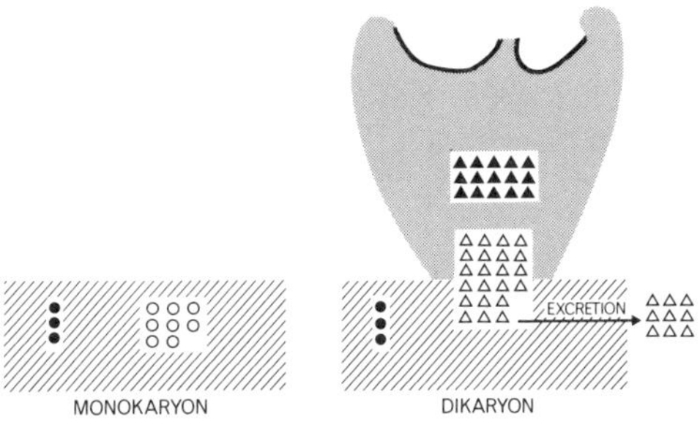

FIGURE 1. Diagrammatic representation of polypeptides specifically synthesized in monokaryons and the derived fruiting dikaryon of *Schizophyllum commune* after four days in surface culture. Each symbol represents a single polypeptide differentially synthesized among 400 analyzed. ●, polypeptide exclusively synthesized in monokaryons; o, polypeptide synthesized in both the monokaryons and the vegetative mycelium of the dikaryon but not in basidiocarps; Δ, polypeptide not synthesized in monokaryons but synthesized in the dikaryon both in vegetative mycelium and basidiocarps; ▲, polypeptide exclusively synthesized in basidiocarps. (Based on data from (19), reproduced from (20)).

all RNA sequences of the monokaryons appeared to be present in the fruiting dikaryon. Significantly, this analysis confirmed the results obtained with RNA:scDNA hybridizations by showing no differences in the class of rare RNA sequences (about 13 000 sequences). The novel RNA sequences in the fruiting dikaryon appeared confined to the class of abundant RNA sequences (calculated as 230 sequences) comprising one third of the cDNA mass. Assuming equal frequency in this class (a gross oversimplification) this would indicate the existence in the fruiting dikaryon of 35 prevalent mRNAs which are absent in the monokaryons, whereas any regulation in the class of rare mRNAs is too low to be detected.

The presence of a number of prevalent mRNAs in the fruiting dikaryon was confirmed by analyzing polypeptides synthesized *in vitro* (21). While this method did not reveal a significant difference between the RNAs from monokaryons and dikaryon after two days of cultivation in surface

TABLE 1
RELATIVE ABUNDANCE OF RNA SEQUENCES IN MONOKARYONS AND FRUITING DIKARYON AS REVEALED BY HYBRIDIZATIONS WITH SPECIFIC CDNA CLONES

Clone	Insert (ntp)	mRNA (nt)	Relative Abundance[a]	
			Monokaryon	Dikaryon
pSc1 (7D5)[b]	620	650	n.d.[c]	100
pSc2 (17B5)[b]	470	1250	n.d.	51
pSc3 (1D10)[b]	530	775	1425	1150
pSc4	570	580	53	1440
pSc5	290	1150	n.d.	67
pSc6	400	770	n.d.	60
pSc7	740	910	n.d.	102
pSc8	430	630	n.d.	25
pSc9	480	900	n.d.	23
pSc10	100	800	23	2

[a] Labelled cloned were hybridized to total RNA on dot blots. No corrections were made for differences in insert lengths and all abundancies are given relative to pSc1 mRNA.
[b] Former designation of these clones (18).
[c] n.d., not detected.

culture, at the fourth day when basidiocarps were present the RNA from the fruiting dikaryon directed the synthesis of 22 specific polypeptides compared to 8 specific for the RNA of the monokaryons.

Analysis with Cloned Sequences

To examine the developmentally regulated prevalent RNA sequences in more detail, cDNA clones complementary to these sequences were isolated (18 and unpublished). cDNA made on poly(A)-RNA from the fruiting dikaryon was cloned into the *Pst*1 site of pBR327 and 2000 bacterial colonies were screened with labelled cDNAs made on the RNAs of the monokaryons and the fruiting dikaryon. A preliminary list of the cDNA clones isolated appears in Table 1. As shown

TABLE 2
RELATIVE ABUNDANCE OF SPECIFIC RNA SPECIES IN 4-DAY OLD SURFACE CULTURES

Culture	pSc1	pSc2	pSc3
Dikaryon -light, air	100[a]	100[a]	100[a]
-light, 5% CO_2	12	-	100
-darkness, air	69	53	183
Dikaryon -mycelium	100[a]	100[a]	100[a]
-basidiocarps	555	69	4
Monokaryon	n.d.[b]	n.d.	60

[a]Values set at 100.
[b]n.d., not detected.

eight clones were isolated corresponding to mRNAs present in the fruiting dikaryon but not or at much lower concentration in the monokaryons. One clone (pSc3) was isolated because it represents a very abundant mRNA present in both the monokaryons and the dikaryon. A few clones (e.g. pSc10) were also found which gave a stronger signal with monokaryotic than with dikaryotic cDNA.

Estimates of the frequencies at which homologous sequences of the isolated dikaryon-specific clones occurred in the cDNA library indicate that the cDNA clones isolated may nearly represent all sequences that are strongly regulated. Assuming equal cloning efficiency they represent more than 5 % of the cDNA mass which was earlier found as the magnitude of the cDNA mass not represented in cDNAs made on RNA from the monokaryons (21). The number of regulated sequences recovered in the cDNA clones is thus much lower than the number of 35 earlier estimated on the basis of the assumption that all sequences in the class of abundant mRNAs occur in equal frequencies. This clearly is not the case, as shown in Table 1.

The dikaryon-specific clones all hybridize to RNA sequences which rise in concentration during basidiocarp formation. While some are particularly abundant in the basidiocarps, other seem to be more abundant in the supporting mycelium (Table 2). However, low hybridization signals were also found with RNAs from two-day old surface

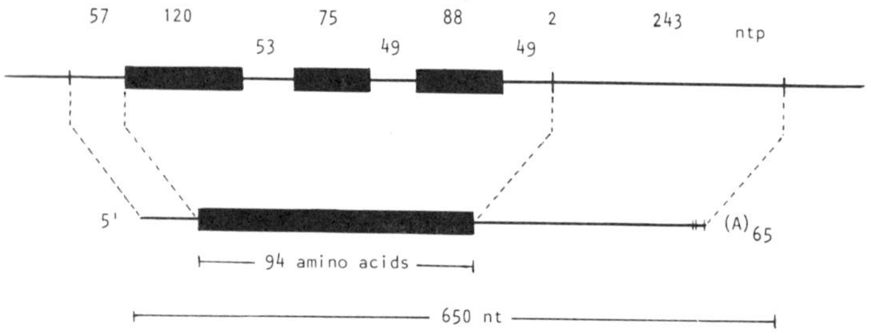

FIGURE 2. Structure of the regulated pScl gene and its mRNA. Dark boxes represent coding sequences. The lengths of the various segments is given in nucleotides. One amino acid codon is generated by splicing the second intron. The stop codon is generated by splicing of the third intron. Data based on (17).

cultures which had not yet started to form basidiocarps. Therefore, the clones were also used to probe RNAs from mycelia grown in shaken culture. All clones isolated as specific for basidiocarp formation also hybridized weakly to RNAs from the dikaryon but not to those of the monokaryons. Thus this represents a real difference between the RNAs from vegetative monokaryons and dikaryon, apparently regulated by the incompatibility genes, but not detectable by the methods of hybridization of total RNA to scDNA or cDNA or *in vitro* translation of RNA. The steep rise in concentration of these mRNAs during basidiocarp formation, however, suggests that they are rather involved in this process than in the monokaryon-dikaryon transition. Possibly the hetero-allelic incompatibility genes permit leaky transcription of the genes coding for these mRNAs with full expression only occurring at the time of basidiocarp formation. This is also suggested by the fact that some of these mRNAs are particularly abundant in the basidiocarps themselves (e.g. pScl mRNA) or are depressed in concentration when the dikaryon is surface-grown in the presence of 5 % CO_2 or in darkness, conditions that prevent the formation of basidiocarps (Table 2).

One of the regulated mRNAs and its gene (corresponding to pScl) have been completely sequenced (17). They code for a small protein of Mr 9842 (Fig. 2). The function of this protein is unknown but the presence of a leader sequence and its low molecular weight suggest that it may be one of the proteins excreted by the fruiting dikaryon. Of particular interest is the presence in the gene of three very small introns which contain an internal sequence with complementarity to the 5'-end of the intron, suggestive for a role in splicing (17). The 9000 ntp long genomic clone on which the pSc1 gene was located did not contain other genes abundantly expressed so that this gene apparently is not a member of a cluster of genes that are similarly regulated (18).

CONCLUDING REMARKS

The present study shows that with current techniques only a small number of genes can be detected which are differentially regulated during basidiocarp formation in *Schizophyllum commune*. All of these genes produce prevalent mRNAs at the time of fruiting. Low concentrations of most of these mRNAs are detectable in the vegetative dikaryon before fruiting but not in the monokaryons. Possibly, the interactions of the incompatibility genes condition fruiting genes for transcription with full expression occurring at the time of fruiting.

With respect to the small number of mRNAs apparently regulated during basidiocarp formation it should be recognized that some mRNAs may be strongly expressed only in certain cell types but are represented in too low concentration in RNA preparations from the whole fungus to be detected. Notwithstanding this provision, the present results contrast with data on extensive gene regulation during development in some other fungal systems (22-26). On the other hand, other reports indicate very limited differential gene expression during development, both in fungi (27-29) and plants (30). The reasons for these discrepancies are not clear. In the present study we took care to maintain steady state growth conditions and to avoid the generation of transcriptionally inactive cells. This minimizes the danger that isolated RNA populations reflect differential decay of mRNAs rather than differential transcription. This may have occurred in some of the studies indicating extensive regulation. In addition, by using the monokaryons as a control we might have eliminated changes in gene expression occurring during growth but not related to morphogenesis.

ACKNOWLEDGMENTS

The author is grateful to his collaborators G.H. Mulder and J. Springer for permission to use some of their unpublished data (Tables 1 and 2).

REFERENCES

1. Chang ST, Hayes WA (1978). "The Biology and Cultivation of Edible Mushrooms". New York: Academic Press.
2. Raper JR (1966). "Genetics of Sexuality in Higher Fungi" New York: The Ronald Press Co.
3. Casselton LA (1978). Dikaryon formation in the higher Basidiomycetes, In Smith JE, Berry DR (eds). "The Filamentous Fungi" vol. 3. London: Arnold, p. 278.
4. Koltin Y (1978). Genetic structure of incompatibility factors. The ABC of Sex. In Schwalb MN, Miles PG (eds) "Genetics and Morphogenesis in the Basidiomycetes". New York: Academic Press, p.31.
5. Raper CA (1978). Control of development by the incompatibility system in Basidiomycetes. In Schwalb MN, Miles PG (eds): "Genetics and Morphogenesis in the Basidiomycetes". New York: Academic Press, p.3.
6. Swamy S, Uni I, Ishikawa T (1984). Morphogenetic effects of mutations at the *A* and *B* incompatibility factors in *Coprinus cinereus*. J Gen Microbiol 130: 3219.
7. Leslie JF, Leonard TJ (1979). Monokaryotic fruiting in *Schizophyllum commune*: Genetic control of the response to mechanical injury. Mol Gen Genet 175:5.
8. Esser K, Saleh F, Meinhardt F (1979). Genetics of fruit-body production in higher Basidiomycetes. II Monokaryotic and dikaryotic fruiting in *Schizophyllum commune*. Curr Gen 1:85.
9. Raudaskoski M, Viitanen H (1982). Effects of aeration and light on fruit-body induction in *Schizophyllum* commune. Trans Brit Mycol Soc 78:89.
10. Anderson MR, Deppe CS (1976). Control of fungal development. I. The effects of two regulatory genes on growth of *Schizophyllum commune*. Develop Biol 53,21.
11. De Vries OMH, Reddingius J (1984). Synthesis of macromolecules and compartment size in monokaryotic and dikaryotic hyphae of *Schizophyllum commune*. Exp Mycol 8:378.
12. De Vries OMH, Hoge JHC, Wessels JGH (1980). Regulation of the pattern of protein synthesis in *Schizophyllum*

commune by the incompatibility genes. Develop Biol 74: 22.

13. Zantinge B, Dons H, Wessels JGH (1979). Comparison of poly(A)-containing RNAs in different cell types of the lower eukaryotic *Schizophyllum commune*. Eur J Biochem 101:251.
14. Zantinge B, Hoge JHC, Wessels JGH (1981). Frequency and diversity of RNA sequences in different cell types of the fungus *Schizophyllum commune*. Eur J Biochem 113: 381.
15. Hoge JHC, Springer J, Zantinge B, Wessels JGH (1982). Absence of differences in polysomal RNAs from vegetative monokaryotic and dikaryotic cells of the fungus *Schizophyllum commune*. Exp Mycol 6:225.
16. Dons JJM, De Vries OMH, Wessels JGH (1979). Characterization of the genome of the Basidiomycete *Schizophyllum commune*. Biochim Biophys Acta 563:100.
17. Dons JJM, Mulder GH, Rouwendal GJA, Springer J, Bremer W, Wessels JGH (1984). Sequence analysis of a split gene involved in fruiting from the fungus *Schizophyllum* commune. EMBO J 3:2101.
18. Dons JJM, Springer J, De Vries SC, Wessels JGH (1984). Molecular cloning of a gene abundnatly expressed during fruiting body initiation in *Schizophyllum commune*. J Bacteriol 157:802.
19. De Vries OMH, Wessels JGH (1984). Patterns of polypeptide synthesis in non-fruiting monokaryons and a fruiting dikaryon of *Schizophyllum commune*. J Gen Microbiol 180:145.
20. Wessels JGH, Dons JJM, De Vries OMH (1985). Molecular biology of fruit-body formation in *Schizophyllum commune*. In: Moore D, Casselton LA, Wood DA, Frankland JC (eds). Developmental Biology of the higher Fungi. Cambridge: Cambridge University Press, p. 485.
21. Hoge JHC, Springer J, Wessels JGH (1982). Changes in complex RNA during fruit-body initiation in the fungus *Schizophyllum commune*. Exp Mycol 6:233.
22. Blumberg DD, Lodish HF (1980). Changes in the messenger RNA population during differentiation of *Dictyostelium discoideum*. Develop Biol 78:285.
23. Timberlake WF (1980). Developmental gene regulation in *Aspergillus nidulans*. Develop Biol 78:487.
24. Gwynne DI. Brandhorst BP (1982). Changes in gene expression during sporangium formation in *Achlya ambisexualis*. Develop Biol 91:263.
25. Johnson SA, Lovett JS (1984). Gene expression during

development of *Blastocladiella emersonii*. Exp Mycol 8: 132.

26. Hoge JHC, Heisterkamp ECP, Dons JJM (1983). Changes in translatable RNA population during basidiospore germination in *Schizophyllum commune*. FEMS Microbiol Lett 17:7.
27. Rozek CE, Timberlake WE (1980). Absence of evidence for changes in messenger RNA populations during steroid-hormone induced cell differentiation in *Achlya*. Exp Mycol 4:33.
28. Clancy MJ, Buten-Magee B, Straight D, Kennedy AL, Partridge RM, Magee PT (1983). Isolation of genes expressed differentially during sporulation in the yeast *Saccharomyces cerevisiae*. Proc Natl Acad Sci USA 80:3000.
29. Percival-Smith A, Segall J (1984). Isolation of DNA sequences preferentially expressed during sporulation in *Saccharomyces cerevisiae*. Mol Cell Biol 4:142.
30. De Vries SC, Springer J, Wessels JGH (1983). Sequence diversity of polysomal mRNA in roots and shoots of etiolated and greened pea seedlings. Planta 158:42.

Molecular Genetics of Filamentous Fungi, pages 207–223

FUNGAL SPORE GERMINATION AND MITOCHONDRIAL BIOGENESIS[1]

Robert Brambl

Department of Plant Pathology, The University of Minnesota
Saint Paul, Minnesota 55108

ABSTRACT The germination of fungal spores is dependent upon function of the mitochondrial respiratory system. The conidia of *Neurospora crassa* contain a functional respiratory membrane that is active at the onset of spore germination. These spores also contain a population of preserved mRNA which can be translated *in vitro* into the subunit peptides of cytochrome oxidase and ATPase, and transcript for a representative subunit of the ATPase can be detected in this population by hybridization with a cDNA probe. The dormant spores of *Botryodiplodia theobromae* contain mitochondria that do not have a functional respiratory membrane; most of the cytochrome components are absent, as are catalytic activities of cytochrome oxidase and ATPase. The subunits of cytochrome oxidase are preserved during dormancy in the separate cellular compartments of synthesis. Upon initiation of germination the cytoplasmically synthesized subunits are rapidly translocated into the mitochondria where they assemble with mitochondrially synthesized subunits to yield a functional enzyme. Mitochondria isolated from dormant spores cannot incorporate cytoplasmic subunits of cytochrome oxidase, but this defect is corrected early in germination. In contrast, few of the subunits of the mitochondrial ATPase are preserved in the dormant spores. Instead, the subunits of this enzyme are translated *de novo* from messenger RNA preserved in the dormant spores to yield an active enzyme immediately after initiation of germination. Studies of the assembly of subunits of cytochrome oxidase and ATPase

[1]This research was supported by Research Grant GM19398 from the National Institute of General Medical Sciences.

in germinating spores of Neurospora have revealed an unusual type of post-translational modification of certain subunits with a derivative of pantothenic acid. This modification apparently is involved in the assembly of the subunit peptides of these enzymes.

INTRODUCTION

The germination of metabolically quiescent, dormant fungal spores into rapidly growing cells characteristically includes the development of a strong aerobic respiratory capacity. This mitochondrial respiration and certain other metabolic activities, notably cytoplasmic protein synthesis, seem to be essential for spore germination. For several years my laboratory has been exploring mechanisms by which activated spores establish a functional respiratory system. We have concentrated our attention upon specific components of the respiratory membrane in order to learn how these paradigmatic components are assembled and function during spore germination.

We have asked whether there are structural or functional differences between mitochondria of dormant and germinated spores and whether requisite enzyme complexes must be assembled for establishment of the respiratory system. If certain enzymes are absent, by what mechanism are they reassembled? Is this reassembly regulated at a transcriptional, translational, or post-translational level? What, if any, interactions exist between the mitochondrial and nucleocytoplasmic genetic systems for the reactivation of the spore respiratory system? Finally, can the germinating spores be used productively as an experimental tool to examine cellular assembly processes such as mitochondrial biogenesis?

In this chapter, I briefly review some of our experimental results, both published and unpublished, that help answer these questions and which provide the rationale for our present and future investigations into mitochondrial biogenesis, enzyme assembly processes, and regulation of nuclear and mitochondrial gene expression.

RESULTS

Conidial Germination in *Neurospora crassa*.

General features. Respiration and oxgen uptake by the germinating conidia of *Neurospora* was considered earlier (1) to be dependent primarily upon the function of an alternate, non-cytochrome pathway of electron transport. This pathway, mediated by an alternate oxidase, was believed to support the respiratory demand of the spores until the cytochrome-mediated electron transport system became functional late in the germination sequence. Subsequent studies have shown, however, that the alternate oxidase, characterized by sensitivity to salicyl hydroxamate, does not normally function during conidial germination and that both respiration and spore germination are dependent upon the cyanide-sensitive, cytochrome-mediated electron transport (2). The alternate oxidase does become functional in the absence of normal respiration when the spores are incubated at high temperature (45°C), but under these conditions growth does not occur beyond emergence of the germ tube (3,4). Incubation of the germinating spores with cycloheximide, an inhibitor of cytoplasmic protein synthesis, sharply reduces the normally accelerating rates of oxygen uptake to the initial, basal level, but it does not completely inhibit the respiration; inhibitors of mitochondrial protein synthesis such as chloramphenicol only slightly reduced oxygen uptake (2). These results imply that development of the respiratory activity in germinating *Neurospora* conidia is not dependent initially upon either type of protein synthesis.

We (10) and others (9) have found that brief exposure to water causes a metabolic activation of the *Neurospora* conidia, with initiation of respiration, the assembly of additional polyribosomes, and the initiation of protein synthesis. Therefore, the conidia used in these studies were harvested from the parent mycelium by use of a paraffinic hydrocarbon (10), rather than with water, to ensure that the spores had not been physiologically activated for germination in advance of experimentation.

Biogenesis of cytochrome c oxidase. The dormant conidia of *Neurospora* contain a complete, and potentially functional respiratory system. This conclusion is based on low-temperature spectrophotometric analyses of the mitochondrial cytochromes which showed that all the cytochrome components of the respiratory membrane were present and on catalytic activity assays of cytochrome *c* oxidase and ATPase (2,5,6).

Cytochrome *c* oxidase of Neurospora consists of seven distinct subunit peptides (2,7,8); the three subunits of highest M_r are products of the mitochondrial genetic system and the four smaller subunits are products of the nucleo-cytoplasmic genetic system. Immunoprecipitation analyses of the mitochondria of radiolabeled dormant spores showed that all seven subunits of cytochrome *c* oxidase were present and apparently assembled into a normal complex (2). It is these components of the respiratory system that are preserved in the dormant spores and which contribute to the respiration that begins when the spores are suspended in a germination medium. The fact that these components are synthesized during sporulation for function during germination also explains why respiration early in germination is partially resistant to inhibitors of protein synthesis.

The synthesis of the subunits of cytochrome *c* oxidase during conidial germination was examined by immunoprecipitation assays of pulse-labeled spores and electrophoresis (2). During the first 60 min of germination only low quantities of the mitochondrial subunits were synthesized and little if any of the cytoplasmic subunits were synthesized. During the next 60 min interval of germination, however, all seven subunits of the enzyme were synthesized and assembled, and during the last 120 min of the 240 min germination period, synthesis of all subunits reached maximum rates. These newly synthesized subunits were not required for respiratory activity in germinating spores, since inhibition of their synthesis with chloramphenicol or cycloheximide did not abolish the basal level of cyanide-sensitive respiration in these cells. Dormant conidia of three mutant strains of Neurospora previously thought to contain little if any assembled cytochrome *c* oxidase were found to contain about 25% to 60% of the catalytic activity found in wild-type spores (2,5), and their respiration during germination is strongly dependent upon cyanide-sensitive respiration (2). These findings reinforce the conclusion that the function of the mitochondrial respiratory system is obligately required for spore germination.

Messenger RNA in dormant spores for enzyme subunits. The dormant conidia of Neurospora contain a population of preserved mRNA. Whether its translation is required for germination is not yet known, and neither is its genetic informational contribution to the germination process (10,11). At least some of this mRNA is preserved in the form of polyribosomes, but it is likely that much of it is preserved in a non-polyribosomal form because upon spore activation with water (in the absence of significant new RNA synthesis) the polyribosome content of the spores increases dramatically (9,10). This preserved mRNA also is polyadenylated (11).

The poly A(+) RNA fraction of dormant spores has been isolated by affinity chromatography and translated *in vitro* with wheat germ or reticulocyte systems (11). The radiolabeled translation products then were reacted with specific antisera to subunits of cytochrome *c* oxidase and ATPase to determine if the preserved mRNA contained the genetic information for the subunits of these two enzymes. In both tests, the antisera specifically precipitated labeled peptides which, upon co-electrophoresis with the enzymes isolated from labeled mitochondria, migrated with the authentic subunits but without exact congruence (11). Presumably, this lack of congruence is due to the translation of the mRNA into precursors of the subunits that contain amino-terminal leader sequences that normally would be removed *in vivo* upon translocation into the mitochondria (12).

Further evidence (11) that the preserved mRNA of dormant conidia contains transcripts for the respiratory enzyme subunits comes from specific hybridization experiments of the RNA fractions of these spores with a cDNA (13) containing the nuclear gene sequence for the proteolipid (subunit 12) of the *Neurospora* ATPase-ATP synthase. Dot blot hybridization assays, employing radiolabeled cDNA in a pBR322 vector and poly A(+) and poly A(-) RNA isolated from dormant spores, demonstrate that these RNA fractions of the cells contain transcripts of the proteolipid gene (11). A comparison of the hybridization with RNA fractions from mycelial cells and from dormant spores shows a higher proportion of transcripts in the poly A(+) RNA fraction than in the poly A(-) fraction in mycelial cells, whereas in the dormant spores a higher proportion of the transcripts appear in the poly A(-) fraction that did not bind to oligo(dT)-cellulose. This observation suggests that the extent of polyadenylation of the same transcript differs in the two types of cells, with the mRNA of the dormant spores possessing the shorter poly A tracts.

Modification of enzyme subunits by pantothenate. Lakin-Thomas and Brody (14) found that five proteins in whole-cell extracts of *Neurospora* contain radioactivity contributed by $[^{14}C]$pantothenate and that two of these appeared to be associated with mitochondria. Further study in our laboratory (15) indicated that there were three proteins in these mitochondria that contained this unusual post-translational modification, identified as 4'phosphorylpantetheine (14), one of which is subunit 6 of cytochrome c oxidase and two of which are subunits 8 and 11 of the ATPase-ATP synthase. In the absence of pantothenate supplementation the enzyme activities in germinating spores of an auxotrophic strain do not increase above the levels of the dormant spores and the enzyme subunits synthesized during germination are not assembled into functional complexes (16). Under cellular pantothenate deprivation, all the subunit peptides of these two enzymes are synthesized and apparently accumulate in the mitochondria, but the subunits are not assembled into complexes with the normal proportions of subunits and the incorporation of ^{55}Fe-labeled heme a into cytochrome c oxidase is sharply reduced. Because the free subunits seem to contain far more radiolabel from exogenous pantothenate than the subunits assembled into complexes, we propose that the 4'phosphorylpantetheine is attached transiently to the unassembled subunits and that it is removed upon assembly of the subunits into the enzyme complexes (16). The failure to modify the three subunit peptides causes an interruption of the assembly pathway of cytochrome c oxidase and ATPase-ATP synthase.

Conidial Germination in *Botryodiplodia theobromae*.

General features. The germination of the spores of *Botryodiplodia*, like that of *Neurospora*, requires function of the mitochondrial respiratory membrane, since the inhibitors cyanide, antimycin A, and oligomycin completely inhibit spore germination (17). Cycloheximide also blocks germination and completely inhibits respiration, but inhibitors of mitochondrial protein synthesis such as chloramphenicol and erythromycin have little effect upon these processes (17). In all these cases direct experiments with radiolabeled amino acids and polarographic measurements of oxygen uptake showed that inhibitors were disrupting the target processes in these spores. The major conclusion drawn from these simple experiments is that cytoplasmic protein synthesis is essential to spore germination and to respiration but that mitochondrial protein synthesis is not required for development of an active respiratory system in the germinating spores.

Fatty acid synthesis also is required for germination and respiration in the spores of Botryodiplodia, since incubation with the antilipogenic drug cerulenin blocks both of these processes (18). Although function of the respiratory system does require fatty acid synthesis, physical assembly of the respiratory system in the Botryodiplodia spores apparently does not require this synthesis. All cytochromes are assembled into the mitochondria of cerulenin-treated spores and, strikingly, this drug specifically increases the mitochondrial concentration of spectrophotometrically detectable cytochrome a by about three-fold over normal levels in spores whose cytochrome c oxidase catalytic activity and respiration are inhibited by cerulenin (18).

The mitochondria of the dormant spores of Botryodiplodia are notable because they do not contain an assembled respiratory membrane. Low temperature (77°K) absorption difference spectrophotometry showed that the mitochondrial fraction of the ungerminated spores contains only cytochrome c and one of two expected cytochromes b, but cytochrome a and cytochrome b_{554} are absent (19). Analysis of pyridine hemochromogen preparations also showed that these mitochondria contained no heme a or protoheme.

Analyses of the catalytic activities of cytochrome c oxidase (8,20) and of oligomycin-sensitive ATPase (6) in the dormant and germinating spores of Botryodiplodia showed that neither of these enzymes could be detected in mitochondria of dormant spores. Nevertheless, in each case the catalytic activities are elaborated rapidly during spore germination to reach a near-maximum activity by 240 min of incubation. Development of cytochrome c oxidase activity is blocked by cycloheximide only if this inhibitor is present in the spore incubation medium at or before 95 min of germination, and after this point the development of the enzyme activity becomes partially resistant to the drug; but the development of enzyme activity early in germination is resistant to inhibitors of mitochondrial protein synthesis (20). A similar pattern exists for the ATPase of these spores in that development of the catalytic activity during germination is sensitive to inhibition with cycloheximide, but insensitive to inhibitors of mitochondrial protein synthesis. These results led us to conclude that the products of mitochondrial protein synthesis required for the assembly of these enzymes must be preserved in the dormant spores and that the enzyme subunits of cytoplasmic protein synthesis must be translated de novo upon initiation of spore germination, probably from the mRNA population we knew to be preserved in these spores (22). Subsequent experiments showed that this conclusion was correct only with respect to the ATPase, and the requirement for cytoplasmic protein synthesis for development of cytochrome c oxidase activity remains unexplained.

Reassembly of cytochrome c oxidase. The initial expectation (above) was that the mitochondrial subunits of cytochrome c oxidase were preserved in the dormant spores but that there were no cytoplasmic subunits of this enzyme present in the spores until initiation of germination at which time they were synthesized. Unexpectedly, however, studies (8) of the biogenesis of the subunits of this enzyme during spore germination showed that low quantities of the three mitochondrial subunits were synthesized during the first 60 min interval and that synthesis of the four smaller cytoplasmic subunits could not be detected. Synthesis of the four cytoplasmic subunit peptides did not begin until after 60 min of spore incubation. Nevertheless, oxygen uptake and cytochrome c oxidase activity both increase during this first 60 min interval, even in the apparent absence of synthesis of the cytoplasmic subunits of the enzyme to complement those predicted to be preserved in the mitochondria.

Botryodiplodia spores were produced on a medium containing radiolabeled amino acids in order to generally label the proteins of the dormant spores. Mitochondria were prepared from these spores and tested with antisera prepared against purified cytochrome c oxidase. The electrophoretic analyses of the immunoprecipitates showed that, as predicted, the mitochondria contained the three mitochondrially synthesized subunits of the enzyme; no other subunits were detectable in these mitochondria. However, a similar assay of the post-ribosomal supernatant fluid of these same spores showed that this fraction contained large amounts of four labeled peptides that were precipitated with this antiserum and which closely resemble the four cytoplasmic subunits of the enzyme in electrophoretic mobility (8). This finding suggested that the two groups of subunit polypeptides of cytochrome c oxidase could be compartmentalized during spore dormancy in two separate cellular fractions. The preserved components of the enzyme are mobilized from these respective compartments of synthesis for assembly into a functional enzyme upon activation of the spores for germination.

Subsequent tests of this hypothesis employed other

approaches. To determine the point when heme _a_ was incorporated into the developing enzyme complex, the spores were incubated with ^{55}Fe to radiolabel protoporphyrin IX which is then incorporated into cytochrome _c_ oxidase as heme _a_. We found that the radiolabel was incorporated into the enzyme at very rapid rates between 0 and 60 min of incubation; subsequently the incorporation rates dropped almost to zero and then increased rapidly again after 180 min when labeled amino acids are incorporated rapidly into the enzyme. This initial burst of incorporation was exactly the pattern one would expect if the enzyme were being assembled from pre-existing subunit peptides. The subunit peptides of cytochrome _c_ oxidase were radiolabeled during sporulation with ^{3}H-amino acids, and immunoprecipitation assays were performed with mitochondria isolated from these spores early in germination in a label-free medium. Since other experiments had shown that the cytoplasmic subunits of the enzyme were not synthesized during the first 60 min of germination, the appearance of radiolabeled cytoplasmic subunits in the enzyme during this early interval indicated that the subunits present in the cytoplasm of the dormant spores must have been translocated into the mitochondria for assembly with the mitochondrial subunits. Finally, assays of mitochondria with specific antisera prepared against any one of the four cytoplasmic subunits of the enzyme showed that these antisera would effectively precipitate the mitochondrial subunits from mitochondria of spores radiolabeled in the first 60 min of germination; these mitochondrial subunits should not have been precipitated with these particular antisera if they were not associated with cytoplasmic subunits that were imported from the cytoplasm at the outset of germination (8).

Reassembly of ATPase-ATP synthase. The mitochondria of dormant spores of _Botryodiplodia_ also do not contain an active oligomycin-sensitive ATPase, and this activity increases in the course of spore germination through processes which are sensitive to cycloheximide but insensitive to inhibitors of the mitochondrial protein synthesis system (6).

To determine if the function of this enzyme in germinating spores depended upon assembly of preserved subunits (like that of cytochrome _c_ oxidase), we radiolabeled sporulating cultures and later isolated the mitochondria from the dormant spores. Immunoprecipitation assays with antisera to the F_1-ATPase and to the proteolipid subunit of the enzyme showed that cytoplasmically synthesized subunits 2 and 12 and mitochondrially synthesized subunit 7 could be found in the mitochondria but that none of the enzyme subunits were detectable in the post-ribosomal, extramitochondrial supernatant fraction. Other subunits of the F_1F_0-ATPase were not present, and those three subunit peptides found in the mitochondria were not assembled into a complex but were present as unassociated peptides (6).

The absence of a normal complement of preserved subunits of the ATPase either in the mitochondria or the cytoplasm of the dormant spores suggested the possibility that they were synthesized de novo upon the onset of germination to lead to development of a functional enzyme. We knew from other studies that the earliest protein synthesis during spore germination must depend upon translation of a population of mRNAs preserved in the dormant spores and that new mRNA synthesis does not begin until approximately 45 min of germination. Consequently, we radiolabeled the spores during several intervals of germination with [^{3}H]leucine, and assayed the mitochondria of the spores with antiserum to the F_1-ATPase. Co-electrophoresis of the immunoprecipitates with enzyme isolated from mature cells was used to determine which of the subunits were synthesized during each of the intervals. We found that during the first 45 min labeling period, readily detectable quantities of all the subunits of the ATPase were synthesized and assembled into the mitochondria (6).

This finding made it likely that the preserved mRNA was indeed being translated into subunits of the ATPase during the first 45 min of spore incubation. To test this expectation, the poly A(+) RNA fraction was isolated from dormant spores and translated in vitro with a reticulocyte system. Assays of the translation products were performed with antiserum to the F_1-ATPase followed by co-electrophoresis with the authentic enzyme. The results showed that this mRNA of the dormant spores clearly could be translated into peptides closely resembling the subunits of this enzyme; the slight discrepancies in electrophoretic mobility could be explained by the translation of the mRNA into peptides with short amino terminal extensions. The precipitation of these peptides from the translation mixture was completely dependent upon the specific antiserum and the presence of a translatable poly A(+) RNA. This result therefore indicates that the preserved mRNA of the spores contains transcripts for the subunit peptides of the ATPase and it strengthens the proposal that assembly of this enzyme during spore germination depends upon de novo translation of this preserved mRNA.

Subunit peptide translocation by mitochondria. The translocation of proteins into the mitochondria occurs through a process that involves initial binding to the outer membrane and movement through the intermembrane space followed by translocation through the inner mitochondrial membrane--usually accompanied by precise proteolytic trimming of the peptide--and installation into the membrane or matrix space.

We have asked if there is a difference between the capabilities of mitochondria isolated from dormant and germinated spores to translocate the cytoplasmically synthesized subunits of cytochrome *c* oxidase and ATPase. Poly A(+) RNA from dormant spores was translated *in vitro* and the specific translation products were identified by precipitation assay with antisera to the two enzymes which showed synthesis of peptides that correspond approximately in electrophoretic mobility to the mature subunits of the two enzymes. Incubation of these translation products with mitochondria from the germinated spores, followed by immunoprecipitation and electrophoretic assay, showed that the mitochondria incorporated subunit peptides of both enzymes and converted them to peptides that are electrophoretically congruent with the subunits of the mature enzymes. (This translocation into the mitochondria was inhibited by respiratory uncouplers, and the peptides associated with the mitochondria were resistant to experimental proteolysis.) Strikingly, however, mitochondria prepared from dormant spores differed in their ability to incorporate subunit peptides of the two enzymes: subunits of the ATPase were readily incorporated and processed, but subunits of cytochrome *c* oxidase were not incorporated. This difference in ability to translocate these two groups of subunits complements the studies of *in vivo* assembly, since the subunits of cytochrome *c* oxidase are excluded from the mitochondria during sporulation to accumulate in their cytoplasmic compartment of synthesis. On the other hand, there is no accumulation of the ATPase subunits in the cytoplasm of the dormant spores. It seems possible that the inability to translocate the subunits of cytochrome *c* oxidase may be related to a stage-specific blockage of transport, perhaps as a programmed step to bring about inactivation--and subsequent reactivation--of the mitochondrial respiratory membrane. Since the subunits of the ATPase are incorporated by the dormant spore mitochondria, it seems unlikely that a general requirement for membrane translocation, such as a membrane electrochemical potential, can account for the inability of these mitochondria to incorporate the subunits of cytochrome c oxidase. This finding also implies that different translocation systems exist for subunits of these two enzymes.

DISCUSSION

Function of the mitochondrial systems for electron transport and ATP synthesis seem to be essential for fungal spore germination (reviewed in ref. 24). The results obtained in our laboratory and summarized in this paper show that the asexual spores of two fungi, *N. crassa* and *B. theobromae* meet this requirement in two different ways. In *B. theobromae* the mitochondria of dormant spores do not contain such respiratory membrane complexes as cytochrome *c* oxidase and ATPase-ATP synthase, but these enzymes are rapidly assembled during spore germination through distinctly different mechanisms. The subunits of cytochrome *c* oxidase are preserved during dormancy in the two compartments of synthesis, the cytoplasm and mitochondria, and upon spore activation the cytoplasmic subunits are recruited from the cytoplasm, translocated into the mitochondria, and assembled with the mitochondrially synthesized subunits and heme *a* to give a functional enzyme. A translocation system in the mitochondrial membrane for import of these subunits into the organelle appears to be inactive in the dormant spore, and this system's inactivity during sporulation may lead to a stage-specific exclusion of the subunits from the mitochondria and their accumulation in the cytoplasm. The subunits of the ATPase-ATP synthase are not all preserved in the dormant spores, and no assembled enzyme can be found in these cells. Upon spore activation, however, subunits of this enzyme are translated *de novo* from a population of mRNAs that are synthesized during sporulation and preserved during spore dormancy. Mitochondria of dormant spores do not exclude the subunits of ATPase from translocation, a process that seems not to regulate assembly of this enzyme. These results also make it likely that two distinct transport mechanisms exist in the mitochondrial membrane for the peptide subunits of these two enzymes.

In contrast, the dormant conidia of *N. crassa* contain mitochondria that are potentially functional in respiration. All the components of this respiratory system are present in these mitochondria, and the initial respiration of germinating spores does not depend upon enzyme subunit assembly or translation of preserved mRNA. Nevertheless, as shown by immunoassay of their translation products, these dormant conidia contain transcripts for the subunit peptides of both cytochrome c oxidase and ATPase-ATP synthase. Whether translation of these preserved transcripts contributes to synthesis of the new respiratory enzymes later in germination is not yet known.

The mitochondria of dormant spores of Botryodiplodia apparently cannot translocate the subunit peptides of one of the enzymes required for mitochondrial function, and this defect is corrected early in germination. What is the basis of this defect and how is it reversed? The translocation system for such peptides likely involves several steps, including binding to the cytoplasmic face of the outer membrane, translocation of the peptide through one or both mitochondrial membranes, and proteolytic removal of amino-terminal leader sequences or other covalent modification. It seems most likely that the exclusion of the cytoplasmically synthesized subunits of cytochrome c oxidase in Botryodiplodia during sporulation must occur at the membrane recognition or membrane translocation steps, and it is at the latter step in particular that we are now attempting to identify a stage-specific, reversible mechanism for exclusion of the cytoplasmic subunits from the mitochondria. The clearest results in analyzing the translocation system of spore mitochondria will come eventually from experiments that combine assays of isolated mitochondrial membranes or membrane components with substrates derived from in vitro translation of defined, single mRNAs (prepared from cloned cDNA sequences) for the subunit peptides. In these experiments one might expect to find either a germination-induced activation or modification of an installed, pre-existing translocation system or a de novo synthesis of the translocation system. Either mechanism could involve the cytoplasmic protein synthesis which we believe to be required for development of this enzyme activity in the germinating spores.

An issue yet to be explored is the relationship of the mitochondrial changes observed during spore germination to the process of spore formation and development of the dormant state. One would like to know how the mitochondria in sporulating cells are modified to exclude certain enzyme subunits and by what mechanism existing complexes of cytochrome c oxidase and ATPase-ATP synthase are disassembled or prevented from further assembly. Does a highly specific proteolysis or a type of post-translational modification, such as that by pantothenate (15,16), play a regulatory role in enzyme assembly and disassembly?

One of the strongest advantages of spore germination as an experimental system to study cell assembly processes and gene expression is that reactivation and metabolism seem to depend primarily upon post-transcriptional regulation before new transcription is initiated. Thus, post-transcriptional regulatory mechanisms can be investigated independently of transcription without use of inhibitors. Furthermore, the germinating spores may prove to be a productive device to examine regulation of transcription and interactions between nuclear and mitochondrial genes with defined, cloned gene sequences to determine what degree of coordination exists between the two genetic systems. If one system or the other can be selectively inhibited with temperature sensitive mutations, chemical inhibitors, or physical manipulations such as heat shock and nutrient deprivation, it should be possible to determine if there are specific consequences upon expression of genes of the opposite genetic system. Such experiments will be especially attractive when it becomes possible to examine regulation of expression of genes in the nucleus and mitochondrion for subunit components of the same enzyme.

The experiments described in this paper shed some small light into potential functions of the preserved mRNA found in both types of dormant spores. In both cases translation of the isolated mRNA *in vitro* and immunoassay show that the preserved mRNA contains transcripts for subunits of cytochrome *c* oxidase and the ATPase-ATP synthase, and in *Botryodiplodia* translation of these preserved transcripts for the ATPase appears to be required for assembly of the enzyme into mitochondria. The only other known function for this mRNA in fungal spores is the preservation in *Neurospora* conidia of transcripts for the major heat shock proteins which may be selectively recruited in spores for translation under conditions of heat shock (4). The spores apparently preserve mRNA for respiratory enzyme subunits that are themselves preserved and functional. These transcripts, like those of the heat shock mRNAs at normal growth temperatures, may remain untranslated early in germination. Even though translation of the preserved mRNA may not be required for the initial respiration of germinating *Neurospora* spores, it could nevertheless contribute to the accelerating respiration of these spores later in germination before new transcripts have accumulated for these subunit peptides. It also is possible that multiple mechanisms of enzyme reassembly are useful as a fail-safe rescue in spores attempting to germinate under marginally favorable circumstances.

The differences between the two types of spores, the soil- and water-borne conidia of Botryodiplodia and the air-borne conidia of Neurospora, are curiously disparate; and one wonders if the differences in the two organisms' habitats and the functions of the two types of spores could help explain such basic differences in spore organization. In Botryodiplodia the water-dispersed conidia may function not only in distribution through space but also through time, surviving unfavorable environments for resumption of growth in plant residue or soil. Such spores, perhaps vulnerable to many on-again, off-again stimuli for germination, may rely upon more complicated cellular rearrangements for dormancy and germination than do the air-borne spores of Neurospora whose primary function is not survival through adverse growth conditions but distribution through space and whose growth into a mycelium may depend upon a rapid response to a single germination stimulus of water and minimal nutrients.

ACKNOWLEDGMENTS

Mark Josephson, Herman Wenzler, Susan Stade, Kristin Peterson, Alice Bonnen, Charles Guy, and Nora Plesofsky-Vig are colleagues whose work and keen abilities have been invaluable in this research and the preparation of this review. I acknowledge with gratitude their contributions and those of our other laboratory colleagues to this research.

REFERENCES

1. Schmidt JC, Brody S (1976). Biochemical genetics of Neurospora crassa conidial germination. Bacteriol Rev 40:1.
2. Stade S, Brambl R (1981). Mitochondrial biogenesis during fungal spore germination: Respiration and cytochrome c oxidase in Neurospora crassa. J Bacteriol 147:757.

3. Michéa-Hamsehpour M, Grange F, Ton-That TC,Turian G (1980). Heat-induced changes in respiratory pathways and mitochondrial structure during microcycle conidiation of Neurospora crassa. Arch Microbiol 125:53.
4. Plesofsky-Vig N, Brambl R (1985). The heat shock response of Neurospora crassa: Protein synthesis and induced thermotolerance. J Bacteriol (In press).
5. Stade S, Brambl R (1981). Cytochrome c oxidase in cytochrome c oxidase-deficient mutant strains of Neurospora crassa. J Biol Chem 256:10235.
6. Wenzler H, Brambl R (1981). Mitochondrial biogenesis during fungal spore germination. Catalytic activity, composition, and subunit biosynthesis of oligomycin-sensitive ATPase in Botryodiplodia. J Biol Chem 256:7166.
7. Josephson M and Brambl R (1980). Mitochondrial biogenesis during fungal spore germination. Purification, properties, and biosyntheis of cytochrome c oxidase from Botryodiplodia theobromae. Biochim Biophys Acta 606:125.
8. Brambl R (1980) Mitochondrial biogenesis during fungal spore germination. Biosynthesis and assembly of cytochrome c oxidase in Botryodiplodia theobromae. J Biol Chem 255:7673.
9. Mirkes, PE (1974). Polysomes, ribonucleic acid and protein synthesis during germination of Neurospora crassa conidia. J Bacteriol 117:196.
10. Bonnen A, Brambl R (1983). Germination physiology of Neurospora crassa conidia. Exptl Mycol 7:197.
11. Hammet, JR, Otto, CM, Russell, PJ, Brambl R (1985). Stored polyadenylated ribonucleic acid and conidial germination in Neurospora crassa. (Submitted).
12. Hay, R, Böhni, P, Gasser S (1984). How mitochondria import proteins. Biochim Biophys Acta 779:65.
13. Viebrock A, Pertz A, Sebald W (1982). The imported preprotein of the proteolipid subunit of the mitochondrial ATP synthase from Neurospora crassa. Molecular cloning and sequencing of the mRNA. EMBO J 1:565.
14. Lakin-Thomas, PL, Brody S (1985). A pantothenate derivative is convalently bound to mitochondrial proteins in Neurospora crassa. Eur J Biochem 146:141.
15. Plesofsky-Vig N, Brambl R (1984). Three subunit proteins of membrane enzymes in mitochondria of Neurospora crassa contain a pantothenate derivative. J Biol Chem 259:10660.

16. Brambl R, Plesofsky-Vig N (1985). Assembly of subunits of cytochrome c oxidase and ATPase-ATP synthase in Neurospora crassa requires pantothenic acid. (Submitted).
17. Brambl R (1975). Characteristics of developing mitochondrial genetic and respiratory functions in germinating fungal spores. Biochim Biophys Acta 396:175.
18. Brambl R, Wenzler H, Josephson M (1978). Mitochondrial biogenesis during fungal spore germination: Effects of the antilipogenic antibiotic cerulenin upon Botryodiplodia spores. J Bacteriol 135:211.
19. Brambl R, Josephson M (1977) Mitochondrial biogenesis during fungal spore germination: Respiratory cytochromes of dormant and germinating spores of Botryodiplodia. J Bacteriol 129:291.
20. Brambl R (1977). Mitochondrial biogenesis during fungal spore germination. Development of cytochrome c oxidase activity. Arch Biochem Biophys 182:273.
21. Brambl R (1975). Presence of polyribosomes in conidiospores of Botryodiplodia theobromae harvested with nonaqueous solvents. J Bacteriol 122:1394.
22. Wenzler H, Brambl R (1978). In vitro translation of polyadenylate-containing RNAs from dormant and germinating spores of the fungus Botryodiplodia theobromae. J Bacteriol 135:1.
23. Brambl R (1985). Mitochondria from dormant spores of Botryodiplodia theobromae are specifically defective in translocation of subunit peptides of cytochrome c oxidase. (Submitted).
24. Brambl R (1981). Respiration and mitochondrial biogenesis during fungal spore germination. In Turian G, Hohl HR (eds): "The fungal spore: Morphogenetic controls," New York: Academic Press, p 585.

Molecular Genetics of Filamentous Fungi, pages 225–238

THE MOLECULAR BIOLOGY OF MICROTUBULES IN ASPERGILLUS[1]

Berl R. Oakley

Department of Microbiology
The Ohio State University
Columbus, OH 43210

ABSTRACT Members of my lab are attempting to determine the mechanisms of microtubule-mediated motility by isolating mutations that affect microtubule function and characterizing these mutations genetically, morphologically and biochemically. We have isolated more than 100 cold-sensitive (cs-) revertants of benA33, a heat-sensitive (hs-) β-tubulin mutation that blocks the functioning of microtubules at high temperatures. Of 78 cs- revertants analyzed genetically, reversion was due to mutations closely linked to benA33 in 44 cases and to mutations unlinked to benA33 in 34 cases. We have mapped 18 of the closely linked revertants precisely with respect to benA33. All map to the right of benA33 and the recombination frequencies with respect to benA33 range from 2.8 X 10^{-5} to 1.2 X 10^{-4}. The close linkage of these mutations to benA33 argues that they are in the benA gene. Of the remaining 34 revertants, 30 have been analyzed. In 10 revertants, cold sensitivity is caused by mutations that coincidentally occcurred during mutagenesis and that have nothing to do with the suppression of benA33. In the remaining 20 cases, cold sensitivity is caused by mutations that cause the suppression of the heat sensitivity conferred by benA33. Of the 20 cs- suppressors of benA33, 16 map to the tubA, α-tubulin, gene and four map to two additional loci we have designated mipA and mipB. Nuclear division and migration are inhibited in many of

[1]This work was supported by grants GM31837 from the NIH and 1-187 from the March of Dimes Birth Defects Foundation.

the revertants at a restrictive temperature, indicating that microtubule functioning is disrupted. Mitochondrial movement is not inhibited, however, indicating that nuclei and mitochondria may move by different mechanisms. We have found that the frequency of non-disjunction is significantly increased in diploids carrying benA33 and in diploids constructed from some of the benA33 revertants. In an effort to identify the non-tubA suppressors of benA33, we have developed improved transformation frequencies and are now attempting to clone these suppressors by complementation.

INTRODUCTION

The central goal of our work is to determine the mechanisms by which organelles move through cells and chromosomes move in the mitotic spindle. We have several more specific aims. We wish to determine how α and β tubulin interact to assemble into microtubules, to identify the proteins in addition to α and β tubulin that are components of microtubules, to determine the identities of proteins that interact with microtubules, and to determine how microtubules and associated structures produce force in chromosomal and organellar movement.

One approach we are taking is to isolate mutations in genes whose products are essential to microtubule function as revertants of a heat-sensitive (hs-) β-tubulin mutation. This approach is based on the work of Jarvik and Botstein (1) who showed in phage P22 that 1) temperature sensitivity is often caused by missense mutations, 2) revertants of temperature-sensitive (ts-) mutations often carry second-site mutations, mutations in genes whose products interact with the protein encoded by the original mutant gene, and 3) these second-site mutations often confer cold sensitivity.

These results led myself and others to suggest that if we were able to isolate a heat-sensitive (hs-) mutation in a gene that encodes a known microtubule protein such as α or β tubulin, we should be able to isolate mutations in other genes whose products interact with this protein as revertants of the original mutant (2,3). We might, moreover, expect some of the reversions to cause conditional lethality e.g. some revertants of hs- mutants might be cold-sensitive (cs-). This approach was initiated in

Aspergillus nidulans by Morris *et al.* (4) who isolated several revertants of weakly hs- β-tubulin mutations and found, among the revertants, a second-site mutation that enabled them to identify an α-tubulin structural gene, the *tub*A gene.

Oakley and Morris (5) later isolated and characterized a more tightly heat-sensitive β-tubulin mutation, *ben*A33. This mutation inhibited nuclear division and movement and caused an increase in the mitotic index at restrictive temperatures but did not block microtubule assembly. This mutation also conferred resistance to a broad spectrum of anti-microtubule agents and the heat sensitivity conferred by this mutation was partially suppressed by antimicrotubule agents. These and other data led Oakley and Morris to conclude that *ben*A33 probably inhibits microtubule function by inhibiting microtubule disassembly.

My lab has isolated and characterized a large number of conditionally-lethal suppressors of *ben*A33. I will not reiterate the details of, nor give the primary data from, these studies which are compendious and are being published elsewhere. Rather, I will take this opportunity to give an overview of the results and tc make some general comments on the utility and limitations of this approach.

RESULTS

Revertants of *ben*A33 carry mutations in the *ben*A gene, the *tub*A gene, and two other genes.

My lab has isolated over 2600 revertants of *ben*A33 and has found 128 cold-sensitive revertants among them. We have analyzed 78 cs- revertants genetically (6,7) and have found that in 44 cases the reversion is due to a mutation that is closely linked to the *ben*A gene. We have analyzed 33 of these 44 revertants with greater precision and have found that reversion is, in each case, due to a mutation less than or equal to 0.1 map unit from *ben*A33 (7). In these cases reversion is, thus, likely to be due to mutations in the *ben*A gene. In 34 revertants, reversion was due to mutations that recombined freely with *ben*A33 and are thus extragenic suppressors of *ben*A33. In 10 revertants we have found that the reversion is genetically separable from cold sensitivity (8). In these cases cold sensitivity is thus due to a mutation that occurred coincidentally during mutagenesis and is unrelated to the reversion. In 20 of the remaining 24 revertants, cold-sensitivity is due to the mutation that causes the reversion from hs- to hs+. We have now analyzed these 20 extragenic suppressors and 16 of them map to the tubA, α-tubulin, locus. Three of them map to another locus on linkage group VIII we have designated mipA (mip for microtubule interacting protein) and the remaining suppressor maps to another locus we have designated mipB. In summary we have isolated large numbers of α- and β-tubulin mutations and smaller numbers of mutations in two additional loci among revertants of benA33.

Phenotypes of extragenic suppressors of benA33 when separated genetically from benA33

Jarvik and Botstein (1) found that many of the cs- extragenic suppressors of ts- mutations in phage P22 retain their cs- phenotype when crossed into a background with a wild-type (wt) copy of the ts- mutation. I have pointed out, however (3), that other phenotypes are possible and among extragenic suppressors of benA33 we have, indeed, found other phenotypes (8). We have found that each of the cs- tubA suppressors of benA33 is lethal in a background with the wt benA gene and each of the three mipA alleles is silent in a wt benA background. Only one extragenic suppressor, mipB, appears to confer cold sensitivity when crossed into a background with a wt benA gene.

I have previously suggested (3) that in some cases extragenic suppressors of tubulin mutations might compensate for the original mutation such that revertants that carry both the original mutation and the suppressor would show no conditional lethality but the suppressor would confer conditional lethality when separated genetically from the mutation it suppresses. With this possibility in mind we have examined revertants in which cold sensitivity is caused by a coincidental cs- mutation. We have found that one of these revertants carries an α-tubulin mutation, which we have designated tubA4, that suppresses the heat sensitivity conferred by benA33 (6,8). The benA33, tubA4 double mutant is cs+, hs+ and resistant to the antimicrotubule agent, benomyl, but tubA4 in a background with a wt benA gene confers super-sensitivity to benomyl and weak cold sensitivity.

In order to determine if tubA4 suppresses benA33 specifically or also suppresses other hs- benA alleles, we have crossed tubA4 into strains carrying benA31 and benA32, two benomyl-resistant, hs- benA mutations previously isolated by B. R. Oakley and N. R. Morris (unpublished). We have found (8) that while tubA4 slightly reduces the benomyl resistance conferred by benA31 and benA32 it does not suppress the heat sensitivity conferred by these two mutations. TubA4 is, thus, an allele-specific suppressor of benA33.

Fine structure mapping of the benA gene

As stated previously, in 33 revertants of benA33 cold sensitivity and reversion from hs- to hs+ are due to mutations that map to within 0.1 map units of benA33. The close linkage of these mutations to benA33 suggests that they are within the benA gene and the fact that they suppress the heat sensitivity conferred by benA33 suggests that they are in regions of the benA gene essential to the functioning of the β tubulin encoded by this gene. Since my lab is interested in determining the mechanisms of microtubule function including the mechanisms of interaction between α and β tubulin, we were interested in mapping these mutations within the benA gene to define genetically some of the essential regions of the gene. Such an analysis requires the examination of many progeny of a revertant x wt cross. Fortunately we have devised a procedure that allows us to select rare recombinants among tens of thousands of parentals. Since the revertants are cold sensitive and the wild-type parent is benomyl sensitive, parental ascospores will not grow on benomyl at low temperatures. A crossover between benA33 and a suppressor of benA33 will produce two recombinants, one of which will carry benA33 without a suppressor. This class of recombinants will be cs+, hs-, and benomyl resistant and can be selected among thousands of parental ascospores on benomyl at low temperatures. By determining the number of viable ascospores and the number of cs+, hs-, benomyl-resistant recombinants, one can determine recombination frequencies between benA33 and intragenic suppressors. By using appropriate outside markers one can determine which side of benA33 the suppressors map to.

We have mapped 17 of our cs- benA mutants and another cs- revertant of benA33 isolated by C. F. Roberts and D. R. Kirsch in the lab of N. R. Morris. After appropriate corrections were made to remove distortions due to different viabilities of the revertants, we have found that all of these mutations map to the right of benA33 with recombination frequencies between 2.5×10^{-5} and 1.2×10^{-4}. These values are very low and suggest strongly that these mutations are within the benA structural gene. In addition, these mutations are not distributed randomly. Many of the mutations map to two clusters with recombination frequencies of 4.8×10^{-5} and 8.0×10^{-5}.

In summary, these are likely to be mutations within the benA, β-tubulin gene that alter the structure of the β tubulin encoded by this gene such that the functioning of the molecule is restored at high temperatures but disrupted at low temperatures. The fact that many of these genes map to two clusters suggests that these are regions of the β-tubulin gene particularly important to microtubule function.

Nuclear division and movement but not mitochondrial movement are inhibited in most revertants of benA33.

One of the reasons for isolating revertants of benA33 was to isolate mutations that disrupted microtubule function. Since many of our revertants are cold sensitive, we expected that microtubule function might be disrupted at low temperatures in many of the revertants. To test this possibility we have germinated conidia of many of the revertants at a restrictive temperature of 25°C. When conidia (uninucleate asexual spores) are incubated in growth medium they germinate, the single nucleus divides repeatedly and the nuclei move from the conidium into the growing germ tube. One can determine if nuclear division and movement have occurred in a germling simply by counting the nuclei and determining whether one or more nuclei have moved from the conidium into the germ tube (5,9). If germlings are stained with the DNA-binding dye 4',6-diamidino-2-phenylindole (DAPI), one can visualize not only nuclei but also mitochondrial genomes. One can thus determine if any of these mutations affect mitochondrial movement as well. We have examined nuclear division, nuclear movement, and mitochondrial movement in revertants carrying intragenic suppressors of benA33, tubA extragenic suppressors of benA33, and mipA and mipB suppressors of benA33. In each of 7 revertants carrying intragenic suppressors of benA33 examined, nuclear division and movement were inhibited at a restrictive temperature of 25°C relative to a wt control at the same temperature and relative to the revertants at a permissive temperature of 37°C (7). The fraction of germlings exhibiting nuclear division at the restrictive temperature ranged from 5% to 23% and in the wt control nuclear division had occurred in 97% of germlings examined. At the permissive temperature nuclear division rates in all revertants were substantially higher than at the restrictive temperature, ranging from 57% to 83%. These values are less than that of the wt control (100%) and presumably reflect the fact that microtubules in these revertants do not function normally even at the permissive temperature.

Nuclear movement was also inhibited at the restrictive temperature in these revertants (7). At the restrictive temperature the fraction of germlings in which at least one nucleus had migrated out of the conidium ranged from 36% to 67% for the various revertants as opposed to 98% for the wt control. Nuclear movement was also somewhat inhibited at the permissive temperature, the fraction of germlings exhibiting nuclear movement ranging from 49% to 71%.

Interestingly there was no inhibition of mitochondrial movement in any of these revertants. At the restrictive temperature the fraction of germlings in which mitochondria had moved out of the conidium ranged from 88% to 92%, (wt control = 86%) and at the permissive temperature values ranged from 95% to 99% (wt control = 97%).

Nuclear division and movement were also inhibited at 25°C in revertants carrying cs- tubA mutations (8). Among seven cs- tubA mutations examined, the fraction of germlings exhibiting nuclear division ranged from 9% to 51% (wt control = 97%) and the fraction exhibiting nuclear movement ranged from 39% to 76% (wt control = 98%). At a permissive temperature of 37°C the fraction exhibiting nuclear division ranged from 42% to 99% (wt control = 100%) and the fraction exhibiting nuclear migration ranged from 30% to 84% (wt control = 99%). Mitochondrial movement was not inhibited at the restrictive temperature, the fraction of germlings exhibiting mitochondrial migration ranging from 89% to 100% (wt control = 86%).

Nuclear division and movement were not strongly inhibited at the restrictive temperature in revertants carrying mipA and mipB. The fraction of germlings exhibiting nuclear division ranged from 82% to 94% and the fraction exhibiting nuclear movement ranged from 84% to 94%.

Nondisjunction rates are altered in diploids carrying *ben*A33 and in diploids constructed from revertants of *ben*A33.

Since microtubules are essential to chromosome segregation in mitosis, we wished to determine if mutations that affected microtubules altered the fidelity of chromosomal segregation. In *A. nidulans* recessive conidial color markers make the detection of nondisjunction comparatively easy. One simply constructs diploids carrying two or more recessive color mutations in different genes. The resulting diploid will have conidia with wt (green) color but if nondisjunction occurs the wt copy of one or both of the color mutations will be lost in many cases and sectors with conidia of other colors will form as the colony grows. We have used this assay to examine the rates of nondisjunction in diploids homozygous or heterozygous for *ben*A33 and in heterozygous diploids constructed from revertants of *ben*A33 and a wt strain (6,10). *Ben*A33 causes a striking increase in the frequency of nondisjunction in heterozygous or homozygous diploids. At 25°C the nondisjunction rate for the *ben*A33/wt diploid is 2.06 times greater than that of a diploid that is isogenic except that it carries two wt *ben*A alleles. The nondisjunction frequency for the *ben*A33/*ben*A33 diploid is even greater, 4.01 times greater than that of the wt/wt diploid. At 37°C the *ben*A33/*ben*A33 diploid will not grow because of its heat sensitivity but nondisjunction occurs in the *ben*A33/wt diploid at a frequency 2.37 times greater than that of the wt/wt diploid.

Nondisjunction frequencies in the revertant/wt diploids are quite variable, ranging at 25°C from below that of the wt/wt diploid to 2.10 times greater than that of the wt/wt diploid. At 37°C results were similar, nondisjunction rates in the revertant/wt diploids ranging from below that of the wt/wt diploid to 2.76 times greater than the wt/wt diploid. These results are not easy to interpret because two mutations that affect microtubule function (*ben*A33 and a suppressor of *ben*A33) are present in each revertant as well as wt alleles of each of these genes. The most significant evidence that the suppressors of *ben*A33 affect the frequencies of nondisjunction is not that nondisjunction frequencies are elevated in many of the revertant/wt diploids relative to the wt/wt diploid but that the frequencies in many of them are lowered relative to the *ben*A33/wt diploid.

Progress in cloning mip genes

Our stategy for identifying the products of these genes is to clone the genes and sequence them. Having isolated and sequenced these genes we can make gene fusions with the mip genes and have them expressed in E. coli or synthesize peptides corresponding to portions of the genes. We can then use these peptides or fusion proteins to induce antibodies that will recognize the native proteins encoded by the mip genes and use the antibodies to purify the native proteins.

At present all of the techniques required for this approach are common laboratory practice except for cloning the mip genes. Since mipA and mipB confer conditional lethality, an obvious approach is to clone the wt copies of these genes by complementation as is commonly done in Saccharomyces cerevisiae and Schizosaccharomyces pombe. The problem with this approach is that to this point no shuttle vectors have been reported for A. nidulans so that transformation frequencies are much lower than for S. cerevisiae or S. pombe. Secondly as Neff et al. (11) have shown for S. cerevisiae and Hiraoka et al. (12) have shown in S. pombe, it may not be easy to clone genes whose products compose microtubules in conventional high copy number shuttle vectors. Since the stoichiometry of the proteins that compose microtubules may be critical, placing many copies of a gene that encodes a microtubule protein in a cell may disrupt the stoichiometry and render the microtubules disfunctional.

To surmount these problems we have attempted to clone genes in A. nidulans, including the mipB gene, by integrative transformation. In order for this approach to be feasible we must have very high frequencies of integrative transformation and we must be able to recover our gene of interest from the transformant. We have systematically approached the problem of increasing integrative transformation frequencies using a plasmid, pDJB1, that carries the orotidine-5'-phosphate decarboxylase gene of Neurospora crassa which complements the pyrG gene of A. nidulans (13). We have carefully modified the transformation procedures of Ballance et al. (13) by carefully varying times and temperatures of incubation in the steps of transformation, by varying protoplasting conditions, by protoplasting very young germlings and, most significantly, by osmotically balancing the polyethylene glycol solutions used in the transformation procedure with 0.6M KCl. By making these improvements we have routinely been able to obtain more than 100 stable transformants per microgram of pDJB1 per 10,000 viable protoplasts (B. R. Oakley and B. L. Mitchell, unpublished). We can calculate, given this transformation frequency and the size of the A. nidulans genome, that if we transform 10^6 viable protoplasts with 100 μg of a gene library that has 4-6 kb inserts we should find one transformant that carries our gene of interest. Given reasonable losses of viable protoplasts during the protoplasting and transformation procedures, we would have to start with 10^7 - 10^8 conidia before protoplasting. The amounts of DNA and numbers of conidia are, thus, large but manageable.

A second requirement for cloning by integrative transformation is that we must be able to recover our gene of interest from the transformant. If integrative transformation occurs, the gene of interest should be linked to an antibiotic resistance marker. We can recover the antibiotic resistance marker by cutting DNA of the transformed cell with an appropriate restriction endonuclease, sizing the fragments, ligating fragments of an appropriate size, and transforming E. coli with the ligated fragments, selecting for antibiotic resistance. A fraction of the plasmids that confer antibiotic resistance in E. coli should carry our gene of interest and these plasmids can be identified by their ability to transform A. nidulans and complement mutations in our gene of interest. We have been able to recover the Neurospora orotidine-5'-phosphate decarboxylase gene from transformed A. nidualans in this manner (B.R. Oakley, B. L. Mitchell and J. E. Rinehart, unpublished).

Given our ability to transform at relatively high frequencies and to recover a transforming gene, we were encouraged to attempt to clone mip genes by integrative transformation. We have constructed a library that contains more than 99.9% of the A. nidulans genome and have transformed a strain that carries mipB with the library. We have obtained two transformants that are no longer cold sensitive and by Southern blot analysis we have determined that they have plasmid sequences integrated into their genomes (C. F. Weil, unpublished). We are now attempting to recover the mipB gene from these transformants.

DISCUSSION

Isolating and characterizing revertants of benA33 has proved a laborious but effective approach for isolating mutations that affect microtubules. It has allowed the isolation of conditionally lethal mutations in genes that encode α tubulin, β tubulin and two other, previously unidentified genes whose products interact with β tubulin and are involved in microtubule function. Although identifying and characterizing these mutations has been time consuming, the effort has been worthwhile because these mutations should be of great value in understanding the structure and function of α and β tubulin, and the mip mutations should permit the identification of two previously unidentified proteins that interact with β tubulin.

In some ways these results are disappointing in that we had hoped to identify more than two mip loci. Thomas et al. (14) have, by contrast, found 16 complementation groups represented among conditionally lethal extragenic suppressors of a conditionally lethal β-tubulin mutation of Saccharomyces cerevisiae. Isolating and characterizing the mipA and mipB genes and identifying the proteins they encode are likely to be lengthy processes, however, and it is unlikely that identifying additional mip loci will be rate-limiting in our work. It is reasonable to expect, moreover, that extragenic suppressors of β-tubulin mutations will be allele specific and the single allele we have tested, tubA4, is indeed, allele specific, suppressing benA33 but not benA31 or benA32. It is likely, therefore, that isolating revertants of β-tubulin mutations other than benA33 would permit the identification of new alleles of mipA and mipB and the identification of new mip loci.

Each category of revertants of benA33 is potentially useful in understanding the mechanisms of microtubule-mediated motility. Understanding the precise way in which each intragenic and extragenic suppressor suppresses the heat sensitivity conferred by benA33 and causes cold sensitivity should reveal something of how α and β tubulin and the proteins encoded by the mipA and mipB genes are involved in the functioning of microtubules. Stated more simply, morphological and biochemical analysis of how these mutations affect microtubule function should allow us to understand how wt microtubules function.

Among the revertants of benA33 we have isolated a large number of mutations that map to the benA and tubA loci. We have shown that many of these mutations disrupt microtubule function at low temperatures. These mutations must necessarily be in regions of the benA and tubA genes that are essential to the functioning of the tubulin molecules encoded by these genes. Our fine structure mapping of the intragenic suppressors of benA33 has given an indication that these mutations are not randomly distributed in the benA gene but tend to be in clusters. It is reasonable that these clusters are in regions of the gene particularly important to the functioning of β tubulin. Sequence analysis of the benA and tubA mutations should allow us to define essential regions of the tubulins, and understanding the amino acid substitutions in these mutants may help us understand how these molecules interact to assemble into microtubules. If our cs- tubA mutants consistently show, for example, the substitution of hydrophilic amino acids for hydrophobic amino acids, we will have good evidence that hydrophobic interactions are involved in the functioning of microtubules. If these data are coupled with biochemical characterization of microtubules in the mutants, we should have a good idea of how such interactions are involved in microtubule functioning.

The mipA and mipB mutations are of particular interest because they are mutations in previously unidentified genes whose products are essential to microtubule function. Identifying the products of these genes will, thus, reveal two new proteins involved in microtubule-based motility. Since identifying the proteins involved in mitosis and microtubule-based organellar translocation has proved difficult, identitication of the products of mipA and mipB would represent a major advance.

ACKNOWLEDGEMENTS

This discussion was based on experiments conducted by Elizabeth Oakley, Janet Rinehart, Brenda Mitchell, Kimberly Kniepkamp, and Clifford Weil. I would like to thank Elizabeth Oakley for invaluable assistance in preparing this manuscript.

REFERENCES

1. Jarvik J, Botstein D (1975). Conditional-lethal mutations that suppress genetic defects in morphogenesis by altering structural proteins. Proc Natl Acad Sci USA 72:2738.
2. Oakley BR (1981). Mitotic mutants. In Zimmerman AM, Forer A (eds): "Mitosis/Cytokinesis," New York: Academic Press, p 181.
3. Oakley BR (1983). Conditionally lethal tubulin mutations of Aspergillus nidulans. J Submicrosc Cytol 15:363.
4. Morris NR, Lai MH, Oakley CE (1979). Identification of a gene for α-tubulin in Aspergillus nidulans. Cell 16:437.
5. Oakley BR, Morris NR (1981). A β-tubulin mutation in Aspergillus nidulans that blocks microtubule function without blocking assembly. Cell 24:837.
6. Oakley BR, Oakley CE, Rinehart JE, Mitchell BL, Kniepkamp KS (1984). Genetic and phenotypic characterization of conditionally lethal mutations in genes that encode microtubule proteins of Aspergillus nidulans. In Seno S, Okada Y (eds): "International Cell Biology 1984," Tokyo: The Japan Society for Cell Biology, p 486.
7. Oakley BR, Oakley CE, Kniepkamp KS, Rinehart JE (submitted for publication). Isolation and characterization of cold-sensitive mutations at the benA, β-tubulin, locus of Aspergillus nidulans.
8. Oakley BR, Oakley CE, Weil CF, Rinehart JE (in preparation). Isolation and characterization of extragenic suppressors of a heat-sensitive β-tubulin mutation of Aspergillus nidulans.
9. Oakley BR, Morris NR (1980). Nuclear movement is β-tubulin dependent in Aspergillus nidulans. Cell 19:255.
10. Oakley BR, Mitchell BL (in preparation). Tubulin mutations alter the frequency of non-disjunction in Aspergillus nidulans.
11. Neff NF, Thomas JH, Grisafi P, Botstein D (1983). Isolation of the β-tubulin gene from yeast and demonstration of its essential function in vivo. Cell 33:211.
12. Hiraoka Y, Toda T, Yanagida M (1984). The NDA3 gene of fission yeast encodes β-tubulin: A cold-sensitive nda3 mutation reversibly blocks spindle formation and

chromosome movement in mitosis. Cell 39:349.
13. Ballance DJ, Buxton FP, Turner G (1983). Transformation of Aspergillus nidulans by the orotidine-5'-phosphate decarboxylase gene of Neurospora crassa. Biochem Biophys Res Commun 112:284.
14. Thomas JH, Novick P, Botstein D (1984). Genetics of the yeast cytoskeleton. In Borisy GG, Cleveland DW, Murphy DB (eds): "Molecular Biology of the Cytoskeleton," Cold Spring Harbor, NY: Cold Spring Harbor Press, p153.

Molecular Genetics of Filamentous Fungi, pages 239–251

IDENTIFICATION AND FUNCTION OF BETA TUBULIN GENES IN ASPERGILLUS NIDULANS[1]

Gregory S. May, James A. Weatherbee, John Gambino
Monica L.-S. Tsang and N. Ronald Morris

Department of Pharmacology, UMDNJ/Rutgers Medical School
Piscataway, NJ 08854

ABSTRACT There are two genes for beta-tubulin in *Aspergillus nidulans*, *ben*A and *tub*C. Although benomyl resistant *ben*A mutants of *A. nidulans* grow in the presence of benomyl, they normally fail to conidiate. To determine whether the benomyl sensitivity of conidiation was caused by a beta tubulin, conidiation resistant (CR) mutants were selected and their tubulins were analyzed by 2D gel electrophoresis. All of the CR mutants lacked beta3-tubulin, suggesting that the sensitivity of conidiation to benomyl was caused by the presence of this tubulin. Two beta tubulin sequences (B5 and B14) were cloned from a genomic library by homology to a chicken beta tubulin cDNA and used to construct shuttle vectors, which also carried a *pyr*4^+ gene from *Neurospora crassa*. A *pyr*G^- strain of *A. nidulans* was transformed to uridine prototrophy with these plasmids. In each case transformation was integrative and homologous to the cloned gene. These (*pyr*4^+, *ben*A^+) transformants were crossed to a *pyr*G^-, *ben*A^- (benomyl resistant) strain and the progeny analyzed for segregation of the plasmid linked *pyr*4^+ marker versus the *ben*A gene. This identified B5 as the *ben*A sequence and B14 as the *tub*C sequence. Disruption of the *tub*C gene by integrative transformation with an internal fragment of the B14 sequence not only caused the beta3-tubulin polypeptide to disappear but resulted in the appearance of the CR

[1]This work was supported by USPHS Grant GM 29228-04 to N. Ronald Morris and by Anna Fuller Fund Postdoctoral Fellowship #600 to Gregory S. May.

phenotype. Ten CR mutants were analyzed genetically by being crossed to a tubC, pyr4+ transformant. All were at the tubC locus.

INTRODUCTION

Tubulin, the major structural protein of microtubules, is evolutionarily one of the most highly conserved of proteins, yet it is also highly polymorphic (1). Tubulin protein polymorphisms were initially found as multiple tubulin bands or spots on one dimensional or two-dimensional protein gels. More recently, tubulin polymorphism has been found to go deeper than the tubulin proteins *per se*. Genetic and molecular biological analysis has shown that all eukaryotes that have been examined, with the possible exception of *Saccharomyces cerevisiae*, have multiple genes for both alpha and beta tubulins (2,3). Although the reason for this diversity of tubulin proteins and genes is not well understood, the multiplicity of tubulin proteins would seem to suggest that different tubulins may serve different functions in different microtubules, and the multiplicity of tubulin genes on the other hand might indicate that regulation of the synthesis of different tubulins was of key importance. In order to analyze this problem we have chosen to work with the fungus *Aspergillus nidulans*, which has several important attributes that make it suitable for analysis of microtubule function and regulation, the most important being its excellent genetic system (4,5) and the fact that it is transformable (6,7,8,9).

A. nidulans has been shown to have both multiple tubulin polypeptides and multiple tubulin genes. There are four known tubulin genes in *A. nidulans*: tubA, which codes for alpha1- and alpha3-tubulins, tubB, which codes for alpha2-tubulin, benA, which codes for beta1- and beta2-tubulins and tubC, which codes for beta3-tubulin (10,11,12,13,14). Alpha1-, alpha3, beta1- and beta2-tubulins are the predominant tubulin proteins found during vegetative growth, whereas alpha2- and beta3-tubulins are less abundant during vegetative growth than the other isomorphs.

Previous work from this laboratory has shown that the benA and tubA gene products are involved in both the mitotic spindle and cytoplasmic microtubules (15,16,17). The evidence is as follows. The benA mutants were originally selected for resistance to benomyl, but some of the benA mutants are also temperature sensitive. Incubation of a temperature sensitive benA mutation (benA33) at restrictive temperature caused cells to be blocked in mitosis with a hyperstable mitotic spindle and also increased the stability of cytoplasmic microtubules (16). Thus we know for a certainty that benA gene products must be used for both spindle and cytoplasmic microtubules. We, however, do not know whether the same protein gene product is involved in both types of microtubules, since the benA gene codes for two polypeptides. Chromosome movement and nuclear migration are also inhibited at restrictive temperature in strains carrying the benA33 mutation, indicating that microtubule depolymerization is necessary for both chromosomal and nuclear movement to occur (16).

TubA1 was isolated as an indirect suppressor of a temperature sensitive benA mutant (15). We have shown that the tubA$^-$ mutation causes both spindle and cytoplasmic microtubules to become relatively "unstable", and we have postulated that it is this "instability" that is responsible for suppressing the microtubule "hyperstability" caused by the benA mutation (15). Thus we know that tubA gene products must also be involved in the mitotic spindle and in cytoplasmic microtubules. However, since tubA, like benA, has two polypeptide gene products, we don't know whether the same polypeptide is involved in both types of microtubules.

No naturally occuring or induced mutations have been found in the tubB gene, so we have no information about the function(s) of alpha2-tubulin; but we have recently obtained information that sheds some light on the function of tubC and its gene product, beta3-tubulin. These experiments are the subject of this report.

RESULTS

Selection of Conidiation Resistant Mutants

BenA mutants, although able to grow in the presence of benomyl, are unable to conidiate in the presence of the drug (13). We suspected that the sensitivity of conidiation to benomyl of benA mutants might be caused by a new benomyl sensitive beta tubulin being used during conidiation that was not used for vegetative growth. This new tubulin might either replace the benA beta tubulin or be synthesized in addition to benA beta tubulin. It should be noted that benA⁻/benA⁺ diploids are sensitive to benomyl. The obvious candidate for the new conidiation specific tubulin was beta3, which, as noted above is a minor component of the Aspergillus tubulins found during vegetative growth.

The direct approach to this problem would be to compare the tubulins synthesized in the mycelium during vegetative growth with those synthesized in the conidiophore (the asexual spore-bearing structure) during normal conidiation. Unfortunately, there is at the moment no convenient way to purify conidiophores away from vegetative mycelium, so that the tubulins of the pure tissues can not be easily compared. As an alternative approach, we decided to select conidiation resistant (CR) mutants of a benA⁻ (normally conidiation sensitive) strain. This was done by looking for spots of conidiation in an otherwise aconidial lawn of benA22, which was grown on medium containing a concentration of benomyl sufficient to inhibit conidiation. Our belief was that some of these mutants should exhibit electrophoretic shifts in the postulated conidiation specific tubulin similar to the electrophoretic shifts exhibited by the beta tubulins of benomyl resistant benA mutants. Our reason for using benA22 as the parental strain was that the 2^+ charge shift in beta1- and beta2-tubulins caused by this mutation reveals beta3-tubulin (which is normally occluded by beta1- and beta2-tubulins on two dimensional gels of tubulins from wild type strains) (12).

Twenty conidiation resistant (CR) mutants were isolated, three spontaneous and seventeen as the result of mutagenesis with ultra violet light. The tubulins from the parental benA22 and benA22/CR mutant strains were then partially purified by DEAE chromatography followed by ammonium sulfate fractionation and compared by two dimensional gel electrophoresis (Fig. 1). All twenty benA22/CR mutants were found to lack the beta3-tubulin polypeptide (13). This does not necessarily indicate that all 20 mutations were null mutations, but could simply mean that the mutant beta3-tubulins failed to purify as normal beta3-tubulin.

In order to determine whether the absence of beta3-tubulin cosegregated with the CR phenotype, two of the CR mutants were crossed to a wild type strain and the progeny analyzed for their ability to conidiate on

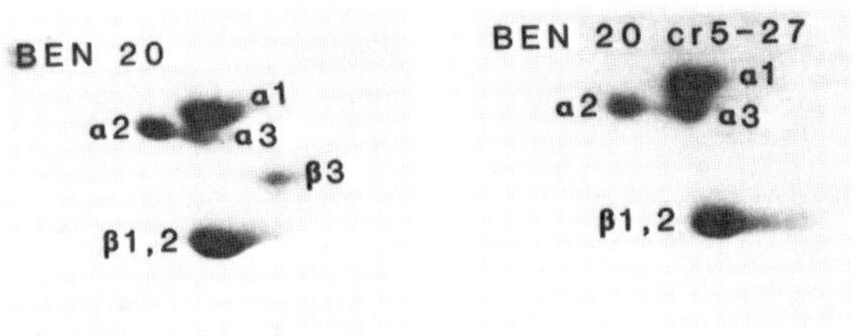

FIGURE 1. Demonstration of the absence of beta3-tubulin from a strain containing the CR^- mutation. Coomassie Blue-stained gels of the tubulins prepared from the parental strain BEN 20 and one of the CR^- mutants (CR5-27).

benomyl and for beta3-tubulin. The absence of beta3-tubulin cosegregated perfectly with the CR phenotype among the 11 segregants tested: six conidiation resistant segregants lacked beta3-tubulin and 5 conidiation sensitive segregants had beta3-tubulin. Although the number of segregants tested was small, this result supported our original suggestion that beta3-tubulin had a special role in conidiation.

Cloning and Characterization of Beta Tubulin Genes and Construction of Plasmids

Two beta-tubulin sequences, B5 and B14, were cloned from an *A. nidulans* Charon 4A library (generously supplied by Dr. William Timberlake) by hybridization with a chicken cDNA beta-tubulin probe (generously supplied by Dr. Donald Cleveland). These clones were confirmed as beta-tubulin by partial sequencing and were shown to be different by restriction mapping and sequence analysis. The sequences also failed to hybridize with each other at high stringency. These sequences were used in the construction of a series of shuttle vectors all of which contained a plasmid gene for ampicillin resistance and the $pyr4^+$ gene (for orotidylate decarboxylase (ODC)) from *Neurospora crassa*. Three such shuttle vectors were made: AIpGM1 (for *Aspergillus* integrative plasmid GM1), which contained all but the 12 amino-terminal amino acids of the B14 sequence, AIpGM4 which contained the B5 sequence and AIpGM6, which contained an internal 750 BP fragment of the B14 tubulin sequence and was designed specifically for a "gene disruption" experiment (see below). Transformation of <u>pyrG</u>$^-$ strains of <u>A</u>. <u>nidulans</u> (which lack ODC activity and are unable to grow in the absence of uridine) was followed by the restoration of uridine prototrophy to transformed strains.

Site Specific Integrative Transformation of Tubulin Clones

AIpGM1 and AIpGM4 transformants were isolated and analyzed by Southern blotting. In the majority of the transformants plasmid was found to have integrated into the genomic DNA at the genomic restriction fragment homologous to its own sequence (Fig. 2). In some instances there were multiple tandem integrations at this site. In

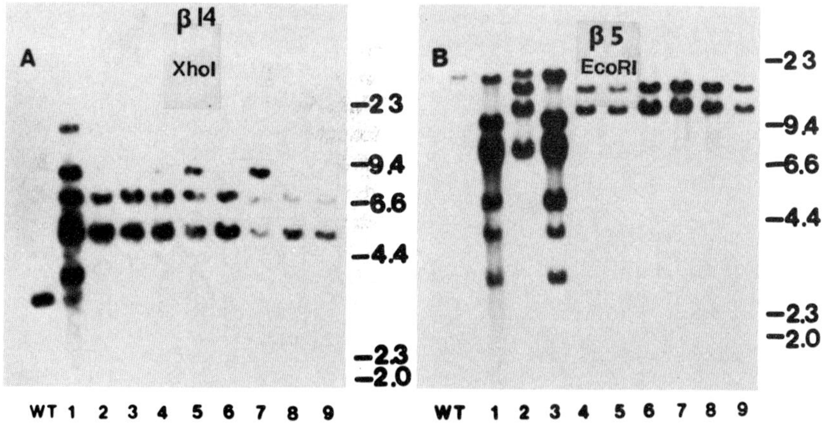

FIGURE 2. Southern analysis of transformant DNA from strains transformed with (A) AIpGM1, the B14 bearing plasmid or (B) AIpGM4, the B5 bearing plasmid. Transformant DNA (2ug) was digested with the indicated restriction endonuclease, fractionated on an agarose gel, transferred to nitrocellulose and probed with either nick translated B14 or B5 DNA. Lane WT was total DNA from the nontransformed wildtype parental strain and lanes 1-9 contain total DNA from tranformants. The position and size in killobases of lambda Hind III markers is indicated at the right.

only a few instances did the plasmids integrate at other sites in the genome, and when this occurred there appeared to be multiple ectopic integration events.

Since each plasmid integrated specifically at its own homologous site in the genome, it was possible to determine which of the two tubulin sequences, B5 or B14, represented the benA gene by mapping the locus of the plasmid, pyr4+ marker with respect to the benA gene. This was done by crossing a benomyl resistant benA−, pyrG− strain with a benomyl sensitive benA+, pyr4+ strain transformed with either AIpGM1 or AIpGM4. The progeny of each of these crosses were then scored for benomyl resistance and for ability to grow in the absence of uridine. The result of these crosses was that the specifically integrated AIpGM4 plasmid, carrying the B5 tubulin sequence, was shown to map at the benA locus, since the benA− and the pyr4+ markers exhibited little recombination in this cross; whereas, the AIpGM1 plasmid, carrying the B14 sequence clearly was integrated at another beta tubulin locus, presumably that of the gene, tubC, that codes for beta3-tubulin, since the two markers segregated freely from each other (Table 1).

TABLE 1
SEGREGATION OF THE PYR4 MARKER
INTEGRATED AT THE B5 OR B14 SEQUENCE

	B14			B5	
	ben^r	ben^s		ben^r	ben^s
pyrG−	114	74	pyrG−	102	6
pyrG+	77	105	pyrG+	2	9

Strains carrying the pyr4 integrated into either the B14 or B5 sequence were cross with a pyrG−, benA^r (benomyl resistant) strain, and phenotype of the progeny was determined. The numbers represent the number of progeny with a given phenotype. The data for the B14 were from crosses with five transformants and progeny from 11 cleistothecia and those for the B5 were from three transformants and 7 cleistothecia.

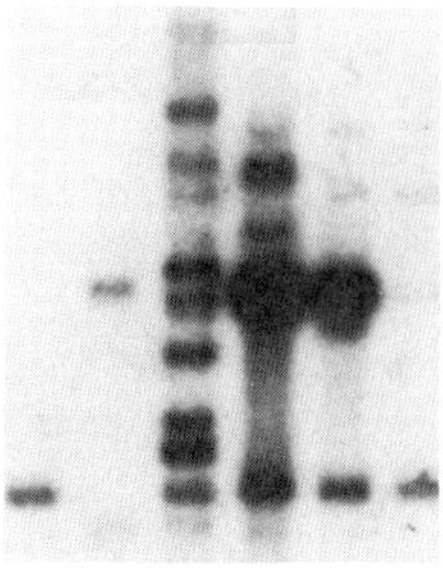

FIGURE 3. Southern analysis of total DNA from five strains transformed with the plasmid AIpGM6. Transformant DNA (2ug) was digested with the indicated restriction endonuclease, fractionated on an agarose gel, transferred to nitrocellulose and probed with nicktranslated B14 sequence. Lane WT was total DNA from the nontransformed wildtype parental strain and lanes 1-5 contain total DNA from the tranformants. It can be seen that the 3.2 Kb Xho I band characteristic of the B14 sequence is absent in only lane 1, indicating a gene disruption. This strain also had the phenotype of a CR^- mutant and lacked beta3-tubulin.

Integrative Transformation Disrupts tubC and Beta3-tubulin

In order to confirm the identity of the B14 sequence as tubC a benA22, $pyrG^-$ strain was constructed and transformed with AIpGM6, the shuttle vector containing an internal 750 base pair B14 fragment. The reason for choosing a benA22 strain as a recipient was, as before, in order that beta3-tubulin could be seen on two dimensional gels without being occluded by the benA beta1- and beta2-tubulins. Transformants were selected for uridine prototrophy. DNA was prepared from each of 5 transformants and analyzed by Southern blotting. AIpGM6, presumably because of its small size, integrated into its homologous beta tubulin sequence in only one of these transformants (Fig. 3).

Because integration of an internal sequence into a gene should disrupt the gene, and consequently the corresponding protein gene product, the tubulin proteins from these five AIpGM6 transformants were partially purified and analyzed by two dimensional gel electrophoresis. Beta3-tubulin protein was missing from the one transformant in which integration was into the homologous tubulin gene but was unaffected in the others (data not shown, see Fig. 4 below). This experiment proves that the B14 sequence integrates into and is homologous with the gene, tubC, that codes for beta3-tubulin. The tubC gene has not yet been mapped, but could be mapped using the tightly linked pyr4+ marker introduced into the genome by the integrated plasmid.

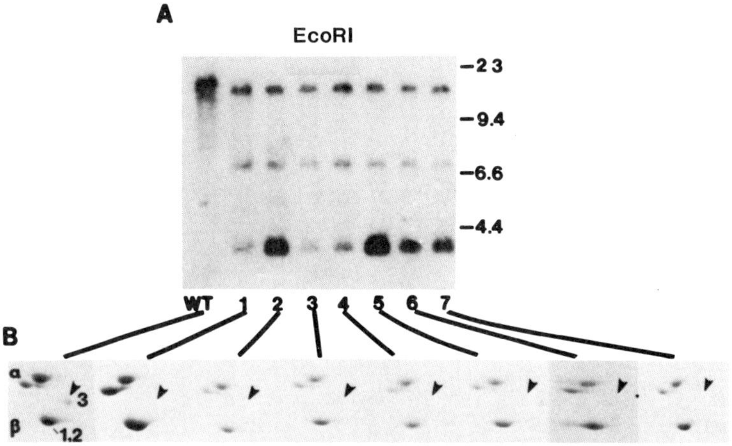

FIGURE 4. (A) Southern analysis of seven of the AIpGM6 transformants that exhibited the CR⁻ phenotype and (B) two-dimensional gels of the tubulins of the same strains. Transformant DNA was digested with the indicated restriction endonuclease, fractionated on an agarose gel, transferred to nitrocellulose and probed with nick translated B14 DNA. Lane WT was total DNA from the nontranformed wildtype parental strain and lanes 1-7 contain total DNA from the transformants. The position and size of lambda DNA markers is indicated at the right. A tubulin enriched fraction was prepared from the parental strain and the seven transformants. Only the tubulin region of each gel is shown and the position of the beta3-tubulin is marked with an arrowhead in each gel.

Since the loss of beta3-tubulin in spontaneous and UV-induced benA22/CR mutations caused conidiation resistance (see above), we thought that disruption of the tubC gene (and beta3-tubulin) by AIpGM6 would also cause conidiation resistance. The tubC site specific AIpGM6 integrant and the other four ectopic AIpGM6 integrants were plated on complete medium containing 4.8 ug/ml of benomyl. Although conidiation of the four AIpGM6 ectopic transformants was inhibited by benomyl, benomyl had no detectable effect on the conidiation of the transformant in which integration was site specific.

In order to determine whether the CR phenotype was always correlated with ablation of the tubC locus we performed a second transformation experiment, which produced 150 $pyr4^{+}$ transformant. Of these only 9 exhibited the CR phenotype. Seven of the CR transformants were analyzed by two dimensional gel electrophoresis, and all were found to lack beta3-tubulin. (Fig. 4). This experiment confirmed our finding with the CR mutations that loss of beta3-tubulin in a benA22 background is responsible for conidiation resistance.

The similarity of the physiological and biochemical phenotypes of the benA22/CR mutations and of the AIpGM6 transformant suggested that at least some of the CR mutations might be in the tubC gene. To test this possibility the $pyrG^{-}$ marker was crossed into each of the CR mutants and these pyrG-, CR strains were in turn crossed to a $pyr4^{+}$, site specific AIpGM1 integrative transformant. The progeny were analyzed for segregation of the CR phenotype with respect to the pyr^{4+} marker integrated at the tubC locus. The seven CR mutations that have been mapped to date all map at the tubC locus.

DISCUSSION

In this report we show by means of integrative transformation experiments that two cloned, beta-tubulin genomic DNA sequences, B5 and B14, correspond to the benA and tubC genes respectively. In addition we prove by a "gene disruption" experiment that beta3-tubulin is the product of the tubC gene. We also show in two different ways that beta3-tubulin appears to function specifically during conidiation. Spontaneous mutations that cause the loss of beta3-tubulin, as well as deliberate disruption of the tubC gene, both cause conidiation resistance in a benA⁻, benomyl resistant background. We interpret this to mean that the normal sensitivity of condidiation to benomyl in benA⁻ strains depends on the presence of the benomyl sensitive beta3-tubulin and that when the beta3-tubulin is absent, conidiation is supported by the benomyl resistant beta1- and/or beta2-tubulin and is therefore resistant to inhibition by benomyl. Because beta3-tubulin confers benomyl sensitivity upon conidiation, it must normally function during conidiation; but, since it can be replaced by beta1- and/or beta2-tubulin, it is clearly not necessary for conidiation per se.

The replacement of beta3-tubulin function by beta1- and/or beta2-tubulin might occur in one of two ways. Both the benA and tubC gene products could function concurrently in wild type conidial development and loss of tubC lead to benA takeover of function. Alternatively, beta3-tubulin might be the only tubulin normally found in conidiophores of wild type cells, and beta1- and/or beta2-tubulin could be induced by its absence. We hope in the near future to be able to determine whether there are changes in beta-tubulin mRNAs or proteins during conidiation, using partially purified conidiophore preparations obtained by scraping conidiophores from conidiating lawns on pour plates.

The evidence that beta3-tubulin functions specifically during conidiation does not tell us what its specific function during conidiation might be. We know that beta3-tubulin is not absolutely required for conidiation because conidia are produced in its absence. This should not be interpreted to mean that beta3-tubulin is an unimportant or dispensible protein. It is quite possible that beta3-tubulin has one or more important conidiation-related functions that have not been detected. For example, strains lacking beta3-tubulin may exhibit less accurate chromosome segregation during conidiophore mitosis than wild type strains. Other subtle malfunctions are also possible and will be experimentally tested in the future.

Although we still do not have a very detailed understanding of the functions of the multiple tubulin genes and multiple tubulins of Aspergillus, this organism because of the rich variety of its microtubule containing structures, its cell and tissue types, its excellent genetics and the recently developed systems for DNA mediated transformation, would appear to be among the most promising resources for future progress in microtubule biology.

ACKNOWLEDGMENTS

We wish to thank Happy Smith for excellent technical support and generally making our lives more pleasant.

REFERENCES

1. McKeithan, TW, Rosenbaum, JL (1984). The biochemistry of microtubules. In Shay, JW (ed): "Cell and Muscle Motility," New York: Plenum Press, Vol.5, p 255.
2. Raff, EC (1984). Genetics of microtubule systems. J Cell Biol. 98:1.
3. Cleveland, DW (1983). The tubulins: from DNA to RNA to protein and back again. Cell 34:330.
4. Pontecorvo, G, Roper, JA, Hemmons, CM, MacDonald, KD, Bufton, AWJ (1953). The genetics of *Aspergillus nidulans*. Adv. in Genetics 5:141.
5. Cove, DJ (1977). The genetics of *Aspergillus nidulans*. In Smith, JE, Pateman, JA (eds): "The Genetics and Physiology of *Aspergillus nidulans*," London: Academic Press Ltd., p 81.
6. Ballance, DJ, Buxton, FP, Turner, G. Transformation of *Aspergillus nidulans* by the orotidine-5'-phosphate decarboxylase gene of *Neurospora crassa*. Biochem. Biophys. Res. Commun. 112:284.
7. Tilburn, J, Scazzocchio, C, Taylor, GG, Zabicky-Zissman, JH, Lockington, RA, Davies, RW (1983). Transformation by integration in *Aspergillus nidulans*. Gene 26:205.
8. Yelton, MM, Hamer, JE, Timberlake, WE (1984). Transformation of *Aspergillus nidulans* by using a *trpC* plasmid. Proc. Natl. Acad. Sci. USA 81:1470.
9. Yelton, MM, Timberlake, WE, van den Hondel, CAMJJ (1985). A cosmid for selecting genes by complementation in *Aspergillus nidulans*: selection of the developmentally regulated *yA* locus. Proc.Natl. Acad. Sci. USA 82:834.
10. Sheir-Neiss, G, Lai, MH, Morris, NR (1978). Identification of a gene for beta-tubulin in *Aspergillus nidulans*. Cell 15:639.
11. Morris, NR, Lai, MH, Oakley, CE (1979). Identification of a gene for alpha-tubulin in *Aspergillus nidulans*. - Cell 16:437
12. Weatherbee, JA, Morris, NR (1984). *Aspergillus* contains multiple tubulin genes. J. Biol. Chem. 259:15452.

13. Weatherbee, JA, May, GS, Gambino, J, Morris, NR (1985). The involvement of a particular species of beta-tubulin (beta3) in conidial development in Aspergillus nidulans. J. Cell Biol. (submitted for publication).
14. May, GS, Gambino, J, Weatherbee, JA, Morris, NR (1985). Identification and functional analysis of beta-tubulin genes by site specific integrative transformation in Aspergillus nidulans. J. Cell Biol. (accepted for publication).
15. Oakley, BR, Morris, NR (1980). Nuclear Movement is beta-tubulin-dependent in Aspergillus nidulans. Cell 19:255.
16. Oakley, BR, Morris, NR (1981). A beta-tubulin mutation in Aspergillus nidulans that blocks microtubule function without blocking assembly. Cell 24:837.
17. Gambino, J, Bergen, LG, Morris, NR (1984). Effects of mitotic and tubulin mutations on microtubule architecture in actively growing protoplasts of Aspergillus nidulans. J. Cell Biol. 99:830.

Molecular Genetics of Filamentous Fungi, pages 253–263

SIZE-CONTROL IN THE CHLAMYDOMONAS REINHARDTII FLAGELLUM

Jonathan W. Jarvik and Michael R. Kuchka

Department of Biological Sciences, Carnegie-Mellon University
Pittsburgh, Pennsylvania 15213

ABSTRACT Genetic analysis of short and long-flagella mutants of Chlamydomonas reinhardtii has identified a number of genes with size-control functions. Flagellar regeneration and dikaryon complementation experiments have provided insights into the process of flagellar size-control, in that they have allowed us to test, and exclude, some otherwise attractive models for size-control. Biochemical analysis of the short flagella mutants, and of certain revertants and pseudorevertants derived therefrom, have implicated some flagellar proteins in the size-control process.

INTRODUCTION

Work in our laboratory concerns flagellar and basal body morphogenesis in the green alga Chlamydomonas. The reader may well wonder whether flagellar morphogenesis is relevent to the topic of this meeting, especially since the filamentous fungi lack cilia, flagella, and even centrioles. Let us acknowledge at the outset that it may not be relevant. And yet a case can be made that the research problem discussed in this communication - flagellar size control - may relate rather closely to certain aspects of filamentous fungal biology. Flagella and hyphae share a fundamental size-control property - their growth is primarily one dimensional. Flagella elongate until a particular size has been reached whereupon growth ceases; for hyphae growth does not cease, but mitosis is coupled to cell growth in such a way that as elongation proceeds an appropriate nucleus-tocytoplasm ratio is maintained (1). Thus flagellar size-control formally resembles cellular mitotic-cycle control, in that both processes require a means of cellular self-assessment of size. The mechanism for this self-assessment is not known in any system, and so progress in understanding size-control in one system - the flagellum - may represent progress towards understanding others. Furthermore, a major aspect of flagellar size control is the control of microtubule assembly, and, as several other communications in this symposium make clear, properly regulated microtubule assembly is essential to fungal mitosis, meiosis, and nuclear migration. Again, what we learn about the control of microtubule assembly in one system may help with understanding it in others.

THE CHLAMYDOMONAS SYSTEM

Several facts about *C. reinhardtii* make it a system well suited for studying flagellar size control. Flagella are not essential for cellular viability, and so a large variety of control-defective mutants can be isolated and analyzed. The organism provides an excellent laboratory system for formal genetic analysis: it is haploid, tetrad analysis is routine, and diploids or dikaryons can be constructed for complementation analysis. Flagellar length can be simply and accurately measured. And pure flagella can be obtained from the organism in high yield for biochemical analysis.

PHENOMENOLOGY OF FLAGELLAR SIZE-CONTROL

In *C. reinhardtii*, at least two flagellar size-control phenomena exist. At the cell population level, flagellar length averages about 12 microns, with a standard deviation of only about 1 micron. At the individual cell level, the two flagella are identical in length - or at least their lengths are indistinguishable in the light microscope. Size-control is homeostatic in nature. If a cell loses its flagella, it immediately commences to regenerate them to their former length. If it loses just one of its two flagella, it typically shortens the other one as the first one regrows; when the two get to equal length they then grow out in concert until the original length is attained (2).

FLAGELLAR STRUCTURE IN THE CONTEXT OF SIZE CONTROL

The flagellar axoneme is very complex in biochemistry and ultrastructure. Most of the complexity derives from lateral elements (dynein arms, radial spokes, etc.) which are attached to a longitudinal frame of outer doublet microtubules. It is clear from combined genetic, ultrastructural, and biochemical studies in C. reinhardtii that most, and possibly all, of these lateral elements are not needed for flagellar growth per se, although flagella without them may be defective for motility. Similarly, the central pair microtubules, and their associated lateral elements, are not needed for flagellar growth (3). On the other hand, flagella without outer doublet microtubules do not grow to more than a few microns in length (4). Thus one aspect of the problem of flagellar size control may reduce to a problem in microtubule assembly. But this is not to trivialize the problem! Flagellar doublet microtubules are continuous with basal body microtubules at their cell-proximal (non-growing) ends, and they are intimately associated with plug-like structures embedded in the flagellar membrane at their distal (growing) ends (5). These plugs are present throughout flagellar growth, and so the structural environment for axoneme microtubule assembly in vivo is fundamentally different from that for microtubule assembly as it has been studied up to now in vitro. Since outer doublet microtubule growth - i.e., tubulin subunit addition - occurs at the distal plug region, the plugs may well be the sites at which size-control is ultimately exerted. But it must be emphasized that even if size control is exerted at the microtubules' distal tips, the tips cannot be the only elements in the size control system, since there would have to be a way for a tip to "sense" how large the flagellum is - and this implies the existence of a signal about the extent of flagellar growth.

Another place at which size-control might be exerted is at the flagellar base. The notion here is that the extent of flagellar growth could be controlled by regulating the entry of unassembled axoneme proteins into the flagellar compartment. In fact, the flagellar membrane is closely apposed to the axoneme in the region of the flagellum/cell junction (the transition region), and electron dense structures in the axoneme's interior which could represent barriers to free diffusion of proteins are present in the transition region. Again, even if assembly is regulated at the base by a "gatekeeper" mechanism, this can only be part of the story - there also must be a signal to indicate how large the flagellum is.

FLAGELLA-LENGTH MUTANTS

Through the analysis of flagellar length mutants we hope to identify genes and gene products involved in the flagellar size-control process. A major virtue of this approach is that we can proceed without any prejudice about the molecular basis of size-control. Instead, we hope that mutants defective for size-control function will lead us to relevant molecules and/or structures.

Chlamydomonas' two flagella are located at the cell anterior; they beat with opposed power and recovery strokes and move the cell forward as though it is doing the breast stroke. As it turns out, flagella length mutants have abnormal motility phenotypes. Short flagella mutants swim slowly - because their flagella exert less force per beat cycle. Long flagella mutants swim erratically - apparently because their power and recovery strokes become uncoordinated. These motility phenotypes have been the basis for isolations of flagellar length mutants (4, 6, 7). Another method for mutant isolation which is not based on motility phenotypes is also available - flow cytometry/cell sorting using laser light scattering. Forward laser light scattering by cells with short flagella is considerably reduced relative to scattering by wild type, and reconstruction experiments have demonstrated that a tenfold enrichment for *shf* cells can be obtained in a single cell-sorting pass. We intend to use this approach to isolate additional short-flagella mutants in the future.

SHORT FLAGELLA MUTANTS

We isolated a number of strains with short but motile flagella by screening clones of mutagenized cells for slow-swimmers (Table 1). We call the mutants *shf* for *sh*ort *f*lagella. They are not defective in phototaxis, backward swimming response, or flagellar agglutination in mating. They regulate flagellar length with a precision comparable to that shown by wild type, but to a shorter mean value. They have substantial cellular pools of flagellar protein, ruling out the trivial possibility that their flagella are short because they synthesize insufficient quantities of flagellar protein (4). Because, with the exception of flagellar length, all flagellar properties of which we are aware are normal, we think it likely that the shf mutants are specifically defective in flagellar size regulation.

TABLE 1
SHORT FLAGELLA MUTANTS

	Mean flagellar length (micrometers)	Standard deviation
Wild type	12.0	1.1
Shf1-236	6.1	1.1
Shf1-253	6.6	1.1
Shf1-277	7.1	1.0
Shf2-158	6.4	1.4
Shf2-1249	6.4	1.1
Shf3-1851	5.7	1.6

SHF MUTATIONS ARE RECESSIVE AND REPRESENT AT LEAST 3 GENES

Gametes of six shf strains were mated to wild type, and the flagellar lengths in the resulting quadriflagellate dikaryons were observed. In all cases the two short flagella elongated to wild type length, indicating that expression of the shf mutations is recessive. shf/shf dikaryons were also constructed to test for complementation. If the mutations are in different cistrons, outgrowth of all four flagella is expected; otherwise there should be no flagellar elongation. Results placed the mutations in three distinct complementation groups which we designate shf-1 (3 alleles), shf-2 (2 alleles) and shf-3 (1 allele). In each of the six shf strains the mutant phenotype is due to a single Mendelian determinant, since shf/shf^+ tetrads all showed 2:2 segregation. shf x shf intercrosses demonstrated that the three shf genes are unlinked. Mapping crosses to marked strains demonstrated that shf-1 is very near the centromere of Chromosome VI (7) and shf-2 is approximately 16 map units from msr-1 on Chromosome I. shf-3 has not yet been mapped.

REVERSION ANALYSIS

The phenotypes of several shf mutants have allowed us to obtain revertants by selecting motile cells. We have sought revertants primarily because some might carry extragenic suppressors and analysis of the suppressors might allow us to identify genes whose products functionally interact with the products of the known shf genes (8). Second, some revertants might be intragenic pseudorevertants carrying new mutations in the shf gene which might have useful new molecular phenotypes - e.g., altered electrophoretic properties for the shf gene product (9), or new functional phenotypes - e.g., temperature-sensitive size-control.

One set of revertant selections took advantage of the fact that shf-1-253 is temperature sensitive for flagellar assembly (cells flagellaless at 34°. Other selections took advantage of the fact that certain shf, shf double mutants are flagellaless (Table 2). Our initial set of revertants consists of 24 independent shf-1-253, shf-2-1249 revertants, 17 independent shf-1-253, shf-3-1851 revertants, and 44 independent shf-1-253 revertants. The analysis of these revertants is far from complete, but already we have found examples of shf-1-253 revertants which carry unlinked extragenic suppressors, and other examples which do not.

TABLE 2
PHENOTYPES OF SHF, SHF DOUBLE MUTANTS

shf-1-236, shf-2-1249	Fla-
shf-1-253, shf-2-1249	Fla-
shf-1-277, shf-2-1249	Shf
shf-1-236, shf-3-1851	Fla-
shf-1-253, shf-3-1851	Fla-
shf-1-277, shf-3-1851	Shf
shf-2-1249, shf-3-1851	Shf

The suppressors are of two types - some dominant, some recessive. Dominance tests for the suppression are made by constructing shf/shf, sup dikaryons; if the two short flagella lengthen we conclude that the suppressor is dominant, and if the two long flagella shorten we conclude that it is recessive. Further, we found that the recessive suppressors are allele specific in that they do not suppress the shf-1-277 mutation, whereas the dominant suppressors work on shf-1-277 as well as shf-1-253. Analysis of the motile revertants of shf-1-253, shf-2-1249 has demonstrated that some of them have either reverted at the shf-2 locus or carry suppressors of shf-2.

PROTEIN COMPOSITION OF SHF FLAGELLA

Axoneme and membrane/matrix fractions of shf flagella have been analyzed by one and two dimensional gel electrophoresis. 3 minor protein spots are missing from 2D gels of shf-1 axonemes. In a revertant carrying an allele-specific extragenic suppressor, an acidic protein of approximately 110K daltons molecular weight is restored whereas the other two proteins are still missing. This protein is thus a good candidate for having a direct role in size-control. The one shf-2 mutant analyzed to date, shf-2-1249 has several dramatic differences in protein composition with respect to wild type. Some of these are in the axoneme, and others are in the membrane. The differences are almost surely relevant to the shf phenotype, since they cosegregate with shf-2 in tetrads. Finally, shf-3-1851 also shows abnormalities in membrane protein composition, some of which are shared by shf-2-1249, and this reinforces our belief that these protein differences are relate in some way to the size-control defect. The challenge now is to make a mechanistic correlation between the flagellar protein-composition defects and the size-control phenotype.

LONG FLAGELLA MUTANTS

The analysis of long-flagella mutants is not so advanced as for the short mutants. Two lf mutants were isolated by McVittie (6) and shown to map to two loci, lf-1 and lf-2; we have isolated 5 others and have recieved a sixth from Dr. P. Lefebvre. With the exception of those at lf-1, all lf mutations show recessive expression in lf/wild type dikaryons. The lf-1 mutations are codominant to wild type, in that the dikaryons persist with two long and two normal flagella (6). (If the cell is deflagellated, however, it regenerates four normal flagella.) We have examined complementation behavior in almost all possible shf/lf dikaryons. Each behaved as expected from the dominance properties of the mutations present - e.g., in a recessive shf/codominant lf dikaryon the short flagella elongated to wild type length while the long flagella did not shorten. Furthermore, just as we have been able to use flagellaless shf, shf double mutants to select motile Shf^{+} pseudorevertants, so we have identified a flagellaless lf, shf double mutant - lf-3-89, shf-1-236 - which can be used to select Lf^{+} revertants.

TESTS OF SIZE-CONTROL MODELS

We have taken advantage of certain mutants to test two general size-control models. The results, summarized below, argue against both. One class of model for flagellar size-control is based upon the notion that if the concentration of some flagellar structural component (e.g., tubulin) in the cellular pool is above some critical value there is flagellar growth and that otherwise there is not; flagellar growth proceeds until the concentration of the component in the pool has dropped to the critical concentration. In this class of model the extent of flagellar assembly is governed by the levels of flagellar constituents, meaning that flagellar size could be controlled at the gene expression level. We tested the model using a mutant - vfl-2 - which has normal cell size but a variable number of flagella per cell (10). vfl-2's flagella are of normal length irrespective of the number of flagella per cell, and so, if the above model holds, pool concentration should be equal in all cells. We determined pool concentrations operationally, by measuring the extent of flagellar regeneration in the absence of new protein synthesis, and found they were not equal in all cells but instead were proportional to flagellar number (e.g., the pool in cells with three flagella was three times the size of that in cells with one flagellum). We obtained a similar result using the uni-flagellated mutant uni-1 (11). Here the precursor pool concentration is approximately half of what it is in wild type, yet the cells have flagella of normal length. Thus it does not appear that flagellar length is regulated by limiting amounts of unassembled flagellar protein in the cellular pool.

A number of other models can be formulated in which flagellar size is intrinsically self limiting at the assembly level, with flagellar growth rate slowing progressively as the flagellum lengthens (12, 13). The fact that the flagellar growth zone is at the distal tip makes such models particularly attractive, since it is easy to imagine ways in which flagellar subunits might have increasing difficulty getting to their assembly sites as the flagellum gets longer and longer. We performed a test of this class of size control model by constructing *shf/wild type* quadriflagellate dikaryons, observing the kinetics of outgrowth of the two short flagella, and comparing them to the kinetics of wild type flagellar regeneration. It turns out that the two half-length flagella grow out to full length rapidly - and with kinetics which closely resemble the growth of wild type flagella from zero to half length. This is not what should happen if the above hypothesis for size-control were correct, however; there the two half-length flagella should elongate to full length with deceleratory kinetics which resemble the outgrowth of wild type flagella from half to full length. We conclude that flagellar length is probably *not* regulated by intrinsic kinetic properties of the flagellar assembly process itself (7).

Results published by Lefebvre *et al* (14) are consistent with the above conclusions. Wild type cells were induced to resorb their flagella by adding a calcium chelator to the medium; then, when the flagella had resorbed most of the way, the cells were deflagellated, calcium was restored, and the kinetics of flagellar outgrowth were determined. It was found that the cells regrew their flagella very rapidly, but to wild type length and no further. Since the flagellar precursor pools in these cells were augmented substantially, both by the resorbed proteins and by new proteins synthesized in response to deflagellation, the fact that cells regenerated to normal length implies that pool size does not regulate flagellar length. And, the cells' very rapid regeneration demonstrates that there is no inherent inverse relation between flagellar length and flagellar growth rate.

ACKNOWLEDGMENT

This research was supported by National Science Foundation Grant PCM-8216337 and NIH Research Career Development Award K04AM00710 to JWJ.

REFERENCES

1. Alberghina L, Sturani E (1981). Control of growth and of the nuclear division cycle in Neurospora crassa. Microbiol Rev 45:99-122.
2. Rosenbaum JL, Moulder J, Ringo D (1969). Flagellar elongation and shortening in Chlamydomonas. J Cell Biol 41:600-619.
3. Luck DJL (1984). Genetic and biochemical dissection of the eucaryotic flagellum (1984). J Cell Biol 98:789-794.
4. Baldwin DA, Kuchka MR, Chojnacki B, Jarvik J (1985). Approaches to flagellar assembly and size control using stumpy and short-flagella mutants of Chlamydomonas reinhardtii. In Mol Biol of the Cytoskeleton (ed. G. Borisy et al.) Cold Spring Harbor Laboratory Press, Cold Spring Harbor, NY, 245-255.
5. Dentler WL (1980). Structures linking the tips of ciliary and flagellar microtubules to the membrane. J Cell Sci 42:207-220.
6. McVittie AC (1972). Flagellum mutants of Chlamydomonas reinhardtii. J Gen Microbiol 71:525-540.
7. Jarvik J, Kuchka M, Reinhart F, Adler S (1984). Altered flagellar size control in shf-1 short flagella mutants of Chlamydomonas reinhardtii. J Protozool 31:199-204.
8. Jarvik J, Botstein D (1975). Conditional-lethal mutations that suppress genetic defects in morphogenesis by altering structural proteins. Proc Nat Acad Sci USA 72:2738-2742.
9. Luck DJL, Piperno G, Ramanis Z, Huang B (1977). Flagellar mutants of Chlamydomonas: studies of radial spoke-defective strains by dikaryon and revertant analysis. Proc Nat Acad Sci USA 74:3456-3460.
10. Kuchka MR, Jarvik J (1982). Analysis of flagellar size control using a mutant of Chlamydomonas reinhardtii with a variable number of flagella. J Cell Biol 92:170-175.
11. Huang B, Raminis Z, Dutcher SK, Luck DJL (1982). Uniflagellar mutants of Chlamydomonas: Evidence for the role of basal bodies in transmission of positional information. Cell 29:745-753.
12. Levy EM (1974). Flagellar elongation; an example of controlled growth. J Theor Biol 43:133-149.

13. Child FM (1978). The elongation of cilia and flagella: a model involving antagonistic growth zones. ICN-UCLA Symposium on Molecular and Cellular Biology, , XII, 351-358.
14. Lefebvre PA, Nordstrom SA, Moulder JE, Rosenbaum JL (1978). Flagellar elongation and shortening in Chlamydomonas V. Effects of flagellar detachment, regeneration and resorption on the induction of flagellar protein synthesis. J Cell Biol 78:8-26.

Molecular Genetics of Filamentous Fungi, pages 265–275

EVOLUTION AND PATTERNS OF EXPRESSION OF THE *PHYSARUM* MULTI-TUBULIN FAMILY ANALYSED BY THE USE OF MONOCLONAL ANTIBODIES[1]

Christopher R. Birkett, Kay E. Foster and Keith Gull

Biological Laboratory, University of Kent,
Canterbury, Kent CT2 7NJ, England

ABSTRACT We have used a panel of seven well defined monoclonal antibodies to probe blots of the myxamoebal and plasmodial tubulin isotypes. Differential reactivity of isotypes to these antibodies suggests a) the plasmodial α1 tubulin isotype is complex. The monoclonal antibody KMP-1 reveals a heterogeneity of α1 sub-types that focus within this one 2D gel spot. b) The β1 group of isotypes appear to be very related tubulins. c) The plasmodial specific β2 tubulin differs significantly from the β1 isotypes in its reactivity to anti-β tubulin monoclonals. Further, the plasmodial specific β2 tubulin is recognised by the normally α tubulin specific monoclonal YL1/2, suggesting that this β tubulin carries an α tubulin-like carboxyl terminus. It may be that the plasmodial specific α2 and β2 tubulin isotypes have arisen in evolution via a recombination event involving the regions of the α1 and β1 structural genes encoding the carboxyl termini of these tubulins.

INTRODUCTION

Microtubules are ubiquitous organelles in eukaryotic cells and are composed of two closely related polypeptides, α and β tubulin, which in most organisms are encoded by a

[1]This work was supported by grants from the Science and Engineering Research Council, the Agricultural and Food Research Council and the Leukaemia Research Fund.

multi-gene family (1). There is now evidence that differential expression of these tubulin multi-gene families or post-translational modifications of the tubulin polypeptides can give rise to the existence of multiple isotypes of α and β tubulin in organisms, tissues or even single cells (1-8).

The slime mould *Physarum polycephalum* provides a useful system in which to study the relationship between tubulin gene expression, tubulin isotypes and microtubular organelles. Defined arrangements of microtubule types exist within the three major cell types of the *Physarum* life cycle: the myxamoeba, the flagellate and the plasmodium. Myxamoebae possess cytoplasmic, centriolar and mitotic spindle microtubules; the flagellate contains cytoplasmic, basal body and flagella axoneme microtubules; whilst the plasmodium forms only spindle microtubules during mitosis and meiosis (9,10).

Recent studies of *Physarum polycephalum* have revealed a cell type dependent expression of tubulin isotypes in this slime mould. The organism is particularly useful since the tubulin isotypes are distinct, electrophoretically separable species (7,8). The myxamoebae express α1 and β1 tubulin isotypes, whilst the plasmodium expresses α1, α2, β1 and β2 tubulin isotypes. Thus the α2 and β2 tubulin isotypes represent plasmodium specific tubulins. Identification of these electrophoretic species as α or β tubulins has been based on extensive evidence. The myxamoebal tubulins α1 and β1 were originally identified on the basis of their self assembly into microtubules and by peptide mapping (11,12). The plasmodial tubulins, α1, α2, β1 and β2 were identified by peptide mapping, immunoprecipitation and by hybrid selection of specific mRNA by cloned tubulin DNA sequences followed by *in vitro* translations (8). Also, all four of these plasmodial tubulins are found in microtubules purified by *in vitro* assembly (7).

A major aim of studies of tubulin gene families has been to link a particular gene with a tubulin polypeptide that has unique properties such as distinct electrophoretic coordinates. In this context the use of mutational analysis of selectable markers has been of great value. For example, in *Drosophila* sterile males which possess a mutation (2) in a particular β tubulin gene (expressed only in the testis) and selection of *Aspergillus* mutants resistant to anti-microtubule drugs (13) are two routes that have proved to be of great value in recognising individual members of these organisms multi-tubulin families at the gene and protein level. The number of α and β tubulin loci in the *Physarum* genome has been examined by studies involving restriction fragment length polymorphisms and the results indicate four unlinked α tubulin sequence loci and at least three unlinked β tubulin sequence loci (14). One of these loci, *bet B* has been shown to be linked to a β tubulin structural gene, *ben D* that encodes a β1 tubulin isotype that is expressed in both myxamoebae and plasmodia (15). However, no other connections have been made between tubulin DNA sequences and any of the electrophoretically defined tubulin polypeptides.

We describe here the use of a novel approach which seeks to use monoclonal antibodies to recognise specific tubulin isotypes expressed within one organism. This approach has general relevance; however it is particularly useful when applied to the *Physarum* system since two of the tubulin isotypes are expressed only in the plasmodium - a syncytium which is unsuited to selectional genetics.

RESULTS

Monoclonal Antibodies

Several monoclonal antibodies to tubulin have been used in this study; two, KMP-1 and KMX-1 were characterised in this laboratory, the others were gifts as follows. DM1B DM3B3 and DM1A were a gift from Dr. S. Blose (Cold Spring Harbor) and were raised against chick brain microtubules. DM3B3 and DM1B react specifically with tubulin, whilst DM1A is an anti-α tubulin monoclonal antibody (16). YOL 1/34 and YL1/2 were a gift from Dr. J. Kilmartin (MRC, Cambridge). Both antibodies were raised against *Saccharomyces* tubulin and are specific for the α subunit of tubulin (17). KMX-1 and KMP-1 were generated during the course of this study using standard hybridoma techniques after the injection of *Physarum* myxamoebal tubulin. KMX-1 showed anti-β tubulin specificity and was typed as an IgG2b immunoglobulin. KMP-1, an IgM antibody, bound to α tubulin on 1D gel blots of *Physarum* myxamoebal tubulin but not at all to mammalian brain tubulin samples. Subsequent examination by Western blotting has confirmed that KMP-1 appears to have a restricted species specificity and does not recognise the α tubulin from various fungi, algae, higher plants, avian, insect and several mammalian sources. In contrast, and more typically for anti-tubulin antibodies KMX-1 did react with the β tubulin from these same sources.

Partially purified preparations of myxamoebal and plasmodial tubulins were produced as detailed previously (7,11). These tubulin preparations were separated on 2D gels (Fig. 1) and then transferred onto nitrocellulose membranes by electroblotting. Separate nitrocellulose blots were then probed with individual members of the panel of monoclonal antibodies. Binding of particular monoclonal antibodies to individual tubulin isotypes was detected by subsequent reaction with peroxidase conjugated second antibodies.

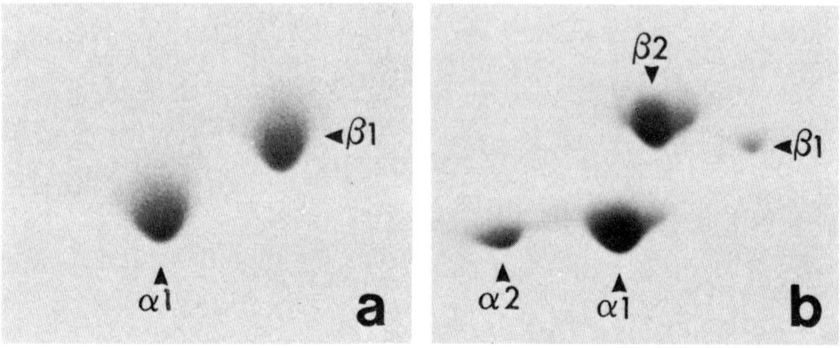

FIGURE 1. Each panel is the tubulin area of a 2D gel stained with coomassie brilliant blue. a) Myxamoebal tubulins. b) Plasmodial tubulins.

The Myxamoebal Tubulins

The panel of monoclonal antibodies was used to probe electroblotted 2D gels of myxamoebal tubulins (Fig. 1). *Physarum* myxamoebal tubulin migrates as two major spots, α1 and β1 tubulin isotypes. For the four anti-α tubulin and three anti-β tubulin antibodies the results were as expected and no cross recognition of the alternative tubulin subunit was detected. YOL 1/34, YL1/2,DM1A and KMP-1 detected only the myxamoebal α1 tubulin isotype whilst KMX-1, DM3B3 and DM1B all detected the myxamoebal β1 tubulin isotype only.

Recently, Burland *et al* (15) have described a myxamoebal mutant BEN 210 that produces a β tubulin with novel electrophoretic mobility, as well as a β tubulin with normal β1 mobility (15) (Figure 2). The mutation *ben-210* lies in the *ben D* locus and cosegregates with a *bet B* β tubulin DNA fragment (14). This and other evidence (14,15) suggests that at least two β tubulin genes are β1 tubulins that possess identical 2D-gel coordinates.

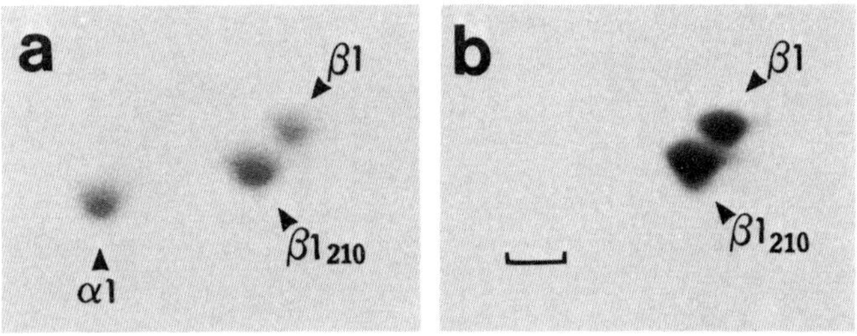

FIGURE 2. a) 2D gel, stained with coomassie brilliant blue of the tubulins purified from the myxamoebae of the mutant BEN 210. b) Equivalent gel after Western blotting and probing with the monoclonal antibody KMX-1

The tubulins from an axenic strain of the BEN 210 mutant were purified, run on 2D gels, electroblotted and the blots probed with the anti-β tubulin monoclonals KMX-1, DM3B3 and DM1B. None of these antibodies was able to distinguish between these myxamoebal β1 tubulin sub-types.

The Plasmodial Tubulins

The 2D gel profile of plasmodial tubulin differs from that of the myxamoebae in that it contains four distinct tubulin isotypes, α1, α2, β1 and β2, of which α1 and β1 isotypes have mobilities similar to those of the myxamoebal α1 and β1 tubulins respectively (Figure 1). The relative abundance of the four plasmodial isotypes is β2>α1>α2>β1. When electroblots of these plasmodial tubulins were probed with the anti-β tubulin monoclonal antibodies, DM1B, DM3B3 and KMX-1 they all recognised the plasmodial β1 tubulin isotype. This was to be expected since it is known that one β tubulin gene, *ben D*, is expressed in both myxamoebae and plasmodia (15). However, all of these antibodies produced a much fainter reaction with the plasmodial β2 tubulin isotype, especially considering the relative proportions of the two isotypes (Figure 1). (There is at least 5 times more of the β2 tubulin isotype than the β1 isotype (7)). All three antibodies recognised the β1 tubulin better than the β2 isotype, however the extent of the distinction varied. DM1B was best able to react with the β2 isotype, whilst DM3B3 gave hardly any reaction. The KMX-1 reactivity to β2 tubulin was intermediate between these two.

The reaction of the anti-α tubulin monoclonal antibodies with the plasmodial tubulins proved to be extremely interesting. The antibodies, YOL 1/34 and DM1A gave similar results in that they detected both α1 and α2 tubulin isotypes and the intensity of the reaction was a reflection of the relative abundance of these plasmodial tubulin isotypes (α1>α2). However, Western blots of plasmodial tubulins probed with the KMP-1 monoclonal antibody revealed that this antibody possesses a limited specificity for the plasmodial α tubulin family. As stated above, KMP-1 recognises the complete α1 tubulin species of the myxamoebae, however it does not detect the complete α1 tubulin species of the plasmodium, nor is it able to detect the plasmodial specific α2 tubulin isotype. KMP-1 appears to recognise an epitope that is lacking from the α2 tubulin isotype and from an α1 tubulin sub-type that focusses near the middle of the plasmodial α1 2D gel spot. Thus, Western blots of the plasmodial tubulins probed with KMP-1 show only detection of the two outer portions of the α1 tubulin 2D gel spot. That this specificity is not an artifact can be seen by counterstaining these same blots with Amido Black and so revealing the presence of unreacted protein in the α2 tubulin position and in the middle of the α1 tubulin 2D gel spot. Also, as stated above, anti-α tubulin monoclonals such as YOL 1/34 can detect all of the plasmodial α1 tubulin 2D gel spot (and the α2 tubulin), and the unreacted species are detected if KMP-1 probed blots are reprobed with YOL 1/34. Evidence from Western blots of single dimension isoelectric focussing gels gave superior resolution and suggest that there are 4 separate α1 tubulin sub-types that comprise the plasmodial 2D gel spot. KMP-1 does not recognise one of the middle pair of plasmodial α1 sub-types. Thus, the KMP-1 monoclonal antibody has revealed that the plasmodial α1 species is complex and contains up to four different electrophoretic sub-types that can be resolved using monoclonal antibodies. One of the plasmodial α1 tubulin sub-types appears similar to the plasmodial α2 tubulin isotype in that it lacks the recognition epitope of KMP-1.

YL1/2 is a well characterised anti-α tubulin monoclonal antibody whose recognition epitope is known in some detail. Wehland *et al* (18) have shown that YL1/2 recognises a linear sequence of amino acids at the extreme carboxyl terminus of α tubulin. Three major sites within the recognition epitope were identified - a negatively charged side chain in the penultimate position followed by an aromatic residue which must carry the free carboxyl group. The third site is a negative charge provided by a carboxylate group on the third residue from the end. When we probed the Western blots of *Physarum* tubulins with YL1/2 we found an intriguing and unique result. With the myxamoebal tubulins YL1/2 detected the α1 tubulin isotype and did not react with the β1 isotype. However, with the plasmodial tubulins, YL1/2 gave only a very weak reaction with the α1 tubulin isotype, did not detect the α2 or β1 tubulin isotypes, yet gave a very strong reaction with the β2 tubulin isotype. Existing information on the YL1/2 epitope suggests that the linear amino acid sequence recognised (usually -Glu-Glu-Tyr) must occur at the extreme carboxyl terminus for positive recognition; it is not sufficient for the sequence to occur within a polypeptide chain (18). In order to check that the YL1/2 recognition epitope on the plasmodial β2 tubulin does occur at the carboxyl terminus we subjected *Physarum* tubulins to mild digestion with carboxypeptidase. This treatment completely removed the recognition epitope from the β2 tubulin isotype. Thus, the plasmodial specific β2 tubulin isotype appears to be a β tubulin which possesses an α tubulin-like carboxyl terminus.

DISCUSSION

All of the monoclonal antibodies used in this study react with their antigenic determinants on α or β tubulin after the tubulin has been denatured. Thus it seems likely that many of these antibodies recognise small sequential amino acid sequences in the tubulin polypeptides. We know that this is, in fact, the case for the YL1/2 monoclonal antibody (18). Consequently, it is likely that a monoclonal antibody which recognises this type of epitope will have the potential for discriminating reactivity based on possibly single amino acid substitutions within the recognition peptide, or modifications due to post-translational events (18). We have been able to show that it is possible to use this discriminatory potential of monoclonal antibodies to discern immunological relationships even between the members of the multi-tubulin family expressed within one organism.

The *Physarum* myxamoebae express two tubulin isotypes, α1 and β1. All of the panel of antibodies that we have used gave expected reactions with these myxamoebal tubulins. All 4 anti-α tubulin antibodies detected only the α1 isotype and all 3 anti-β tubulin antibodies detected only the β1 isotype. Mutational analysis of the myxamoebal β tubulins has revealed that at least two β tubulin genes are expressed in the *Physarum* myxamoeba (15). The three anti-β monoclonal antibodies cannot discriminate between the β1 sub-types separated in the BEN 210 mutant. It is likely that these β1 tubulins are extremely similar with only a few minor amino acid substitutions. In *Chlamydomonas* the two β tubulin genes code for identical proteins (19).

The differential expression of the *Physarum* multi-tubulin family is centred on the appearance in the plasmodium of two novel isotypes - β2 and α2. However, the KMP-1 monoclonal has not only provided us with information regarding the α2 isotype but it is also able to reveal the complexity of the plasmodial α1 tubulin isotype. KMP-1 does not detect the full repertoire of α1 sub-types which focus in this plasmodial α1 2D gel spot. One of these four plasmodial α1 subtypes appears similar to the α2 isotype in that it does not possess the KMP-1 recognition epitope. Schedl *et al* (14) have shown that there are at least four α tubulin loci in the *Physarum* genome, so allowing for the possibility that many of these immunologically defined sub-types may be the products of different α tubulin genes. However, the fact that the monoclonal antibody KMP-1 can discriminate between α tubulin sub-types that focus within one 2D gel spot provides evidence for the power of this immunological approach to analysis of the complexities of tubulin isotype expression.

The fact that all three of anti-β tubulin monoclonal antibodies used in this study reacted only weakly or hardly at all with the plasmodial β2 tubulin isotype is an intriguing finding. It suggests that this plasmodial specific β2 tubulin isotype may differ significantly from the β1 tubulins. This view is reinforced by the unique finding that a well characterised anti-α tubulin monoclonal antibody YL1/2 is able to recognise this plasmodial β2 tubulin. YL1/2 has previously been considered to be a selective probe for α tubulins only. Knowledge of the YL1/2 recognition epitope suggests that the *Physarum* β2 tubulin is a β tubulin that carries an α tubulin-like carboxyl terminus. The YL1/2 antibody does not recognise the other plasmodial specific tubulin isotype, the α2 tubulin. It appears possible that there is some relationship between these two phenomena. Comparison of the carboxyl termini of mammalian and other tubulins indicates a high level of conserved sequence between the α and β tubulins (Figure 3). It may well be that in *Physarum* the novel α2 and β2 tubulin isotypes appeared during evolution by a single recombination event involving the structural genes for the α1 and β1 tubulins.

ALPHA - Gly - Glu - Gly - Glu - Glu - Glu - Gly - Glu - Glu - Tyr

BETA - Gly - Glu - Phe - Glu - Glu - Glu - Gly - Glu - Glu - Asp - Glu - Ala

FIGURE 3. The carboxyl termini of α and β tubulins from porcine brain showing the regions of homology and the YL1/2 epitope on α tubulin. The data are taken from references 20 and 21.

The attraction of this model is that a single recombination event involving the regions coding for the extreme carboxyl termini of the polypeptides could have resulted in the production of two novel tubulins - α2 and β2. The exact nature of this postulated evolutionary route for the *Physarum* multi-tubulin family should be revealed with future analysis at the molecular level.

ACKNOWLEDGMENTS

We thank Marianne Wilcox for expert technical assistance and Dr. S. Blose and Dr. J. Kilmartin for generous gifts of monoclonal antibodies. We also thank Dr. T.G. Burland for supplying us with the axenic strain of the BEN 210 mutant.

REFERENCES

1. Raff EC (1984). Genetics of microtubule systems. J Cell Biol 99:1.
2. Kemphues KJ, Raff EC, Raff RA, Kaufman TC (1980). Mutation in a testis specific β tubulin in *Drosophila*: analysis of its effects on meiosis and map location of the gene. Cell 21:445.
3. Brunke KJ, Collis PS, Weeds DP (1982). Post-translational modification of tubulin dependent on organelle assembly. Nature (London) 297:516.
4. McKeithan TW, Lefebvre PA, Silflow CA, Rosenbaum JL (1983). Multiple forms of tubulin in *Polytomella* and *Chlamydomonas*: evidence for a precursor of flagellar α tubulin. J Cell Biol 96:1056.
5. Russell DG, Miller D, Gull K (1984). Tubulin heterogeneity in the trypanosome, *Crithidia fasciculata*. Mol Cell Biol 4:779.
6. Hussey PJ, Gull K (1985). Multiple isotypes of α and β tubulin in the plant *Phaseolus vulgaris*. FEBS Letts 181:113.
7. Roobol A, Wilcox M, Paul ECA, Gull K (1983). Identification of tubulin isoforms in the plasmodium of *Physarum polycephalum* by *in vitro* microtubule assembly. Eur J Cell Biol 33:24.
8. Burland TG, Gull K, Schedl T, Boston RS, Dove WF (1983). Cell type dependent expression of tubulins in *Physarum*. J Cell Biol 97:1852.
9. Havercroft JC, Quinlan R, Gull K (1981). Characterisation of microtubule organising centre from *Physarum polycephalum* myxamoebae. J Ultrastruct Res 74:313.
10. Havercroft JC, Gull K (1983). Demonstration of different patterns of microtubule organisation in *Physarum polycephalum* myxamoebae and plasmodia using immunofluorescence microscopy. Eur J Cell Biol 32:67.
11. Roobol A, Pogson CI, Gull K (1980). *In vitro* assembly of microtubule proteins from myxamoebae of *Physarum polycephalum*. Exp Cell Res 130:203.
12. Clayton L, Quinlan RA, Roobol A, Pogson CI, Gull K (1980). A comparison of tubulins from mammalian brain and *Physarum polycephalum* using SDS-polyacrylamide gel electrophoresis and peptide mapping. FEBS Lett 115:301.
13. Sheir-Neiss G, Lai MH, Morris NR (1978). Identification of a gene for β tubulin in Aspergillus nidulans. Cell 15:639.

14. Schedl T, Owens J, Dove WF, Burland TG (1984). Genetics of the tubulin gene families of *Physarum*. Genetics 108:143.
15. Burland TG, Schedl T, Gull K, Dove WF (1984). Genetic analysis of resistance to benzimidazoles in *Physarum*: differential expression of β tubulin genes. Genetics 108:123.
16. Blose SH, Meltzer DI, Feramisco JR (1984). 10nm filaments are induced to collapse in living cells microinjected with monoclonal and polyclonal antibodies against tubulin. J Cell Biol 98:847.
17. Kilmartin JV, Wright B, Milstein C (1982). Rat monoclonal antitubulin antibodies derived by using a new non-secreting rat cell line. J Cell Biol 93:576.
18. Wehland J, Schroder HC, Weber K (1984). Amino acid sequence requirements in the epitope recognised by the α tubulin specific rat monoclonal antibody YL1/2. EMBO J 3:1295.
19. Youngblom J, Schloss JA, Silflow CD (1984). The two β tubulin genes of *Chlamydomonas reinhardtii* code for identical proteins. Mol Cell Biol 4:2686.
20. Ponstingl H, Krauhs E, Little M, Kempf T (1981). Complete amino acid sequence of α tubulin from porcine brain. Proc Natl Acad Sci USA 78:2757.
21. Krauhs E, Little M, Kempf T, Hofer-Warbinek R, Ade W, Poustingl H (1981). Complete aminoacid sequence of β tubulin from porcine brain. Proc Natl Acad Sci USA 78:4156.

Molecular Genetics of Filamentous Fungi, pages 277–294

ASPECTS OF THE NEUROSPORA GENOME

David D. Perkins

Department of Biological Sciences, Stanford University
Stanford, California 94305

ABSTRACT Cytological, molecular and genetic features are reviewed and comparisons are made with other ascomycetes. Neurospora chromosomes are large enough during ascus development to permit conventional analysis by light microscopy. This has been complemented by studies of the synaptonemal complex. DNA content is somewhat larger than in Saccharomyces, but the genetic map is shorter. Over 500 gene loci and 220 chromosome rearrangements have been mapped in the seven linkage groups. Chromosome ends are preferentially involved in rearrangements. Insertional translocations are common. Nontandem duplications for many specific chromosome segments can be produced at will. Whole-arm translocations enable centromeres to be manipulated. Rearrangements that involve the nucleolus organizer chromosome allow ribosomal RNA gene dosage to be altered. In one translocation having a divided organizer, retranslocation occurs spontaneously by recombination between blocks of rRNA genes in the two positions. Mendelian chromosomal elements called Spore killers map to a region where recombination is blocked when a killer chromosome is heterozygous. Mating type alleles are stable except for induced null mutation, though examples of mating type reversal are known in other filamentous ascomycetes. Strategies for identifying instabilities are suggested by studies with unstable insertion elements in Ascobolus. Chromosome complements are similar in heterothallic, pseudo-homothallic and homothallic Neurospora species and in related genera of the Sordariaceae. The number of chromosomes is the same in several other Pyrenomycete families but is different in Discomycetes and Plectomycetes.

INTRODUCTION

The filamentous euascomycetes differ from Saccharomyces and other hemiascomycetes in many ways. Some of these features are advantageous for the investigator, others are disadvantageous. Among the advantages are the larger asci, ascospores, and nuclei, and the pigmentation of ascospores. Certain species with large pigmented ascospores are especially useful for genetic analysis of recombination (e.g. 1), unstable genes (2), chromosome rearrangements (3, 4), and meiotic drive (5). The size advantage can be appreciated from Emerson's drawing (Figure 1) which shows asci of representative fungi to scale (6). When asci are larger, nuclei are often larger and chromosomes may become more extended and easier to see. In many species with large asci, individual chromosomes of the complement can be recognized on the basis of their distinctive morphology, using light microscopy. This is notably true of Neurospora and its Pyrenomycete relatives.

THE CHROMOSOME COMPLEMENT: CYTOLOGICAL ASPECTS

I shall begin by describing the chromosomal genome of Neurospora, and then go on to consider molecular, genetic, and comparative aspects.

Fungal cytogenetics was initiated by Barbara McClintock, who adapted to Neurospora the staining and squash techniques she had developed in maize (7). She showed that the chromosomes are most fully extended at the pachytene stage in the young ascus. Her orcein method, with minor modifications, is still superior for revealing pachytene detail, though haematoxylin or a fluorescent dye may be preferred for other purposes.

The pachytene chromosome complement of Neurospora is shown in Figure 2. Individual chromosomes can be identified by their length and chromomere morphology. Chromosome 7, the smallest, can be readily distinguished even though its DNA content is substantially less than the chromosome of *Escherichia coli*. For reviews of the cytogenetics and chromosome cytology of Neurospora, see references 8, 9, and 10.

The Synaptonemal Complex. Information from light microscopy has been complemented by electron microscope studies of the synaptonemal complex, the tripartite pairing

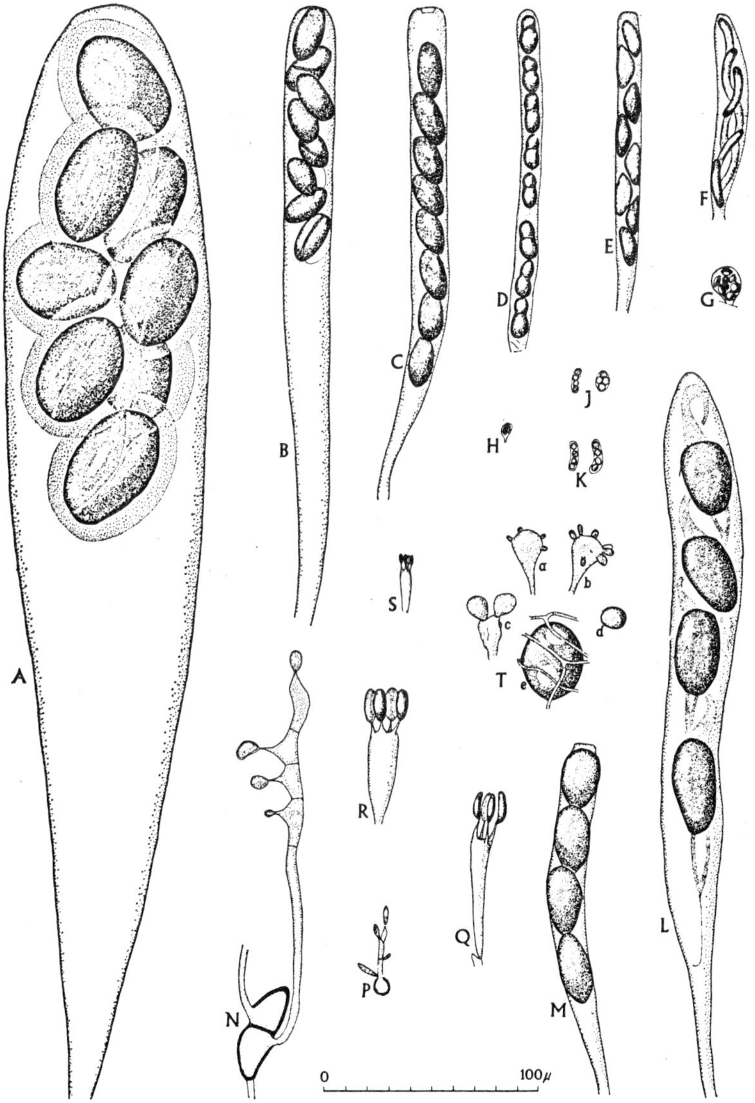

FIG. 1. Relative sizes of meiospores of some ascomycetes and basidiomycetes which have been used in genetic studies: A, *Ascobolus immersus;* B, *Ascobolus magnificus;* C, *Neurospora crassa;* D, *Venturia inaequalis;* E, *Bombardia lunata;* F, *Glomerella cingulata;* G, *Aspergillus nidulans;* H, *Ophiostoma multiannulata;* J, *Saccharomyces cerevisiae;* K, *Schizosaccharomyces pombe;* L, *Podospora anserina;* M, *Neurospora tetrasperma;* N, *Puccinia graminis;* P, *Ustilago maydis;* Q, *Cytidia salicina;* R, *Coprinus fimetarius;* S, *Schizophyllum commune;* T, *Cyathus stercoreus.* Reproduced with permission from Emerson (1966) (reference 6).

structure that is found universally in recombinationally competent eukaryotes during the first meiotic division (reviewed in 11). The typical railroad track appearance of conventionally stained pachytene chromosomes in fungi is due to alignment of the homologues on opposite sides of the 0.2 micrometer wide synaptonemal complex. The complex itself isn't visible in conventional cytological preparations. Gillies first observed a complete fungal synaptonemal-complex karyotype by reconstructing the pachytene nucleus of Neurospora from thin sections, observed by electron microscopy (12). This type of analysis has been extended to numerous other fungi (11). Features are revealed that cannot be seen by conventional microscopy. For example, all chromosome ends except the one that bears the nucleolus organizer are seen to be attached to the nuclear envelope in Neurospora and Sordaria

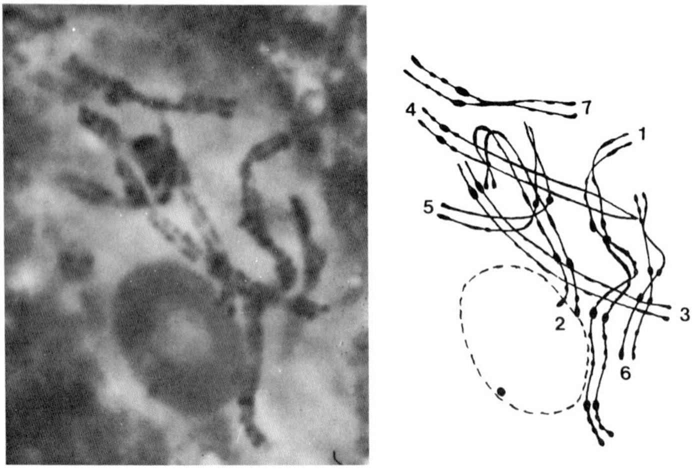

FIGURE 2. The seven bivalents of Neurospora crassa at pachytene. All chromosome segments are readily visible in the microscope but not all can be seen in a single photograph because the focal depth is very shallow at the high magnification that must be used. Orcein staining, X 4000. Prepared and photographed by E. G. Barry and reproduced from (8) with permission of the publisher.

(12, 13). Recombination nodules are seen in association with the complex. The nodules correspond in frequency and position to chiasmata and to reciprocal crossing-over events (13, 14). Synaptonemal complex reconstructions enable the course of chromosome pairing to be followed and to be compared in normal genomes and in rearrangement heterozygotes (13, 14, 15).

SOME MOLECULAR ASPECTS

Molecular characterization of Neurospora reveals a haploid genome of 27,000 kilobases -- less than twice the yeast genome. Neurospora chromosomal DNA is mostly unique-sequence, but with 2% foldback and 8% repetitive sequences (16). Most of the repetition is accounted for by approximately 200 ribosomal RNA genes, arranged tandemly in the nucleolus organizer region (16, 17, 18). 5S ribosomal RNA genes are not clustered but are dispersed through the genome as single copies (18, 19, 20). Chromatin is organized in typical nucleosome units (21). Genes specifying histones H3 and H4 are known to be physically linked and to be present in single copy (22).

The DNA content of individual Neurospora chromosomes, estimated from proportional cytological lengths at pachytene (Figure 3 in reference 8), ranges from about 6200 kb for the largest chromosome down to about 2540 kb for the smallest. The smallest chromosome, number 7, thus contains less DNA than E. coli, but 20% more than the biggest yeast chromosome (23). In certain translocation strains of Neurospora, two short chromosome arms (IIIL and VL) are combined into a single tiny chromosome that can be seen distinctly even though it is considerably smaller than chromosome 7. These and other small chromosome constructs derived from existing Neurospora rearrangements are expected to have DNA contents in the same size range as the Saccharomyces chromosomes. It should be possible to achieve physical separation of intact chromosomal DNA from the small Neurospora chromosomes by pulsed field gradient gel electrophoresis, a technique which has already been used successfully with yeast (24, 25).

The Neurospora genetic maps have been estimated to total about 1000 map units (centimorgans) (8). This gives on average 27 kilobases per map unit, compared with 2 or 3 kb per map unit in Saccharomyces. The greater physical content per genetic map unit will thus make 'walking' with cloned overlapping segments seem more laborious in

Neurospora. But the difference is largely illusory. Total map length is five times greater in Saccharomyces, although DNA content is less. Genes located at equal physical distance in the two organisms are thus farther apart on the yeast genetic map. Calculations in anticipation of walking should take into account the likelihood of regional differences in the density of genetic events for the same physical distance (26). Crossing over per unit physical length is disproportionate in centromere regions or in distal regions in yeast, Drosophila and maize (27, 28). Crossovers may be localized, as in Podospora (29). In Neurospora, *rec* genes can change recombination dramatically for the same interval (see e.g. 30).

GENETIC APPROACHES

Over 500 gene loci and 220 chromosome rearrangements have been mapped to the seven linkage groups in *Neurospora crassa* (31, 32, 8, 33). As in other fungi and in eukaryotes generally, functionally related genes are generally dispersed rather than being clustered.

Mapping has not been limited to conventional gene markers or to conventional methods based on crossing over frequencies. Genetic polymorphisms are abundant in *Neurospora crassa* and other outbreeding Neurospora species, as first demonstrated electrophoretically at the protein level (34). At the DNA level, restriction-fragment length polymorphisms appear to be ubiquitous in strains from nature. For example, Metzenberg has compared a strain from Texas with a laboratory standard strain derived from Louisiana. The strains are very similar in phenotype and crosses between them are completely fertile. Yet nucleotides differ at thousands of sites throughout the genome. The resulting restriction fragment differences can be used as markers for determining linkages of cloned DNA segments in the genetic maps (35). The method has been used to map 23 5S ribosomal RNA genes (20), 11 genes that are expressed differentially during conidiation (36), and the beta-tubulin gene (37).

The genetic maps show not only gene loci but also the breakpoints of selected chromosome rearrangements. Rearrangements that have been mapped include reciprocal and insertional translocations, inversions and translocations in which one breakpoint is at a chromosome tip, interstitial inversions that include a centromere, and nontandem (displaced) duplications, both interstitial and terminal (8). The most precisely mapped breakpoints belong to insertional and terminal translocations, because when either of these is crossed with a normal-sequence strain, meiotic recombination results in progeny that contain duplications for precisely defined chromosome segments. These can be used to determine gene order very efficiently by tests of duplication coverage.

Diploids and disomics are very unstable in Neurospora, for reasons that are not understood (38). An alternative is provided by the segmental duplications derived from insertional or terminal rearrangements. Duplications (partial diploids) obtained in this way have been used to determine dominance and vary gene dosage (see, for example 39). Duplication-generating rearrangements have made it possible to identify and map numerous vegetative (heterokaryon) incompatibility genes, because the incompatibility is expressed not only in heterokaryons but also in heterozygous duplications (40, 41).

Telomere regions are involved in rearrangements more often than would be expected by chance. Terminal translocations have been used to place eight of the 14 telomeres on the genetic maps. Not only telomeres, but also centromeres can be mapped genetically and cytologically, using appropriate chromosome rearrangements. Duplication-generating rearrangements make it easy to establish whether included markers are in one arm or the other. Information on centromeres is also provided by whole-arm (Robertsonian) reciprocal translocations, in which each breakpoint is immediately adjacent to a centromere. One pair of whole-arm translocations with breakpoints on opposite sides of the same two centromeres has been used to substitute one centromere for another in Neurospora (42). In the closely related species Sordaria macrospora, a majority of ultraviolet-induced translocations have been shown to involve the reciprocal interchange of entire chromosome arms (43). When a translocation of this type is crossed with a normal-sequence strain, nondisjunction of centromeres is common, resulting in asci with four inviable ascospores.

Rearrangements that involve the nucleolus organizer chromosome of Neurospora have been used to obtain meiotic segregants in which the number of rRNA genes is doubled or halved (44, 45). Magnification or demagnification then ensues so as eventually to restore the number of copies to its original chromosomal level (17, 46). Restriction-fragment length polymorphisms in the nontranscribed spacers are being used to investigate the mechanism of rDNA demagnification (47, 48).

In one rearrangement strain, part of the nucleolus organizer is translocated so that nucleoli are produced in two places -- one terminal and the other interstitial in another chromosome (45). Meiotic recombination between the two segments occasionally results in retranslocation so as to restore an essentially normal gene sequence (49).

Two other local regions of interest are the Spore killer region of linkage group III and the mating-type region of I.

Spore Killer. Two chromosomal factors called Spore killer (symbolized $Sk\text{-}2^K$ and $Sk\text{-}3^K$) have been found in nature in *Neurospora intermedia* and have been analyzed after introgression into *N. crassa* (5, 50). I shall use $Sk\text{-}2^K$ as an example.

When a killer strain is crossed with the standard wild type or with any other strain that is sensitive to killing ($Sk\text{-}2^K$ X $Sk\text{-}2^S$), only four ascospores survive in each ascus. The survivors are killers. But when killer is crossed by killer ($Sk\text{-}2^K$ X $Sk\text{-}2^K$), all eight ascospores survive. Thus $Sk\text{-}2^K$ does not kill itself.

$Sk\text{-}2^K$ shows 1 : 1 Mendelian segregation, 100% first division segregation, and linkage to centromere markers of linkage group III. Killing does not depend on whether $Sk\text{-}2^K$ is used as male or as female parent.

No difference between $Sk\text{-}2^K$ and $Sk\text{-}2^S$ has been detected in the vegetative phase. Forced heterokaryons containing nuclei of both types in the same cytoplasm are vigorous and healthy. Likewise, meiosis and ascus differentiation are normal in heterozygous crosses where killing will occur. Abnormalities are seen only after $Sk\text{-}2^S$ nuclei are sequestered by being enclosed in an ascospore wall. Sensitive nuclei which would otherwise die are rescued, however, if a killer nucleus is enclosed within the same ascospore wall. This can occur erratically, in disomics, or regularly, when giant ascospores are formed by certain developmental mutants.

When $Sk\text{-}2^K$ is crossed with a non-killer strain, crossing over is blocked in a 30-unit region that extends to both sides of the III centromere. The recombination block appears not to be due to an inversion (51).

Another Spore killer of independent origin, $Sk\text{-}3^K$, resembles $Sk\text{-}2^K$ in its mode of killing, its map location, and its effect on recombination. But Sk-3^K differs from Sk-2^K in specificity of killing and resistance. Sk-3^K does not kill itself, but is sensitive to killing by Sk-2^K, and vice versa. Mutual killing occurs in crosses between Sk-2^K and Sk-3^K, so that all ascospores are killed in most of the asci.

Chromosomally based Spore killing systems are also known in Neurospora sitophila and in other Pyrenomycetes (Podospora, Gibberella). The fungal Spore killers resemble meiotic drive systems in higher plants and animals, which are typically manifested postmeiotically by differential survival of spores or gametes. (For references see 5; 52.) The molecular basis of the delayed-action killer systems remains obscure. Preliminary observations suggest that Spore-killer-like differences may exist mutually between closely related Neurospora species; if so, they might be expected to contribute to the achievement of reproductive isolation.

Mating Type. A mating system that is based on two alleles at a single locus is characteristic of the ascomycetes. Mating types in other Neurospora species are homologous to those in Neurospora crassa. (This may be true also for closely related genera [53].)

The mating types A and a are expressed in two ways in Neurospora crassa. Two strains must be of opposite mating types in order to cross with one another, but they must be of the same mating type in order to be compatible in a heterokaryon during vegetative growth. Neighboring loci seem to have no relation to either function. Mutations affecting meiosis, recombination, and fertility map elsewhere in the genome. The only variants obtained at the mating type locus have been null mutations (with the exception of one mutant which retains mating activity and specificity while losing the vegetative incompatibility function). (For references see 54, 55, 31).

Although several truly homothallic eight-spored Neurospora species have been described (see 56), no interconversion of one mating type to another is known in Neurospora, and there has been nothing to suggest the existence of silent cassettes or a switch mechanism like those described in Saccharomyces (57) or Schizosaccharomyces (58). However, the possibility of mating-type conversions is not out of the question for Neurospora, because other filamentous ascomycetes are known in which one mating-type allele regularly gives rise to the opposite mating type. I am aware of three examples: Sclerotinia trifoliorum (59), Chromocrea spinulosa (60), and Glomerella cingulata (61). All three differ from Saccharomyces in that the mating-type change is unidirectional. In every eight-spored ascus of Sclerotinia, for example, four spores produce self-sterile cultures and four produce cultures that are self-fertile. One mating type of a two-allele system is apparently changing to the opposite mating type during growth, so as to produce a mixed culture, while changes in the reverse direction do not occur.

Genetic Instability. A nitrous-acid induced adenine mutant of Neurospora has been described which showed autonomous instability (62). Prototrophic revertants retained the instability, mutating again to adenine auxotrophy with high frequency. The strain is still available.

Findings with yeast (62a) and with the filamentous ascomycete Ascobolus immersus (2) may suggest strategies for identifying instabilities and transposable elements in Neurospora. Two unstable systems in Ascobolus were discovered in crosses between wild strains from nature. They have several features in common with transposable elements in plants. Mutability was first detected at loci controlling ascospore color. Reversions occur at a well defined brief stage of the life cycle, which is just after ascospore germination for one mutable system and just after fertilization for the second. Mutability is greatly increased at an abnormal temperature in the first system. In a strain that contains a mutable allele at one locus, new mutations occur at other loci, and many of these show the same type of instability as the original, although they may specify an entirely different phenotype. For the second system, ability to induce reversion and ability to undergo induced reversion can be retained or lost independently. Ability to induce is expressed only when the inducing strain is used as female and the entering male nucleus is inducible.

These findings in Ascobolus, reviewed in (2), suggest that a search for transposable elements in other filamentous fungi might focus on transition stages in the life cycle, that conditions such as temperature might be varied outside the norm, and that strains from nature might be examined. Wild-collected Neurospora strains are available that might be used for such a search (63, 64, 33).

COMPARISONS WITH RELATED EUASCOMYCETES

Ten Neurospora species are known. These are reproductively isolated to various degrees, and they fall into three categories with respect to their breeding systems: heterothallic (4 species), pseudohomothallic (1 species), and homothallic (5 species). Neurospora tetrasperma, the pseudohomothallic species, has evolved a precise method of delivering two haploid nuclei of opposite mating type into each of four ascospores. The result is a self-fertile heterokaryon which has no need for an external mating partner in order to complete the sexual cycle. The true-homothallic Neurospora species produce eight-spored asci in which each haploid spore is self-fertile. The vegetative mycelium is devoid of conidia in all five homothallic species, which must therefore depend on ascospores for dispersal. In spite of these differences, the chromosome complements of all ten species are virtually indistinguishable. Behavior of nuclei and chromosomes during karyogamy and meiosis is also very similar (65, 10).

The genera Sordaria, Podospora, and Gelasinospora are classified in the same family with Neurospora -- the Sordariacae. They are markedly different from Neurospora in morphology, physiology, and life style. The ascospores of Podospora and Sordaria pass through the gut of an animal and germinate to produce mycelia that grow on dung. Neurospora ascospores are heat-activated, and burned vegetation is a preferred substrate. Despite the morphological differences and adaptations to markedly different ecological situations, chromosome complements are very similar in all the genera (13, 15, 66, 67). All have seven haploid chromosomes of graded size and a single nucleolus organizer located terminally, or nearly so, on the second longest chromosome. The mating type locus is in the longest chromosome in Podospora, just as in Neurospora (66). More critical comparisons with Neurospora will soon be possible in Sordaria macrospora. This species, being homothallic, has no mating-type. But numerous auxotrophic markers are being characterized and mapped. Many linkages appear to have been conserved. Already 11 homologous or probably homologous loci have been mapped to corresponding chromosomes in Sordaria and Neurospora (68).

The evidence cited is admittedly fragmentary, but it suggests that not only within the genus Neurospora but also in different related genera, species with very different life styles make do with what is essentially the same outfit of chromosomes and gene loci -- the same hardware.

If so, what has changed is apparently the software -- the way the physically similar genomes have been programmed to cope with very different developmental and environmental situations.

Failure to scramble their genomes isn't due to any inherent chromosomal stability, because in the laboratory following ultraviolet mutagenesis and a selective mutant-screening procedure, at least 10% of surviving Neurospora cells contain new chromosome rearrangements (5). Without selective screening, 6 X 10^{-4} of UV-treated survivors in Sordaria macrospora contain detectable rearrangements (4). Rearrangements must be occurring continually in nature. Yet detectable rearrangements fail to become established, at least in the heterothallic species of Neurospora. Only three or four translocations have been detected among thousands of wild-collected Neurospora strains, and these occurred as minority components in widely separated local populations (33).

Two possible explanations come to mind for the failure of new rearrangements to become established. (a) Most aberrations reduce fecundity when heterozygous, because a portion of progeny are aneuploids. A minority sequence will thus be selected against in an outbreeding population. Founder effects are minimized in a heterothallic species where the units of dispersal are haploid cells. (b) Deviations from the established wild-type sequence may be at a selective disadvantage. Most, but not all, newly identified rearrangements of laboratory origin appear to be phenotypically normal in Neurospora. Deleterious changes may be present, however, that are too subtle for the eye to detect. I am not aware of any systematic search for minor effects of rearrangements in strains that appear phenotypically normal.

For Pyrenomycetes outside the Sordariaciae, cytological information is spotty and often unreliable, and genetic information is usually rudimentary. Other well known Pyrenomycetes include saprophytes such as Xylaria (n = 7) (69) and plant pathogens such as Cochliobolus (Helminthosporium) (n = 8) (70), Gibberella (Fusarium) (n = 7?) (71), Venturia (n = 7) (72, 73), Endothea, Ceratocystis, and Glomerella. These all represent different families, and it would be of interest to know how much restructuring of the genome has occurred. The similar chromosome numbers leave open the possibility that there has been little karyotypic change.

The filamentous euascomycetes as a whole are classified into three major groups, Pyrenomycetes (perithecium with ostiole), Discomycetes (open apothecium), and Plectomycetes (closed cleistothecium). Differences in chromosome number are definitely seen between the groups. In the Discomycetes, n = 11 for *Ascobolus immersus* (74, correcting 75), and n = ca 23 for *Neotiella rutilans* (76); in the Plectomycetes, n = 8 for *Aspergillus nidulans* (77, 78).

We don't yet have molecular data from which to establish a reliable phylogenetic tree or to estimate a time scale for divergence within the ascomycetes. Have various taxons been separated for 500 million years? 2000 million? Obtaining 17S ribosomal RNA sequences so as to construct a dendrogram of relationships would be a worthy undertaking.

ACKNOWLEDGMENTS

I am grateful to colleagues for providing unpublished information and for permission to cite their work in advance of publication, as indicated under References, and I thank Sterling Emerson for permission to reproduce Figure 1. Research in our laboratory was supported by Public Health Service Grants AI-01462 and K6-GM-4899.

REFERENCES

1. Rossignol J-L, Nicolas A, Hamza H, Kalogeropoulos A. Recombination and gene conversion in Ascobolus. In: Brooks Low K (ed) "The Recombination of Genetic Material," Academic Press. (In preparation).
2. Decaris B, Francou F, Kouassi A, Lefort C, Rizet G (1981). Genetic instability in *Ascobolus immersus*: Modalities of back-mutations, intragenic mapping of unstable sites, and unstable insertion. Preliminary biochemical data. Cold Spring Harbor Symp. Quant. Biol. 45:509-517.
3. Perkins DD (1974). The manifestation of chromosome rearrangements in unordered asci of Neurospora. Genetics 77:459-489.
4. Arnaise S, Leblon G, Lares, L (1984). A system for the detection of chromosomal rearrangements using *Sordaria macrospora*. Mutat. Res. 125:33-42.
5. Turner BC, Perkins DD (1979). Spore killer, a chromosomal factor in Neurospora that kills meiotic products not containing it. Genetics 93:587-606.

6. Emerson S (1966). Mechanisms of inheritance. 1. Mendelian. In Ainsworth GC, Sussman AS (eds): "The Fungi: An Advanced Treatise. Vol. 2. The Fungal Organism," New York: Academic Press, p. 513.
7. McClintock, B (1945). Neurospora. I. Preliminary observations on the chromosomes of Neurospora crassa. Am. J. Bot. 32:671-678.
8. Perkins DD, Barry EG (1977). The cytogenetics of Neurospora. Adv. Genet. 19:133-285.
9. Perkins DD (1979). Neurospora as an object for cytogenetic research. Stadler Genet. Symp. 11:145-164.
10. Raju NB (1980). Meiosis and ascospore genesis in Neurospora. Eur. J. Cell Biol. 23:208-223.
11. Wettstein D von, Rasmussen SW, Holm PB (1984). The synaptonemal complex in genetic segregation. Annu. Rev. Genet. 18:331-413.
12. Gillies CB (1972). Reconstruction of the Neurospora crassa pachytene karyotype from serial sections of synaptinemal complexes. Chromosoma 36:117-130.
13. Zickler D (1977). Development of the synaptonemal complex and the "recombination nodules" during meiotic prophase in the seven bivalents of the fungus Sordaria macrospora Auersw. Chromosoma 61:289-316.
14. Gillies CB (1979). The relationship between synaptinemal complexes, recombination nodules and crossing over in Neurospora crassa bivalents and translocation quadrivalents. Genetics 91:1-17.
15. Zickler D, Leblon G, Haedens V, Collard A, Thuriaux P (1984). Linkage group-chromosome correlations in Sordaria macrospora: Chromosome identification by three dimensional reconstruction of their synaptonemal complex. Curr. Genet. 8:57-67.
16. Krumlauf R, Marzluf GA (1980). Genome organization and characterization of the repetitive and inverted repeat DNA sequences in Neurospora crassa. J. Biol. Chem. 255:1138-1145.
17. Rodland KD, Russell PJ (1982). Regulation of rRNA cistron number in a strain of Neurospora crassa with a duplication of the nucleolus organizer region. Biochim. Biophys. Acta 697:162-169.
18. Free SJ, Rice PW, Metzenberg RL (1979). Arrangement of the genes coding for ribosomal ribonucleic acids in Neurospora crassa. J. Bacteriol. 137:1219-1226.
19. Selker EU, Yanofsky C, Driftmeir K. Metzenberg RL, Alzner-DeWeerd B, RajBhandary UL (1981). Dispersed 5S RNA genes in N. crassa: structure, expression and evolution. Cell 24:819-828.

20. Metzenberg RL, Stevens JN, Selker EU, Morzycka-Wroblewska E (1985). Identification and chromosomal distribution of 5S RNA genes in Neurospora crassa. Proc. Natl. Acad. Sci. U.S.A. 82 (In press).
21. Noll M (1976). Differences and similarities in chromatin structure of Neurospora crassa and higher eucaryotes. Cell 8:349-355.
22. Woudt LP, Pastink A, Kempers-Veenstra AE, Jansen, AEM, Mager WH, Planta RJ (1983). The genes coding for histone H3 and H4 in Neurospora crassa are unique and contain intervening sequences. Nucleic Acids Res. 11:5347-5360.
23. Fangman WL, Zakian VA (1981). Genome structure and replication. In Strathern JN, Jones EW, Broach JR (eds): "The Molecular Biology of the Yeast Saccharomyces. Life Cycle and Inheritance," Cold Spring Harbor Laboratory, p 27-58.
24. Carle GF, Olson MV (1984). Separation of chromosomal DNA molecules from yeast by orthogonal-field-alteration gel electrophoresis. Nucleic Acids Res. 12:5647-5664.
25. Schwartz DC, Cantor CR (1984). Separation of yeast chromosome-sized DNAs by pulsed field gradient gel electrophoresis. Cell 37:67-75.
26. Rabelais F (1532). La cause pourquoy les lieues sont tant petites en France. In "Pantagruel," Chap. 15. Genève: Droz, 1965 edition, p. 126.
27. Mortimer RK, Schild D (1981). Genetic mapping in Saccharomyces cerevisiae. In Strathern JN, Jones EW, Broach JR (eds): "The Molecular Biology of the Yeast Saccharomyces. Life Cycle and Inheritance," Cold Spring Harbor Laboratory, p 11-26.
28. Carlson WR (1977). The cytogenetics of corn. 6. Distribution of crossovers in the cytological chromosomes. In Sprague GF (ed) "Corn and Corn Improvement," Madison: American Society of Agronomy, p. 256.
29. Marcou D, Masson A, Simonet J-M, Piquepaille G (1979). Evidence for non-random spatial distribution of meiotic exchanges of Podospora anserina: Comparison between linkage groups 1 and 6. Mol. Gen. Genet. 176:67-79.
30. Catcheside DG, Corcoran D (1973). Control of non-allelic recombination in Neurospora crassa. Aust. J. Biol. Sci. 26:1337-1353.
31. Perkins DD, Radford A, Newmeyer D, Björkman M (1982). Chromosomal loci of Neurospora crassa. Microbiol. Rev. 46:426-570.

32. Perkins DD (1984). Neurospora crassa genetic maps, May 1984. Neurospora Newsl. 31:46-54.
33. Perkins, DD et al. (Unpublished).
34. Spieth PT (1975). Population genetics of allozyme variation in Neurospora intermedia. Genetics 80:785-805.
35. Metzenberg RL, Stevens JN, Selker EU, Morzycka-Wroblewska E (1984). A method for finding the genetic map position of cloned DNA fragments. Neurospora Newsl. 31:35-39.
36. Berlin V, Yanofsky C (1985). Isolation and characterization of genes differentially expressed during conidiation in Neurospora crassa. Mol. Cell. Biol. 5:849-855.
37. Orbach MJ. (Personal communication).
38. Smith DA (1973). Unstable diploids of Neurospora and a model for their somatic behavior. Genetics 76:1-17.
39. Metzenberg RL, Chia W (1979). Genetic control of phosphorus assimilation in Neurospora crassa: dose-dependent dominance and recessiveness in constitutive mutants. Genetics 193:625-643.
40. Perkins DD (1975). The use of duplication-generating rearrangements for studying heterokaryon incompatibility genes in Neurospora. Genetics 80:87-105.
41. Mylyk OM (1975). Heterokaryon incompatibility genes in Neurospora crassa detected using duplication-producing chromosome rearrangements. Genetics 80:107-124.
42. Perkins DD, Raju NB, Barry EG. Substitution of centromeres from nonhomologous chromosomes of Neurospora, using whole-arm reciprocal translocations. (In preparation).
43. Leblon G, Zickler D, Lebilcot S. In Sordaria macrospora ultraviolet induced reciprocal translocations involve mainly a limited number of sites located in the centromeric regions. (In preparation).
44. Perkins DD, Raju NB, Barry EG (1980). A chromosome rearrangement in Neurospora that produces viable progeny containing two nucleolus organizers. Chromosoma 76:255-275.
45. Perkins DD, Raju NB, Barry EG (1984). A chromosome rearrangement in Neurospora that produces segmental aneuploid progeny containing only part of the nucleolus organizer. Chromosoma 89:8-17.
46. Rodland KD, Russell, PJ. Magnification of rRNA gene number in a Neurospora crassa strain with a partial deletion of the nucleolus organizer. (In preparation).

47. Rodland KD, Russell PJ (1983). Segregation of heterogeneous rDNA segments during demagnification of a Neurospora crassa strain possessing a double nucleolar organizer. Curr. Genet. 7:379-384.
48. Rodland KD, Russell PJ. (Personal communication).
49. Perkins DD, Raju NB, Metzenberg RL, Barry EG. Restoration of normal chromosome sequence by crossing over between displaced segments in crosses homozygous for a translocation dividing the nucleolus organizer region of Neurospora. (In preparation).
50. Raju NB (1979). Cytogenetic behavior of Spore killer genes in Neurospora. Genetics 93:607-623.
51. Campbell JL, Turner BC Recombination block in the Spore killer region of Neurospora. (In preparation).
52. Kathariou S, Spieth PT (1982). Spore killer polymorphism in Fusarium moniliforme. Genetics 102:19-24.
53. Olive LS, Fantini AA (1961). A new, heterothallic species of Sordaria. Am. J. Bot. 48:124-128.
54. Griffiths AJF (1982). Null mutants of the A and a mating type alleles of Neurospora crassa. Can. J. Genet. Cytol. 24:167-176.
55. Griffiths AJF, DeLange AM (1978). Mutations of the a mating type in Neurospora crassa. Genetics 88:239-254.
56. Frederick L, Uecker FA, Benjamin CR (1969). A new species of Neurospora from soil of West Pakistan. Mycologia 61:1077-1084.
57. Nasmyth KA (1982). Molecular genetics of yeast mating type. Annu. Rev. Genet. 16:439-500.
58. Egel R (1984). The pedigree pattern of mating-type switching in Schizosaccharomyces pombe. Curr. Genet. 8:205-210.
59. Uhm JY, Fujii H (1983). Heterothallism and mating type mutation in Sclerotinia trifoliorum. Phytopathology 73: 569-572.
60. Mathieson MJ (1952). Ascospore dimorphism and mating type in Chromocrea spinulosa. Ann. Bot. 16:449-466.
61. Wheeler HE (1950). Genetics of Glomerella. VIII. A genetic basis for the occurrence of minus mutants. Am. J. Bot. 37:304-312.
62. Barnett WE, DeSerres FJ (1963). Fixed genetic instability in Neurospora crassa. Genetics 48:717-723.
62a. Williamson VM (1983). Transposable elements in yeast. Int. Rev. Cytol. 83:1-25.
63. Perkins DD, Turner BC, Barry EG (1976). Strains of Neurospora collected from nature. Evolution 30:281-313.

64. Perkins DD (1983). Variation in natural populations of Neurospora. Proc. 15th Int. Cong. Genet., Abstracts of Contributed Papers, 451.
65. Raju, NB (1978). Meiotic nuclear behavior and ascospore formation in five homothallic species of Neurospora. Can. J. Bot. 56:754-763.
66. Simonet J-M, Zickler D (1979). Linkage group - chromosome correlation in Podospora anserina. Mol. Gen. Genet. 175:359-367.
67. Lu BC (1967). The course of meiosis and centriole behaviour during the ascus development of the ascomycete Gelasinospora calospora. Chromosoma 22:210-236.
68. Leblon G. (Personal communication).
69. Rogers JD (1975). Xylaria polymorpha. II. Cytology of a form with typical robust stromata. Can. J. Bot. 53:1736-1743.
70. Guzman D, Garber RC, Yoder OC (1982). Cytology of meiosis I and chromosome number of Cochliobolus heterostrophus (Ascomycetes). Can. J. Bot. 60:1138-1141.
71. Raju NB. (Personal communication).
72. Day PR, Boone DM, Keitt GW (1956). Venturia inaequalis (Cke.) Wint. XI. The chromosome number. Am. J. Bot. 43:835-838.
73. Julien JB (1958). Cytological studies of Venturia inaequalis. Can. J. Bot. 36:607-613.
74. Rizet G, Lefort C, Decaris B, Francou F, Kouassi A (1979). Unstable ascospore color mutants of Ascobolus immersus. Mol. Gen. Genet. 175:293-303.
75. Zickler D (1967). Analyse de la méiose du champignon discomycète Ascobolus immersus Pers. Compt. Rend. Acad. Sci. Paris, Sér. D 265:198-201.
76. Westergaard M, Wettstein D von (1966). Studies on the mechanism of crossing over. III. On the ultrastructure of the chromosomes in Neottiella rutilans (Fr.) Dennis. Compt. Rend. Trav. Lab. Carlsberg 35:261-286.
77. Elliott CG (1960). The cytology of Aspergillus nidulans. Genet. Res. 1:462-476.
78. Boothroyd ER (1978). A light-microscopic study of some translocations in Aspergillus nidulans. Can. J. Genet. Cytol. 20:325-328.

Molecular Genetics of Filamentous Fungi, pages 295–307

DISPERSED MULTIPLE COPY GENES FOR 5S RNA: WHAT KEEPS THEM HONEST?[1]

Robert L. Metzenberg, Eric U. Selker[2], Ewa Morzycka-Wroblewska[3], and Judith N. Stevens

Department of Physiological Chemistry
University of Wisconsin
Madison, WI 53706

When we think casually about a "haploid" organism, we generally think in terms of one copy of each gene, with the allele at every locus being completely exposed to the ordeal of natural selection. But every eucaryote contains at least some genes which face the world together as multiple, non-allelic copies. Where the copy number is large, the ability of natural selection to weed out newly-arisen, poorly-functional variants must be correspondingly small. One might expect that a high degree of redundancy would make it impossible to eliminate new mutations by natural selection, and that the copies would diverge rapidly. In fact, this is not the case. Multiple-copy sequences usually remain identical within an organism while diverging between organisms. This phenomenon has been called "concerted evolution". Various mechanisms to account for within-individual homogeneity have been proposed, including unequal crossingover, transposition, gene conversion, and RNA-mediated conversion (1). All of these mechanisms involve either repeated expansion and contraction of the family of genes with loss of variability during the contraction phase, or "correction" of one sequence by another,

[1]This work was supported by NIH grant GM-08995 and by a Helen Hay Whitney Foundation fellowship to E.U.S.
[2]Present Address: Department of Biology and Institute of Molecular Biology, University of Oregon, Eugene, Oregon 97403
[3]Present Address: Department of Molecular Biology and Genetics, Johns Hopkins University School of Medicine, Baltimore, Maryland 21205.

with the number of copies remaining precisely the same. It may be pointed out that "selection" requires death of organisms, "expansion-contraction" requires replacement of one or a few genes, and "correction" may require only replacement of a small number of nucleotides.

Unequal crossingover, an expansion-contraction mechanism, will result in the loss of unique-sequence DNA unless the multiple-copy genes are arranged tandemly. In some lower eucaryotes (e.g., Saccharomyces cerevisiae), all four of the ribosomal RNA species are encoded in tandem repeat units which include all or nearly all of the copies in the genome (2,3). In most higher eucaryotes, the three larger rRNAs, (18S, 5.8S, and 26S) are encoded in one or more tandem repeat units present in the nucleolus organizer(s), and the 5S rRNA is encoded in one or more tandem repeat units of its own. In Neurospora, the 100-odd copies of genes for 5S rRNA are not encoded with the other three rRNAs (which are arranged in tandem repeats), nor is it in any kind of simple repeat unit. A great variety of different-sized restriction fragments of Neurospora DNA hybridize to 5S RNA (4,5), and study of cloned genomic DNA showed that, with one known exception (Selker et al., this volume) DNA fragments in the size range of 5-10 kilobase pairs (kbp) do not contain more than one 5S gene. Intuitively, we believed that some mechanism must exist to keep many or all the copies identical. To see whether this was so, we isolated about two dozen clones of Neurospora DNA which hybridized to 5S RNA, and examined their 5S regions. From sequence information, it was apparent that a bare majority of the clones (13/24 of the total, or 13/21 if three obvious pseudogenes are excluded) contain an invariant, or almost invariant series of 120 base pairs. This sequence, which we call α, shows obvious similarity to the 5S rRNA of other eucaryotes; but it was equally striking that the 5' and 3' sequences flanking each of these genes are very different. While α is clearly the predominant 5S sequence among the clones, it is not the only kind. Others differ from α and from one another in 10% or more of the nucleotides, and we call these the β isotype, the γ, etc. The number of nucleotide differences found between contrasting isotypes within a single strain of Neurospora crassa are greater than the number of differences between frogs and humans. Yet the nucleotide sequence of each of these isotypes can be expressed as an essentially identical secondary structure (Fig. 1).

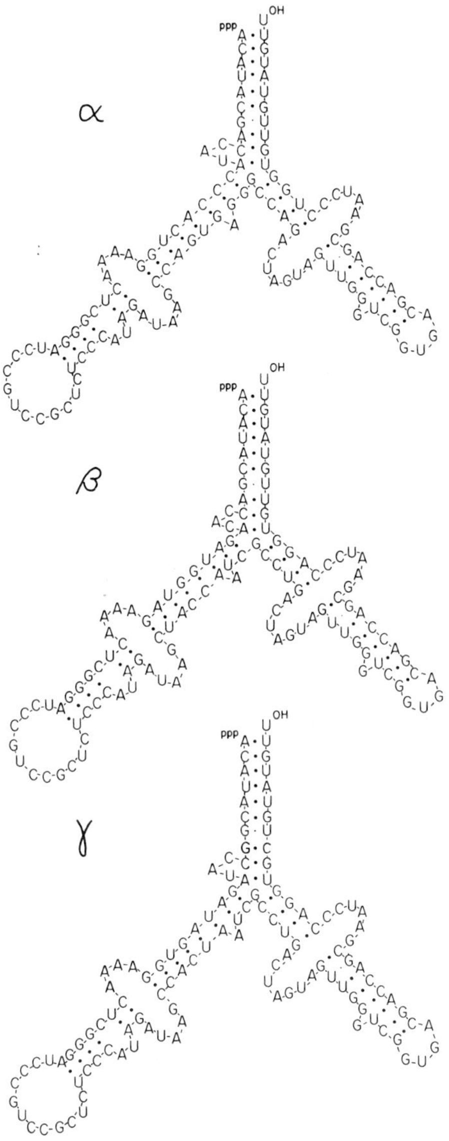

FIGURE 1. Secondary structures of 5S RNA's

This structure is in agreement with the one previously proposed by Nishikawa and Takemura (6) for the structure of 5S RNA from Torulopsis utilis, and is in disagreement with a large number of other suggested structures.

Having found a variety of 5S gene sequences, we wanted to know whether this variety is expressed in vivo and, if so, whether the various 5S RNA's actually make their way into ribosomes. It was therefore reassuring to find that bulk 5S RNA consists largely of molecules identical in sequence with the α isotype 5S genes recognized from cloned DNA fragments. Early results from sequence studies indicated that at least some of the minor isotype genes were transcribed, and that phantom bands corresponding to these transcripts appeared in sequencing ladders (5). To gain more insight into the nature of these minor transcripts, we developed a method of separating 5'-end-labelled 5S RNA isotypes so that they could be sequenced individually (7). Bulk 5S RNA was purified by eletrophoresis through a fully denaturing gel (7M urea) and through a non-denaturing gel, both of which allow the various isotypes to run as a single band. The bulk 5S RNA was then fractionated on long "sequencing" gels under partially denaturing conditions (2M urea). This presumably leaves the 5S RNA's with the most intact hydrogen bonds in a compact, mobile form and those with fewer hydrogen bonds in an extended form of lesser mobility. Whether or not this is the correct explanation, the result is that as many as a dozen bands can be partly or completely resolved from bulk 5S RNA (Fig. 2). Sequence analysis of these showed that the most prominent band represented α isotype 5S RNA, and that the second most prominent band represented β. In addition, an isotype called γ' was seen, which differed by a single nucleotide from the sequence predicted from our previously-reported γ gene (5). Two minor bands were due to RNA's which apparently differ from the α isotype at single positions. No bands corresponding to δ, ζ, or η isotypes, known from their cloned DNA fragments, have been identified. It is possible that these are not transcribed, or that they give rise to very unstable transcripts, or that their transcripts are sufficiently different in properties from those of bulk 5S RNA that they are separated from it by the fully-denaturing gel or the non-denaturing gel that precede the analytical gel. The existence of an assortment of 5S RNA isotypes in the cell left open the possi-

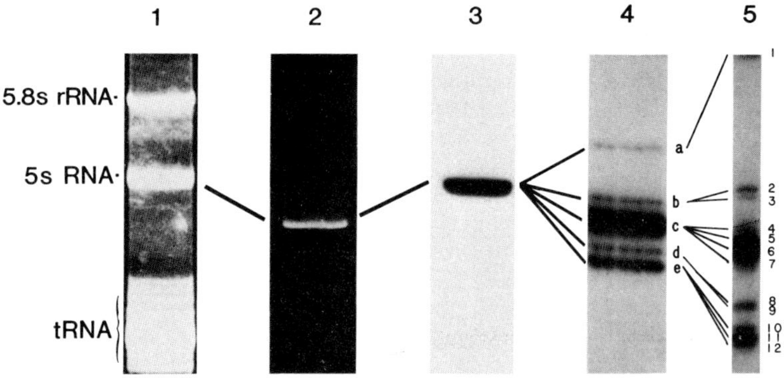

FIGURE 2. Purification of bulk 5S RNA by gel electrophoresis (lanes 1-3) and resolution into various isotypes in 2M urea gels (lane 4=40 cm gel, lane 5=90 cm gel).

bility that only the majority type, α, was present in ribosomes. However, isolation of 5S RNA from ribosomes instead of from whole cells showed almost the identical spectrum of bands in partially denaturing gels.

The evidence, then, indicates that ribosomes themselves are heterogeneous owing to the heterogeneity of their 5S RNA molecules. We may ask whether this is a mere accident of intra-strain genetic drift, or whether the different classes of ribosomes are specialized for distinct functions. If the differences were merely tolerated as unimportant accidents, we might expect that a comparison between different strains of Neurospora crassa would reveal differences in positions of minor bands, or proportions, or both; comparison between different species within the genus Neurospora, and between these and related genera would show even much greater differences. To investigate this question, we examined 5S RNA from distantly-related laboratory strains of N. crassa, and from other filamentous fungi, including N. tetrasperma, N. sitophila, N. discreta, N. intermedia, and from Gelasinospora tetrasperma, Sordaria brevicollis, Aspergillus nidulans, and Cochliobolus heterostrophus, as well as the fission yeast Schizosaccharomyces pombe and the budding yeast

Saccharomyces cerevisiae. The results were clear: every one of these species contains two or more isotypes of 5S RNA (a fact already known for A. nidulans (8) and for S. cerevisiae (9,10). Furthermore, in all the Neurospora species and in G. tetrasperma, an α and a β are present, and in virtually identical proportions, though the G. tetrasperma also contains a β-like 5S RNA isotype which was not seen in the Neurospora species. S. brevicollis contains α 5S RNA identical with that of the Neurospora species and G. tetrasperma, but does not contain β; it does, however, contain at least one other 5S RNA which has not been characterized.

The persistence of these major and minor isotypes across lines of species and even genera suggests that they are not merely tolerated but are of selective value to the organisms. The nature of the functional differences is a subject for easy speculation, but facts are few. Perhaps the most obvious possibility -- that different isotypes predominate during different phases of the life cycle of the organism -- is not borne out by analysis (5), though it remains possible that a particular isotype which is present in all cell types is only functionally important in one of them. Another possibility is that the different isotypes, having secondary structures of different stability, predominate in the cell at different temperatures of growth. Again, the proportions of the isotypes show only an unimpressive change, if any, during growth at 4°C. vs. 30°C (Selker, E., unpublished data). However, it is possible that different 5S RNA's could buffer against changes in translational efficiency at extreme temperatures without the proportions of those isotypes being regulated by temperature. Other hypotheses come all too easily, for example, that the different isotypes are important in different parts of each cell cycle, or that they are used in translation of different classes of mRNA. We do not know whether the ribosomes within a polysome are homogeneous with respect to 5S isotype. Further insight into function may come either from fractionation of whole ribosomes or polysomes according to 5S isotype, or perhaps by mutational extirpation of certain minority isotype 5S genes and examination of the resulting phenotype.

We have only the beginnings of understanding of how the transcription of 5S RNA is accomplished in Neurospora. RNA polymerase III from this organism has been partly characterized (11) and it is clear that the same enzyme preparation actively transcribes both α and β isotypes of 5S RNA, as well as some of the more minor isotypes and at least some precursors of tRNAs(11); Selker et al, unpublished. It is not entirely clear what sequences specify the initiation of transcription. The 5S genes active in frog oocytes are transcribed by a polymerase which takes its cue for initiation from nucleotides which are 50 nucleotides and more downstream from the site of initiation (12). If the same principle should apply to the Neurospora 5S genes, a single kind of RNA polymerase III must accept more or less equally the quite-different internal regions of genes of different isotypes. Examination of the 5'-flanking sequences of the Neurospora 5S genes, all of which are different, might seem to favor the idea that they contain no information for initiation. However, there is one short region upstream from the 5S genes which appears to be conserved in all but one of eight genes which can be shown to be transcribed in vitro (13). Nucleotides -29 to -26 are, respectively, TATA; the single exception (gene 52) has CATA in this position. The "TATA" box is always flanked with a T or an A in position -30, and with a G or an A in position -25. We have made Baℓ31 deletions in plasmids carrying a β 5S gene, starting from a fixed point far upstream from the 5S gene and extending various distances toward the gene (Morzycka-Wroblewska et al., unpublished). Some of the deletions removed the TATA box, while others left it intact. Testing these plasmids which had suffered Baℓ31-induced deletions for in vitro transcription by Neurospora RNA polymerase III showed an informative correlation: those deletions which removed the TATA box abolished transcription or reduced it drastically, and those which fell short of the TATA box did not. At least in vitro, the TATA box appears to be necessary for efficient transcription, and it is a reasonable guess that it is necessary in vivo as well. Recent work indicates that a TATA box (or TACTA box) is important for transcription by RNA polymerase III in a number of organisms (13-15).

The question of whether the 5S genes carry most or all of their punctuation signals in the transcription unit bears on theories of how the numerous copies of the 5S genes were dispersed in the first place, and how they are kept homogeneous within each isotype. Genes in which all signals are included within the transcript itself are, by definition, self-contained, and a reverse-transcript of such an RNA might be able to integrate anywhere in the genome and be functional immediately. On the other hand, if a TATA box is needed in the 5' upstream region for initiation of transcription, and poly-T is necessary on the transcript-like strand for termination, then initial dispersal by reverse transcription of 5S RNA seems less likely. The fact that such regions do seem to be necessary suggests that dispersal probably occurred by other mechanisms. Transposition must be considered not only as a possible mechanism of initial dispersal, but as an ongoing mechanism for purification of multiple gene families by expansion and contraction. If 5S genes are regularly jumping into new places in the genome and the number of 5S genes kept roughly constant by deletion of 5S sequences, we would predict that sequences similar to the flanking sequences of a 5S gene in our reference strain would be present in a distantly-related strain, but that the 120 nucleotide 5S sequence itself would have been lost. Operationally, if 5S genes and their flanking sequences were isolated from the reference strain on the basis of their ability to hybridize with 5S RNA, and these clones were used to isolate clones from an "exotic" or distantly-related strain on the basis of homology of flanking sequences, the clones from the "exotic" strain should not contain a 5S sequence. Isolation and sequencing of several such DNA fragments from the reference strain (Oak Ridge) and distantly-related strain (Beadle and Tatum 25) showed that in each case, the flanking sequences of the latter contained a 5S gene in exactly the same position as in the reference strain. Furthermore, the transcribed regions in each pair of alleles had been kept exactly the same except for a single nucleotide difference in one of them (one mutation among 360 nucleotides). In contrast, the allelic flanking sequences had diverged markedly in both the 5' and 3' regions, though much less so than between non-allelic genes in the same strain (13). It appears that transposition of 5S genes, if it is currently happening at all, is too infrequent to account for the degree of homogeneity of 5S genes within each isotype.

While expansion and contraction of the family of 5S genes of each isotype by some statistical equality of transposition and deletion could be ruled out, expansion and contraction by frequent unequal crossingover could not be rigorously excluded. Our information from DNA sequencing and genome blots indicated that the 5S genes are imbedded in many different kinds of flanking sequences, but it remained possible that many others were imbedded in identical flanking sequences. Such tandem arrays would have been seen on genome blots probed with 5S RNA merely as bands more intense than most, but these would have been hard to distinguish from bands from fragments of unrelated sequence but with fortuitously similar molecular weights. If two clones from non-allelic genes in such a tandem array were isolated, all but one of them would have been discarded as duplicates. It was also clear that, at the distance range of cloned restriction fragments (viz., 5-10 kbp) the 5S genes were not generally arranged in tandem; however, nothing could be said about possible linkage on the scale of ordinary genetic maps. The genome of Neurospora contains about 27,000 kbp of DNA divided among seven chromosomes (16,17) and it remained possible that the hundred 5S genes could lie within a few thousand kbp on one chromosome arm without any two of them appearing on the same restriction fragment. For expansion-contraction by unequal crossingover to be strictly excluded, it seemed necessary to show that indispensible, unique-sequence DNA was interspersed between various 5S genes. We therefore decided to map as many 5S genes as was practical by standard genetic techniques -- that is, to find crossover distances. While the fact of "correction" meant that allelic differences between 5S genes were not available, the failure of "correction" of flanking sequences of these genes had generated a high degree of flanking sequence polymorphism. The existence of such restriction fragment length polymorphisms (RFLPs) allowed us to map these genes with respect to conventional markers affecting nutrition, color, antibiotic resistance, and morphology (18). The map resulting from these crosses is shown in Fig. 3.

The map of these 5S genes shows that they are not distributed randomly. Significantly more of the α genes are on Linkage Group II than would have been expected on the basis of chance, and significantly too many of the β genes are on Linkage Group IV. But more important to the present argument is that genes of every isotype are widely scattered, if not randomly disturbed, and that a very large number of indispensible single-copy genes must lie between them. However, we cannot exclude the possible existence of small clusters of tandemly-arranged 5S genes. If unequal crossingover plays any role in maintaining homogeneity of 5S genes, it must be limited to

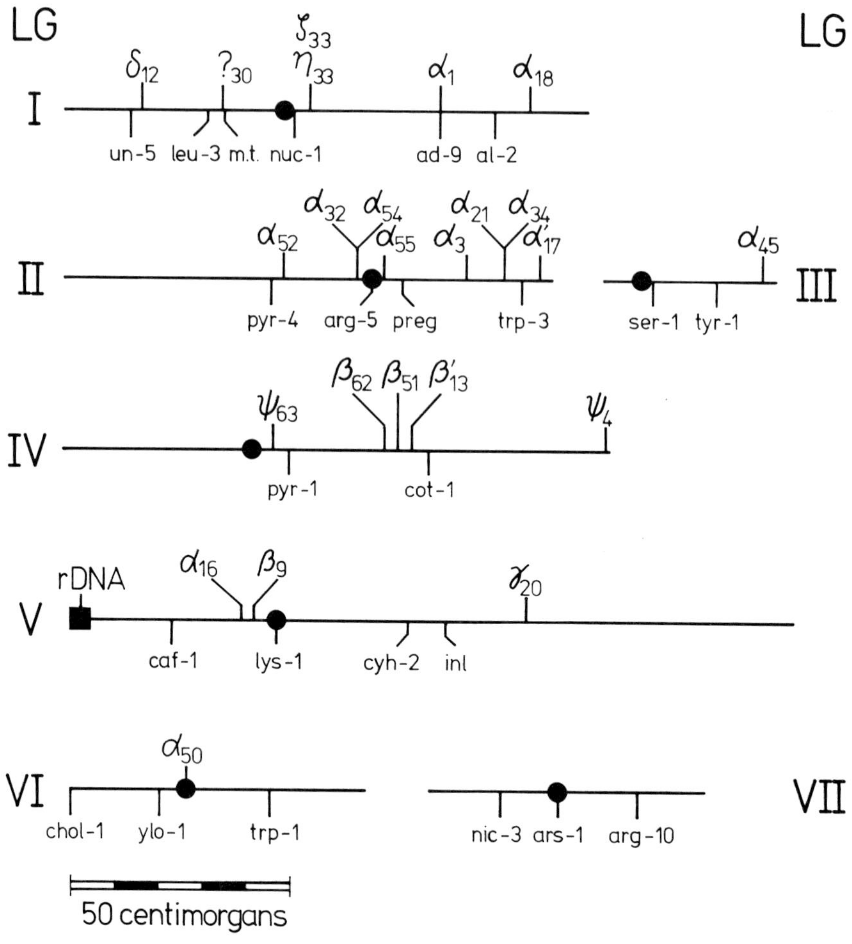

FIGURE 3. Map of some 5S genes and reference markers.

working within such scattered clusters -- if they exist.

With the two expansion-contraction models of maintenance of homogeneity apparently in conflict with the facts, we have considered the two pure "correction" models: classical gene conversion, and RNA-mediated gene conversion. While classical conversion is in many respects an attractive model, it runs into difficulties in the current case. In particular, it is difficult to see why the limits within which "correction" seems to occur are the same as the unit of transcription. On the other hand, an explanation of this oddity flows automatically from the hypothesis of RNA-mediated conversion. A hybrid model also fits the facts: "correction" could involve DNA-DNA contacts as in classical conversion, but transcription would in some way be necessary to bring about these contacts. The question, then, would center around the role of the 5S RNA molecule. Is it a true intermediate in information transfer during "correction," or is it a mere by-product of a mechanism for establishing DNA-DNA contact between two 5S genes of the same isotype? It appears that study of the end result of "correction" cannot answer such questions, and that investigation of the mechanism itself will be necessary.

REFERENCES

1. Dover GA, and Flavell RB (1984). Molecular coevolution: DNA divergence and the maintenance of function. Cell 38:622.
2. Nath K, and Bollon AP (1977). Organization of the yeast ribosomal gene cluster via cloning and restriction analysis. J. Biol. Chem. 252:6562.
3. Petes TD, and Botstein D (1977). Simple Mendelian inheritance of the reiterated ribosomal DNA of yeast. Proc. Nat. Acad. Sci. U.S.A. 74:5091.
4. Free SJ, Rice PW, and Metzenberg RL (1979). Arrangement of the genes coding for ribosomal ribonucleic acids in Neurospora crassa. J. Bacteriol. 137:1219.
5. Selker EU, Yanofsky C, Driftmier K, Metzenberg RL, Alzner-DeWeerd B, and RajBhandary UL (1981). Dispersed 5S RNA genes in N. crassa: structure, expression, and evolution. Cell 24:819.

6. Nishikawa K, and Takemura S (1974). Structure and function of 5S ribosomal ribonucleic acid from Torulopsis utilis. II. Partial digestion with ribonucleases and derivation of the complete sequence. J. Biochem. 76:935.
7. Selker EU, Stevens JN, and Metzenberg RL (1985). Heterogeneity of 5S RNA in fungal ribosomes. Science 227:1340.
8. Bartnik E, Strugala K, and Stepien PP (1981). Cloning and analysis of recombinant plasmids containing genes for Aspergillus nidulans 5S rRNA. Curr. Genet. 4:173.
9. Piper PW, Bellatin JA, and Lockheart A (1983). Altered maturation of sequences at the 3' terminus of 5S gene transcripts of a Saccharomyces mutant that lacks a processing endonucleases. EMBO J. 2:353.
10. Piper PW, Lockheart A, and Patel N (1984). A minor class of 5S rRNA genes in Saccharomyces cerevisiae X2180-1B, one member of which lies adjacent to a Ty transposable element. Nucleic Acids Res. 12:4083.
11. Tyler BM, and Giles NH (1984). Accurate transcription of homologous 5S RNA and tRNA genes and splicing of tRNA in vitro by soluble extracts of Neurospora. Nucleic Acids Res. 12:5737.
12. Sakonju S, Bogenhagen DV, and Brown DD (1980). A control region in the center of the 5S RNA gene directs specific initiation of transcription. I. The 5' border of the region. Cell 19:13.
13. Morzycka-Wroblewska E, Selker EU, Stevens JN, and Metzenberg RL (1985). Concerted evolution of dispersed Neurospora crassa 5S RNA genes: pattern of sequence conservation between allelic and nonallelic genes. Molec. Cell. Biol. 5:46.
14. Rubacha A, Sumner W III, Richter L, and Beckingham K (1984). Conserved 5' flank homologies in dipteran 5S RNA genes that would function on 'A' form DNA. Nucl. Acids. Res. 12:8193.
15. Morton DG, and Sprague KU (1984). In vitro transcription of a silkworm 5S RNA gene requires an upstream signal. Proc. Natl. Acad. Sci. U.S.A. 81:5519.
16. Krumlauf R, and Marzluf GA (1980). Genome organization and characterization of the repetitive and inverted repeat DNA sequences in Neurospora crassa. J. Biol. Chem 255:1138.
17. Perkins DD, and Barry EG (1977). The cytogenetics of Neurospora. Adv. Genet. 19:133.

18. Metzenberg RL, Stevens JN, Selker EU, and Morzycka-Wroblewska E (1985). Identification and chromosomal distribution of 5S rRNA genes in Neurospora crassa. Proc. Nat. Acad. Sci. U.S.A. 82:2067.

Molecular Genetics of Filamentous Fungi, pages 309–319

RAPID EVOLUTIONARY DECAY OF A NOVEL PAIR OF 5S RNA GENES[1]

Eric U. Selker[2], Judith N. Stevens
and Robert L. Metzenberg

Department of Physiological Chemistrv
University of Wisconsin
Madison, Wisconsin 53706

ABSTRACT: In contrast to the case in most organisms, 5S RNA genes are generally dispersed in Neurospora crassa. Of 23 cloned EcoRl fragments which hybridized to 5S RNA, only one contains two 5S genes. We designate these tightly linked 5S genes zeta (ζ) and eta (η). ζ and η are found in tandem directly repeated 794 base pair elements which differ from each other by numerous C-T and G-A differences. We propose that numerous transition mutations occurred in the ζ-η region as a result of deamination of methylated cytosine (C) residues. This region is extensively methylated, whereas several other Neurospora 5S regions examined are not. Some Neurospora strains have one instead of two 5S genes at the ζ-η locus and in these the homologous DNA is not methylated. We suggest that tandem duplications may be targets for methylation in Neurospora.

INTRODUCTION

Why should members of a repeated gene family be tandemly arranged in one organism and dispersed in another? The genes

[1]This work was supported by NIH grant GM-08995 and by a fellowship from the Helen Hay Whitney Foundation to E.U.S.
[2]Present Address: Department of Biology and Institute of Molecular Biology, University of Oregon, Eugene, Oregon 97403

for 5S RNA, the smallest of the structural RNAs found in ribosomes, represent such a case. In most eukaryotes they are found clustered in arrays of hundreds or thousands of tandemly arranged identical or nearly identical repeated units. However, in Neurospora crassa and several other filamentous fungi, the 5S genes are dispersed (1). In fact, in Neurospora we know that 5S genes are broadly scattered on at least six of the seven chromosomes (2). A tandemly arranged gene family can amplify or deamplify by unequal crossing over, and this process tends to homogenize the gene family as variants are occasionally lost. Other processes of "concerted evolution" not limited to tandemly repeated genes may act to keep gene families homogeneous in dispersed gene families (3). Nevertheless, at least in the case of the dispersed 5S genes of Neurospora, heterogeneity is observed. Most of the Neurospora 5S genes have identical 5S RNA structural regions of the type we call alpha (α). Other 5S gene types having 5S RNA regions differing in sequence by 10% or more are also present and are active (1,4). Furthermore, unlike the situation with tandemly arranged 5S genes, the various Neurospora 5S genes, even of a single type, are found embedded in unique flanking sequences. The dispersed arrangement of repeated genes apparently permits sequence divergence. This may allow members of the family to acquire new functions and/or to come under different regulation. Whatever the consequences may be of the dispersed arrangement of the Neurospora 5S genes, the cause of the initial dispersal and the reasons that the genes have remained dispersed may be totally unrelated. For example, it is conceivable that new dispersed 5S genes are continually created by insertion into the genome of reverse transcripts of 5S RNA. Assuming that natural selection acts to preserve an optimal number of 5S RNA genes, we would expect tandem arrays (which are inherently unstable in that they can lose members by an unequal crossing over) to contract and finally disappear as new dispersed genes were generated. Results of a recent analysis of an unusual pair of Neurospora 5S genes points to another possible reason why dispersed genes may be favored in this organism.

MATERIALS AND METHODS

Southern Hybridization Experiments

DNA was isolated from Neurospora (74-OR8-1a) as previously described (5). For the study of the effect of 5-azacytidine, conidia were germinated at a concentration of approximately 5 x 10^6/ml in Vogel's medium (6) containing 2% sucrose and the indicated concentration of 5-azacytidine. Cultures were harvested after 18 hours of growth at 30°C with shaking (300 RPM). One μg samples of DNA were digested with BamH1 (10 u) with or without EcoR1 (10 u) for three hours and the resulting digests were fractionated on 1% agarose gels. Samples (approximately 0.1 ng) of the appropriate digested plasmid DNAs were fractionated on the same gel as controls. The DNA was transferred to membranes by blotting and hybridized with RNA or DNA probes using standard techniques.

Computer Analyses

Programs of the University of Wisconsin Genetics Computer Group were employed. In the sequence comparison by the graphic matrix technique (7), the zeta-eta sequence (8) was compared with itself 10 nucleotides at a time in every possible register. For the plots of nucleotide composition, the frequency of A+G or A+T residues in successive 50 nucleotide segments were calculated. The 50 nucleotide "window" was moved along the sequence in 3 nucleotide increments.

RESULTS AND DISCUSSION

Zeta and eta - a novel tandem pair of Neurospora 5S genes

We have characterized 23 distinct EcoR1-generated cloned fragments of Neurospora DNA which hybridize to 5S RNA. All but one include single 5S RNA genes or pseudogenes. The exceptional fragment contains a pair of 5S regions which we designate zeta (ζ) and eta (η) (8). ζ and η lie on Linkage Group I which is also known to contain α 5S RNA genes and a delta (δ) 5S RNA gene (2). Complete sequence analysis of the ζ-η region revealed that these genes are oriented in the

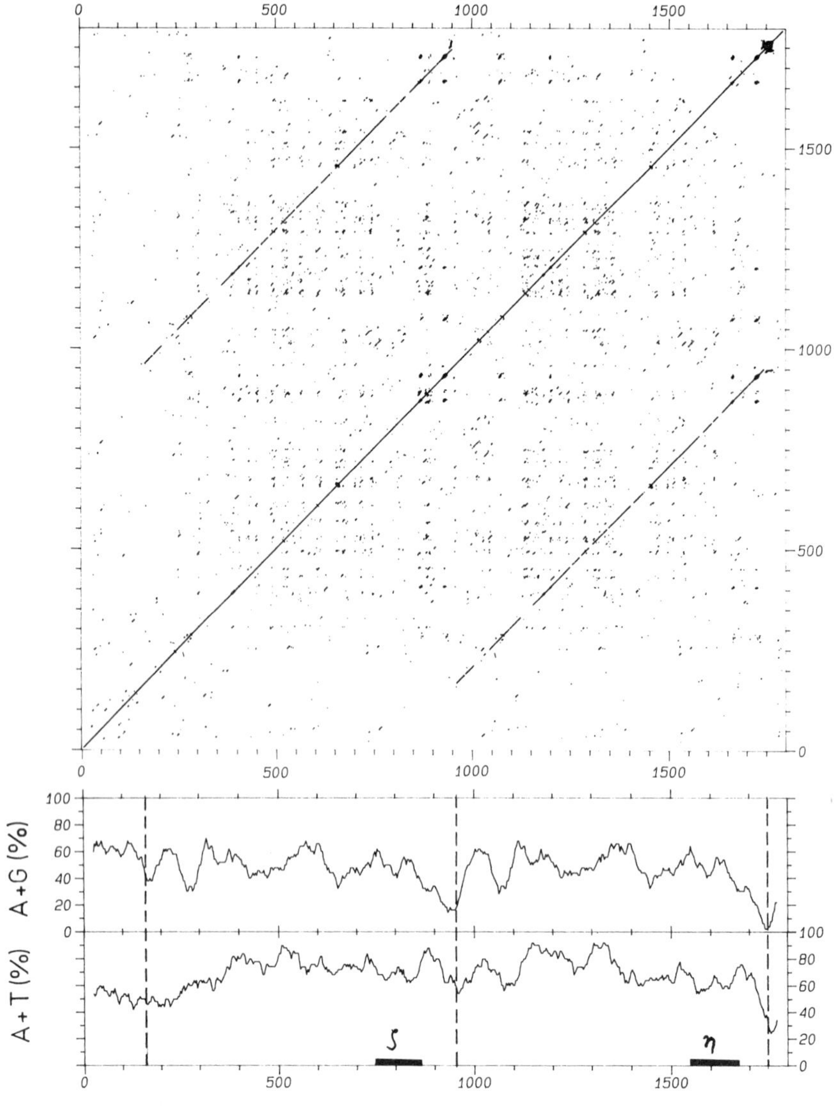

Figure 1a (top). Self comparison of the ζ-η region using the graphic matrix technique. Dots were printed in the matrix at positions where at least 8 out of 10 nucleotides match. Repeated regions appear as lines parallel to the central diagonal line resulting from each base matching itself. b (bottom). Nucleotide composition of the ζ-η region. The ζ and η 5S regions are indicated (heavy lines).

same direction and that both their coding and flanking sequences are similar (8). The ζ and η 5S regions are approximately 16% different in sequence and their flanking sequences are about 14% different. The homology extends from about 600 nucleotides on the 5' side of the 5S genes to almost 100 nucleotides on the 3' side and includes all of the DNA in between ζ and η. This is illustrated by the graphic matrix self comparison (Figure 1a). Thus ζ and η are in tandemly repeated elements, reminiscent of 5S RNA genes in other organisms. Exactly 794 basepairs separate homologous positions in the ζ and η repeats. Unlike the situation in other organisms, however, the repeated elements have diverged. Remarkably, all of the 113 differences between the ζ and η repeats are due to transition mutations. Thus the positions of purines and pyrimidines in the duplicated elements are identical, as illustrated by the plot of purine composition along the sequence (Figure 1b). Also of interest is the fact that the duplicated DNA is unusually A+T rich. It is 70%

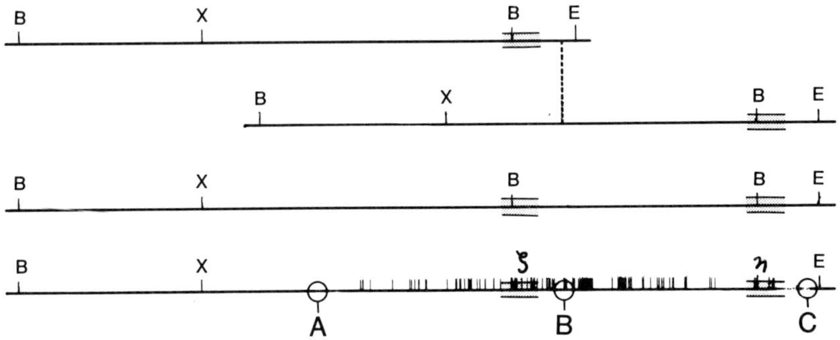

Figure 2. Hypothetical generation of the present day ζ and η regions by an unequal crossover followed by the accumulation of transition mutations. The positions of the 5S regions (stippled areas) and restriction sites (X = Xhol, B= BamH1, E = EcoR1) are shown to scale based on the sequence of the duplicated region (8). The fine vertical lines represent transition mutations which gave rise to the sequence differences observed between the ζ and η repeated elements (8). The circled segments span the boundaries of the duplication and the duplication junction. The structures of these regions are shown in Figure. 3.

A+T compared with 47% A+T in the surrounding DNA and in the genome overall (Figure 1b).

Comparison of the nucleotide sequences of the ζ and η 5S regions with other Neurospora 5S RNA and 5S DNA sequences suggests that transition mutations occurred at numerous positions besides those where the present day ζ and η repeats differ. In addition, it appears that the transitions occurred in the direction G→A (not A→G) and C→T (not T→C). This conclusion is consistent with the A+T rich nature of the ζ-η region. This region most likely evolved by a tandem duplication followed by the occurrence of numerous polarized transition mutations as diagrammed in Figure 2. The fact that there are no transversion differences between the ζ and η repeat units suggests that the duplication event was relatively recent. The regions which must have aligned if the duplication resulted from an unequal crossover event can be identified by inspection of the duplication junction. Figure 3 illustrates that these regions had little homology. It may be significant that one of the partners (C in Figure 3) in the presumptive DNA breakage and reunion event apparently had a strong purine/pyrimidine strand asymmetry. On the strand shown, 20 consecutive C residues immediately follow the recombination position.

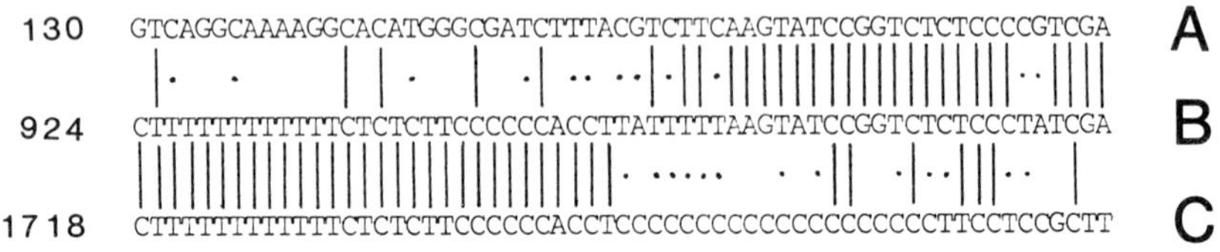

Figure 3. Regions involved in breakage-joining event which generated the ζ-η duplication. Ancestral sequences similar to the present-day left (A) and right (C) boundary regions of the duplication were joined by a crossover event yielding a novel junction similar to the present-day junction sequence (B). Vertical lines indicate positions of identity between the pairs of sequences and dots mark other positions where both members of the pair are either purines or pyrimidines.

Molecular Basis of the Numerous Transition Mutations

In *Escherichia coli* it is known that $\frac{C}{G} \rightarrow \frac{T}{A}$ mutations occur by deamination of methylated C residues (9). Spontaneous deamination of cytosine is frequent in DNA (10) and converts CG pairs into UG pairs. Normally the U residues are recognized and excised by uracil-DNA glycosylase and the base pair is restored by repair or replication. Deamination of 5mC yields 5mU (T) which is not removed by uracil-DNA glycosylase. As a result of mismatch repair or replication this TG pair can become a TA pair thus completing the $\frac{mC}{G} \rightarrow \frac{T}{A}$ conversion.

If C methylation was responsible for the numerous transition mutations observed in the ζ-η region, 5mC must not have been limited to positions 1 or 2 nucleotides preceding G residues, as appears to be the case in other eukaryotes examined. It is known that about 2% of the C residues in Neurospora DNA are methylated, but their distribution is not known (D. Swinton and S. Hattman, personal communication). On the assumption that the process resulting in the transition mutations in the ζ-η region may be still in progress, we examined remaining C residues in this region for methylation. To do this, we assessed the sensitivity of restriction sites to cleavage by restriction enzymes which fail to cut if C residues in their recognition sequences are methylated. The enzymes *BamH1*, *Sau3A*, *Alu1*, *HpaII*, and *Msp1*, were employed. Protection against cleavage was observed. Furthermore, the drug 5-azacytidine, which has been shown to prevent C methylation in many systems, effectively prevented the inhibition of cleavage. An example is shown in Figure 4 using the enzyme *BamH1* which has sites in both ζ and η (see restriction map in Figure 2). DNA from tissue grown under normal conditions could not be efficiently cut at the *BamH1* sites in the ζ and η regions but could be cut at the *BamH1* site upstream of the duplicated DNA. In contrast, DNA from tissue grown in the presence of 5-azacytidine could be cleaved at the *BamH1* sites in ζ and η. We conclude that C residues in the ζ-η region are methylated, consistent with the hypothesis that the numerous transition mutations resulted from deamination of 5mC residues.

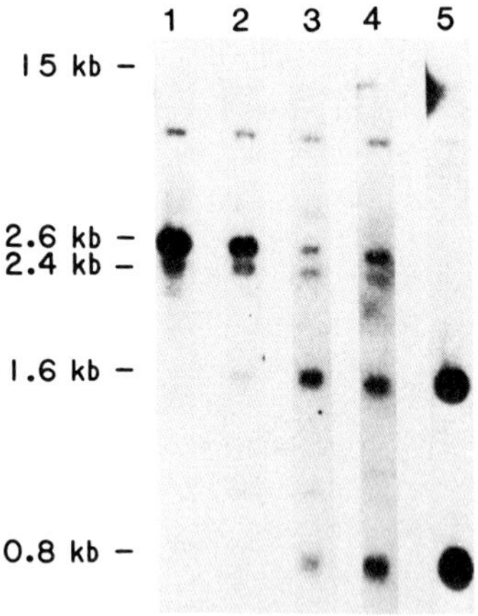

Figure 4. Methylation at BamH1 sites in the ζ-η region. DNA from Neurospora grown under normal conditions (lane 1) or in the presence of 3 μM (lane 2), 12 μM (lane 3) or 48 μM 5-azacytidine (lane 4) was digested with EcoR1 and BamH1, fractionated by electrophoresis, blotted to a membrane and hybridized with a radioactive probe specific for the ζ-η region. Lane 5 shows hybridization to BamH1 digested pJS33, a bacterial plasmid containing the ζ-η region. The band at 0.8 kb represents a DNA fragment generated by digestion of the BamH1 sites in both ζ and η. The band at 1.6 kb represents a fragment extending from the BamH1 site in ζ, upstream to the next BamH1 site. The bands at 2.4 kb and 2.6 kb represent fragments extending from this upstream BamH1 site to the BamH1 site in η or the EcoR1 site 0.2 kb further, respectively (see Figure 2).

Is Methylation a General Feature of Neurospora 5S genes?

To answer this question, we assessed the state of methylation in six standard Neurospora 5S RNA genes (five α genes and one β gene), one possible pseudogene (δ) and one clear-cut pseudogene (ψ4). The results are presented in Figure 5. No inhibition of cleavage was observed, unlike the case in the ζ and η 5S regions.

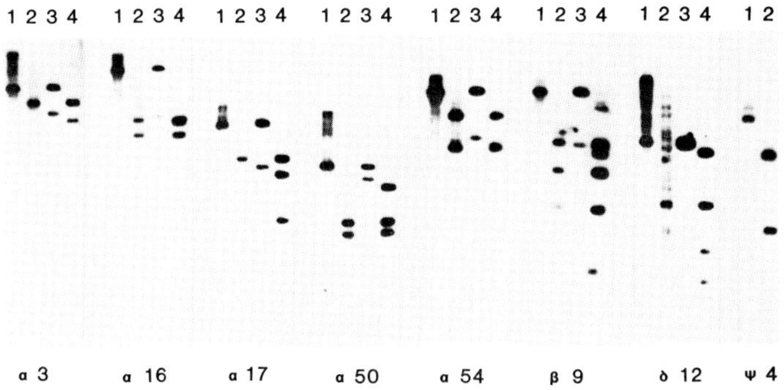

Figure 5. Assessment of methylation at BamH1 sites in eight Neurospora 5S RNA regions. The 5S DNA clones used were generated with EcoR1 except for ψ4 which is a Pst1 clone. One μg of genomic DNA (lanes 1 and 2) or 0.2 ng of cloned DNA of the indicated 5S regions (lanes 3 and 4) were digested with EcoR1 (lanes 1 and 3) or EcoR1 and BamH1 (lanes 2 and 4) except in the case of ψ4 in which case Pst1 (lane 1) and Pst1 and BamH1 (lane 2) were employed. The DNAs were fractioned by electrophoresis, transferred to nitrocellulose membranes and probed with the indicated 5S RNA regions (2). In every case the fragments detected in the BamH1 + EcoR1 (or BamH1 + Pst1) double digests of genomic and cloned DNA match, indicating that the BamH1 sites are not methylated. Additional bands in the genomic DNA digests are due to cross hybridization with other 5S genes; the extra bands in the lanes having the cloned DNA is due to hybridization with the vector (pBR322). The Pst1 fragment detected with ψ4 is also of the same size as the cloned fragment (not shown).

CONCLUSIONS

In contrast to the situation in most eukaryotes, the 5S genes of Neurospora are dispersed. The tandem pair ζ and η are the only known exception and these 5S regions are heavily methylated, have undergone numerous transition mutations, and are probably inactive. Some Neurospora strains have one instead of two 5S genes at the ζ-η locus and in these, the homologous DNA is not methylated (unpublished). In a cross of a strain having the methylated, duplicated ζ-η region and a strain having unique, unmethylated DNA in the homologous region, progeny having the duplication showed methylation at the ζ-η locus whereas progeny lacking the duplication did not. This observation together with the fact that the duplicated and methylated regions seem to coincide, tempts us to speculate that the methylation signal may be the duplication itself. Tandem duplications might be recognized in the genome by the occasional occurrence of single stranded loops which would form whenever the two strands of the tandem repeats underwent unequal pairing as diagrammed in Figure 6. In tandemly repeated DNA, the propensity for mispairing might be a function of the nature or size of the repeat units or of the number of units in the array. Small gene families such as the Neurospora 5S gene family or tRNA gene families may be most functional and/or best preserved in dispersed rather than tandemly repeated arrangements.

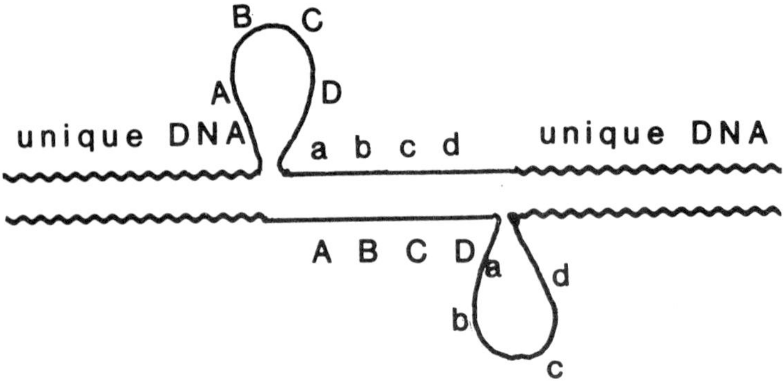

Figure 6. Single stranded loops formed by unequal pairing of DNA strands in hypothetical tandem repeat.

ACKNOWLEDGMENTS

We thank Barbara Mann and Jeanne Selker for critical reading of the manuscript. We also thank Nobuyo Maeda and Michael Slater for introducing us to the local computer system.

REFERENCES

1. Selker EU, Yanofsky C, Driftmier K, Metzenberg RL, Alzner-DeWeerd B, and RajBhandary UL (1981). Dispersed 5S RNA genes in N. crassa: structure, expression and evolution. Cell 24, 819-828.
2. Metzenberg RL, Stevens JN, Selker EU, Morzycka-Wroblewska E, (1985). Identification and chromosomal distribution of 5S RNA genes in Neurospora crassa. Proc. Natl. Acad. Sci. USA, 82,2067-2071.
3. Morzycka-Wroblewska E, Selker EU, Stevens JN, and Metzenberg RL, (1985). Concerted evolution of dispersed Neurospora crassa 5S RNA genes: pattern of sequence conservation between allelic and nonallelic genes. Mol. Cell. Biol. 5, 46-51.
4. Selker EU, Stevens JN, and Metzenberg RL, (1985). Heterogeneous 5S RNA in fungal ribosomes, 227, 1340-1343.
5. Stevens JN, and Metzenberg RL, (1982). Preparing Neurospora DNA: some improvements, Neurospora Newsletter, 29, 27-28.
6. Davis RH, and DeSerres FJ, (1970). Genetic and microbial research techniques for Neurospora crassa, Meth. Enzymol., 17A, 47-143.
7. Maizel Jr., JV and Lenk RP, (1981). Enhanced graphic matrix analysis of nucleic acid and protein sequences, Proc. Nat. Acad. Sci USA, 78, 7665-7669.
8. Selker EU, and Stevens JN, manuscript submitted.
9. Duncan BK, and Miller JH, (1980). Mutagenic deamination of cytosine residues in DNA, Nature, 287, 560-561.
10. Lindahl T, and Nyberg B, (1974). Heat induced deamination of cytosine residues in deoxyribonucleic acid, Biochem., 13, 3405-3410.

Molecular Genetics of Filamentous Fungi, pages 321–332

DNA METHYLATION IN NEUROSPORA: CHROMATOGRAPHIC AND ISOSCHIZOMER EVIDENCE FOR CHANGES DURING DEVELOPMENT[1]

Peter J. Russell*, Karin D. Rodland*[2], Jim E. Cutler*[3], Eliot M. Rachlin+, and James A. McCloskey+

Biology Department, Reed College, Portland, Oregon 97202* and Department of Medical Chemistry, College of Pharmacy, The University of Utah, Salt Lake City, Utah 84112+

ABSTRACT: 5-methylcytosine was detected in the genomic DNA of the fungus *Neurospora crassa* by the use of stable isotope dilution gas chromatography - mass spectrometry (GC/MS). The 5-methylcytosine content of genomic DNA was 0.34 mol% in conidia (1.28% of Cs methylated) and 0.43 mol% in stationary-phase mycelia (1.62% of Cs methylated). Isoschizomer analysis of the ribosomal RNA genes (the rDNA) using the methylation-sensitive restriction enzyme HpaII and the methylation-insensitive enzyme MspI (both of which recognize the sequence CCGG) confirmed the presence of 5-methylcytosine in both conidial and mycelial rDNA. The methylated cytosines are not randomly distributed throughout all the CCGG sequences within the rDNA repeat unit, but are clustered specifically within the nontranscribed spacer region of the repeat unit. At least one of these sites appears to be differentially methylated in conidia and mycelia.

[1]This work was supported by the following grants: NIH GM 32922, GM22488, GM21584, CA09038, and American Cancer Society NP-441.

[2]Present address: Department of Cell Biology and Anat--omy, Oregon Health Sciences University, Portland, OR 97201.

[3]Present address: Department of Microbiology, Montana State University, Bozeman, Montana 59717.

INTRODUCTION

The role of DNA methylation in the regulation of eukaryotic gene expression is not yet clear. The 'higher' eukaryotes, particularly vertebrate animals, contain appreciable levels of 5-methylcytosine (5-mC), primarily in the sequence CpG (1-3). In these organisms many workers have observed a strong negative correlation between 5-mC levels and gene expression (4-14), and treatment with 5-azacytidine, a potent inhibitor of cytosine methylation, has been associated with increases in gene expression (4,8,15-19). However, in a number of systems studied there has been no consistent correlation between the level of gene expression and the relative 5-mC content (8,20-24). For a recent review of the evidence on both sides, see Cooper (4), Doerfler (25) and Riggs and Jones (8).

Efforts to elucidate a general role for changes in DNA methylation have been hampered by the difficulty in detecting extremely low levels of 5-mC. Invertebrates generally have much lower levels of 5-mC than vertebrates, and Drosophila has no detectable 5-mC (25,26). Relatively few studies on the presence of 5-mC in fungi have been reported. Yeast, for example, is reported to contain a relatively low amount of 5-mC (27). One recent systematic study on the incidence and patterns of cytosine methylation in fungi using restriction and nearest-neighbor analysis of DNAs isolated from undifferentiated cells indicates that fungi are a heterogeneous group with respect to cytosine methylation in that the amount of 5-mC ranged from 0 to approximately 0.5 per cent of the total cytosines depending upon the species being examined (28). The rDNA of Schizophyllum commune has also been reported to contain 5-mC (29). However, the levels of 5-mC reported in fungi approach the resolution limits of the typical analytical procedures used (restriction analysis and HPLC) thereby raising questions about the accuracy of the figures obtained. The use of stable isotope gas chromatography - mass spectrometry (GC/MS) has lowered the detection limit of 5-mC from approximately 3% with HPLC to less than 0.2% of the total genome (30). We applied this technique to genomic DNA isolated from both conidia and stationary-phase mycelia from the filamentous fungus Neurospora crassa to determine the amount of 5-mC in this organism and to investigate whether different developmental stages exhibited different 5-mC contents. Since 5-mC is typically found in CpG sequences in eukaryotes, CCGG sequences (recognition sites for the restriction enzymes MspI and HpaII) within the repeat unit containing the major ribosomal RNA genes (the rDNA) were located by restriction enzyme mapping. Isoschizomer analysis with the enzymes MspI and HpaII was then used to determine the regions of DNA methylation within the rDNA of conidia and mycelia.

RESULTS

Analysis of 5-Methylcytosine Content

DNA was isolated from dormant conidia and stationary-phase mycelia of wild-type _Neurospora crassa_ strain 74-OR23-1_A_ as previously described (31). The DNA was lyophilized and 1 μg was hydrolyzed by heating in 95-97% formic acid for 60 min at 180 C. The dried hydrolysates were converted to their _tert_-butyldimethylsilyl derivatives which were analyzed for 5-methylcytosine content by stable isotope dilution gas chromatography-mass spectrometry using the method described by Crain and McCloskey (30). 5-mC was quantified relative to thymine (T), and the percentage of 5-mC in the total genome was calculated from the known abundance of thymidine in the DNA of _Neurospora crassa_ (32). Figure 1 represents the results of this analysis. The 5-mC level of conidia was 0.34 mol%, while that of stationary-phase mycelia was 0.43 mol%.

Identification of CCGG Sequences within the rDNA Repeat Unit

The mapping strategy used to identify the location of CCGG sequences within the rDNA repeat unit involved partial MspI digestions of end-labeled DNA fragments representing the entire rDNA repeat unit. Figure 2 is a map of the rDNA repeat unit in _Neurospora_ (33), with the three EcoRI fragments used in these experiments clearly indicated: Fragments I and II contain mostly 17S and 25S rRNA coding sequences, respectively; and fragment III contains mostly nontranscribed spacer sequences. The EcoRI fragments were isolated from the pBR322-based rDNA clones pMF2 and pRW529 (Figure 2); pMF2 contains a PstI insert within which are EcoRI fragments I and II, and pRW529 is an EcoRI insert of fragment III. Each EcoRI fragment was purified from the appropriate clone and end-labeled with ^{32}P using a 'fill-in'

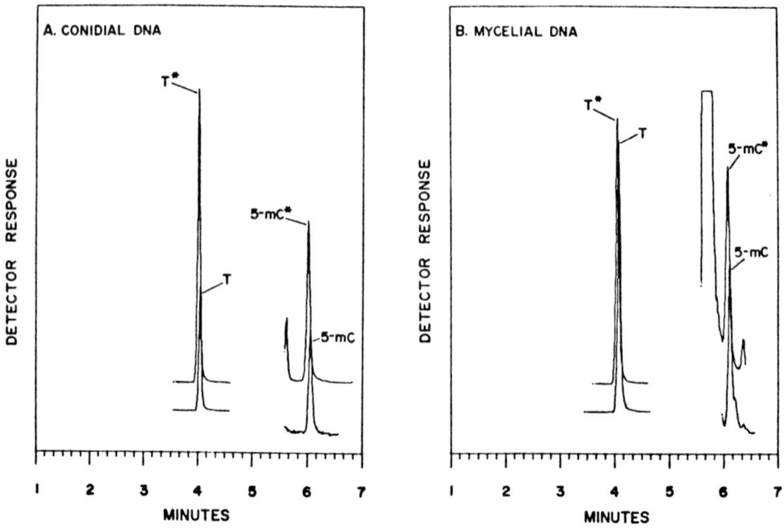

FIGURE 1. Elution profile of thymine (T) and 5-methylcytosine (5-mC) in DNA from Neurospora conidia (A) and mycelia (B) analyzed by stable isotope dilution gas chromatography-mass spectroscopy. T* and 5-mC* denote the labeled internal standards $[^2H_3]$ thymine and $[^{15}N,^2H_3]$-5-methylcytosine, respectively. 5-mC was quantified relative to the thymine peak.

method (34). The end-labeled fragments were purified by electrophoresis in 1% agarose gels, electro-eluted onto DEAE-cellulose, and then cut with an appropriate restriction enzyme to separate the labeled ends. The resulting six fragments, representing the 3' and 5' ends of the EcoRI fragments I, II, and III, were separated by electrophoresis in 1% agarose gels and then purified from the gels by DEAE-cellulose chromatography as before.

Each of the six fragments was subjected to partial digestion with the restriction enzyme MspI, using 0.8 units of MspI per microgram of DNA and incubating at 37 C for 30 min. The resulting fragments were separated by electrophoresis in either 1% agarose or 10% polyacrylamide to obtain good resolution of both large and small fragments. A map of the CCGG sequences within the rDNA was constructed from analysis of the sizes of the fragments produced.

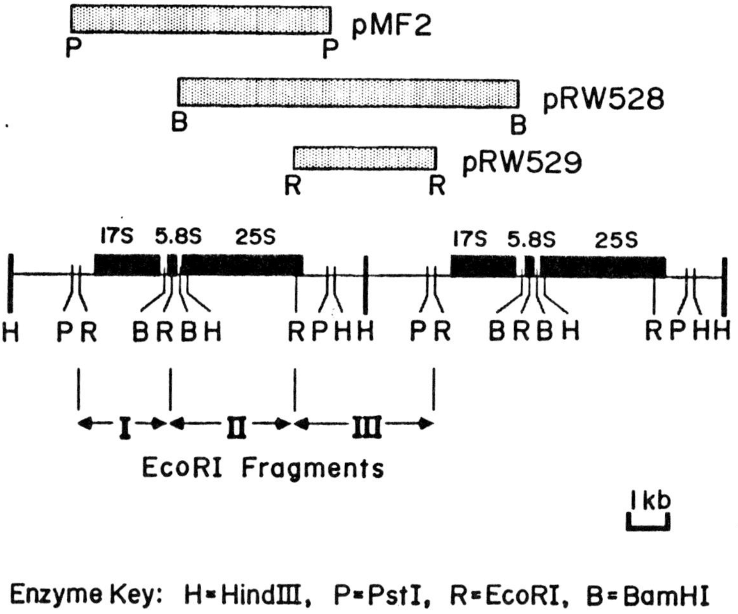

FIGURE 2. Restriction map of the rDNA repeat unit in Neurospora crassa, strain 74-OR23-1A, showing the extents of the three fragments I, II, and III that are produced by digestion with EcoRI. These EcoRI fragments are contained within the recombinant, pBR322-based clones pMF2 and pRW529.

Figure 3 is a restriction digest map of the entire rDNA repeat unit of Neurospora crassa, including the location of MspI sites (CCGG sequences).

Isoschizomer Analysis of rDNA Methylation in Conidia and Mycelia.

The restriction enzymes MspI and HpaII are isoschizomers since they both recognize and cleave the sequence 5'-CCGG-3' in DNA. If the internal C is methylated, however, MspI will cleave the DNA but HpaII will not. The reverse is the case if the external C is methylated. Isoschizomer analysis with the restriction enzymes MspI and HpaII was used to determine whether any of the CCGG sequences identified in cloned rDNA was methylated in genomic DNA, and also to identify any differences in the methylation patterns of conidial and mycelial DNA. DNA was isolated from conidia and exponential-phase mycelia as previously described (31), and subjected to complete digestion with either MspI or HpaII, using a three-fold excess of enzyme and a 4-5 h incubation at 37 C. The resulting fragments were separated by electrophoresis in 1% agarose gels and transferred to hybridization membranes (GeneScreen Plus), then hybridized with ^{32}P-labeled probes as previously described (35,36). The three probes used were the EcoRI fragments I, II, and III described earlier. Each membrane was probed sequentially with each of the three probes; the membranes were stripped between each hybridization. The bands visualized by autoradiography of the membranes are shown in Figure 4. Note that all of the fragments detected with EcoRI fragments I and II are smaller than 1.5 kilobase pairs (kbp) in size. This correlates well with the estimated fragment sizes for this region based on the restriction map. However, probing with EcoRI fragment III results in many bands, including a number of large bands ranging from 1.59 to 4.51 kbp. These bands are larger than predicted by the restriction map, indicating the presence of significant methylation.

DISCUSSION

The results from the GC/MS analysis clearly indicate that *Neurospora crassa* contains significant amounts of 5-methylcytosine. These results contradict those of Cedar et al (37), based on digestion of total genomic DNA with HpaII and MspI, but agree with the HPLC results of J. Silver (personal communication) which indicated 1-3% of Cs methylated (0.27 - 0.81 mol%). Thus, while the isoschizomer method is adequate for detecting 5-mC in the heavily methylated DNA of vertebrates, it is not sufficiently sensitive to detect the low levels of methylation present in *Neurospora* and other fungi. Detection of 5-mC in fungi is very difficult, as the amount present is very close to the sensitivity limit of HPLC. The application of GC/MS to analysis of DNA from other fungi should provide more accurate quantitative data on the presence of modified nucleotides and should contribute to demonstrating the generality of DNA methylation among eucaryotes.

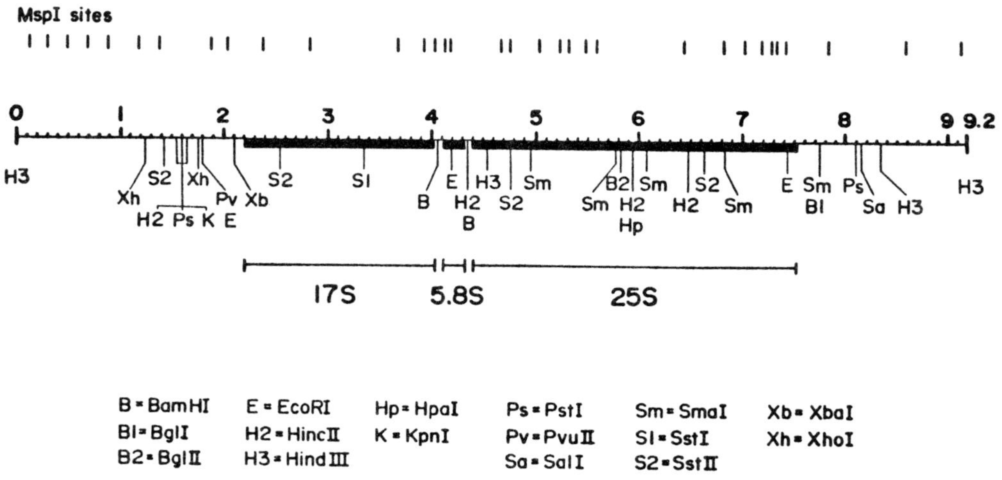

FIGURE 3. A restriction enzyme map of the rDNA repeat unit of *Neurospora crassa* strain 74-OR23-1*A* indicating the presence of MspI sites.

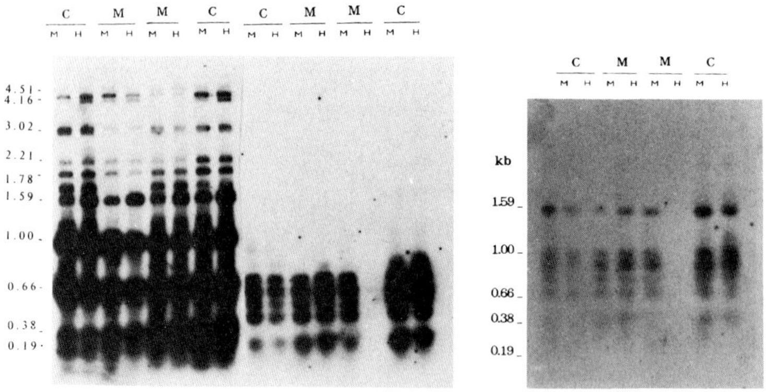

FIGURE 4. Autoradiographic visualization of rDNA fragments produced by digestion of conidial (C) and exponential-phase mycelial (M) genomic DNA with either MspI (M) or HpaII (H) followed by hybridization with EcoRI fragments I, II, and III.

The GC/MS method is sufficiently sensitive to detect a difference of 0.09 mol% in the 5-mC levels of conidial and stationary-phase mycelial DNA. This finding suggests that changes in DNA methylation may be correlated with developmental events in Neurospora. The finding of increased 5-mC levels in stationary-phase mycelia might seem to suggest that increased gene activity is correlated with decreased methylation in Neurospora. However, this method reveals little about changes in DNA methylation at specific loci, and the effect on expression of specific genes. Recent evidence suggests that the significance of cytosine methylation is highly site specific, with the most sensitive areas lying within a region 500-700 bp 5' from the site of transcription initiation (38). Structural analysis, either by the use of methylation-sensitive and insensitive isoschizomers or by Maxam-Gilbert sequencing reactions, provides more specific information about the localization of 5-mC with possible regulatory functions.

The restriction mapping analysis of the Neurospora rDNA repeat unit revealed the presence of some CCGG sequences in the functionally interesting area immediately 5' to the 17S rRNA coding sequence. Southern hybridization analysis of fragments produced by digestion with MspI and HpaII shows that several large fragments are produced in this region, indicating the presence of significant methylation on both cytosines in CCGG recognition sequence. (Recall that MspI will not cut if the external C is methylated, while HpaII will not cut if the internal C is methylated.) Of particular interest is the 1.59 kb fragment which hybridizes to both EcoRI fragments I and III, indicating that it is immediately 5' to the 17S rRNA coding sequence. This fragment gives a very intense hybridization sample in DNA samples cut with HpaII, and a much fainter signal in samples cut with MspI. Since the nucleolus organizer region of Neurospora consists of 185-200 copies of the rDNA repeat unit arranged in a tandem array (36,39, 40), this pattern might have been generated if a specific CCGG sequence was methylated in some, but not all, repeat units. This pattern was observed in both conidial and mycelial samples, and does appear to be developmentally correlated. Comparisons of the overall pattern obtained from conidia and mycelia does reveal the presence of a 4.16 kbp fragment which is always evident in conidia and either missing or very faint in mycelia. In addition, the 3.02 kbp fragment is consistently more intense in conidial rDNA than in mycelial rDNA. These results imply that one or more specific CCGG sites may be methylated or demethylated in correlation with a particular developmental stage.

In sum, the results of this study clearly indicate that methylation of cytosine residues does occur in the filamentous fungus Neurospora crassa. Substantial differences in the total 5-mC of conidia and stationary-phase mycelia are demonstrated by the chromatographic analysis. Detailed study of the CCGG sequences within the rDNA repeat unit of Neurospora strongly suggests the certain CCGG sequences may experience variations in methylation which may be functionally related to gene transcription.

ACKNOWLEDGMENTS

The authors are grateful to Sheryl Wagner and Joanne Pratt for expert technical assistance, and to Peer B. Jacobson for performing some preliminary experiments.

REFERENCES

1. Doerfler W (1981). DNA methylation - a regulatory signal in eukaryotic gene expression. J Gen Virol 57:1.
2. Doskocil J, Sorm F (1962). Distribution of 5-methylcytosine in pyrimidine sequences of deoxyribonucleic acids. Biochim Biophys Acta 55:953.
3. Grippo P, Iaccarino M, Parisi E, Scarano E (1968). Methylation of DNA in developing sea urchin embryos. J Mol Biol 36:195.
4. Cooper DN (1983). Eukaryotic DNA methylation. Hum Genet 64:315.
5. Doerfler W (1984). DNA methylation and its functional significance: studies on the adenovirus system. Curr Top Microbiol Immunol 108:79.
6. Mandel JL, Chambon P (1979). DNA methylation: organ specific variations in the methylation pattern within and around ovalbumin and other chicken genes. Nucleic Acids Res 7:2081.
7. McGhee G, Ginder GD (1979). Specific DNA methylation sites in the vicinity of the chicken beta-globin genes. Nature 280:419.
8. Riggs AD, Jones PA (1983). 5-Methylcytosine, gene regulation, and cancer. Adv Cancer Res 40:1.
9. Rogers J, Wall W (1981). Immunoglobulin heavy chain genes: demethylation accompanies class switching. Proc Natl Acad Sci USA 78:7497.
10. Shen ST, Maniatis T (1980). Tissue-specific DNA methylation in a cluster of rabbit β-like globin genes. Proc Natl Acad Sci USA 77:6634.
11. Van der Ploeg LHT, Flavell RA (1980). DNA methylation in the human gamma-delta-beta globin-locus in erythroid and nonerythroid tissue. Cell 19:947.
12. Vedel M, Gomez-Garcia M, Sala M, Sala-Trepat JM (1983). Changes in methylation pattern of albumin and α-fetoprotein genes in developing rat liver and neoplasia. Nucleic Acids Res 11:4335.
13. Weintraub H, Larsen A, Groudine M (1981). α-Globin switching during the development of chicken embryos: expression and chromosome structure. Cell 24:333.
14. Wilks AF, Cozens PJ, Mattay IW, Jost JP (1982). Estrogen induces a demethylation at the 5' end region of the chicken vitellogenin gene. Proc Natl Acad Sci USA 79:4252.

15. Compere SJ, Palmiter RD (1981). DNA methylation controls the inducibility of the mouse metallothionein-1 gene in lymphoid cells. Cell 25:233.
16. Taylor SM, Jones PA (1979). Multiple new phenotypes induced in 10T1/2 and 3T3 cells treated with 5-azacytidine. Cell 17:771.
17. Mohandas T, Sparkes RS, Shapiro LJ (1981). Reactivation of an inactive human X chromosome: evidence for inactivation by DNA methylation. Science 211:393.
18. Venolia L, Gartler SM, Wassman ER, Yen P, Mohandas T, Shapiro LJ (1982). Transformation with DNA from 5-azacytidine-reactivated X chromosomes. Proc Natl Acad Sci USA 79:2352.
19. Groudine M, Eisenman R, Weintraub H (1981). Chromatin structure of endogenous retroviral genes and activation by an inhibitor of DNA methylation. Nature 292:311.
20. McKeon C, Ohkubo H, Pastan I, de Crumbrugge B (1982). Unusual methylation pattern of the α2(1) collagen gene. Cell 29:203.
21. Cate RL, Chick W, Gilbert W (1983). Comparison of the methylation patterns of the two rat insulin genes. J Biol Chem 258:6645.
22. Bower DJ, Errington LH, Cooper DN, Morris S, Clayton RM (1983). Chicken lens δ-crystallin gene expression and methylation in several non-lens tissues. Nucleic Acids Res 11:2513.
23. Ott M-O, Sperling L, Cassio D, Levilliers J, Sala-Trepat J, Weiss, MC (1982). Undermethylation at the 5' end of the albumin gene is necessary but not sufficient for albumin production by rat hepatoma cells in culture. Cell 30:825.
24. Folger K, Anderson JN, Hayward MA, Shapiro DJ (1983). Nuclease sensitivity and DNA methylation in estrogen regulation of Xenopus laevis vitellogenin gene expression. J Biol Chem 258:8908.
25. Doerfler W (1983). DNA methylation and gene activity. Ann Rev Biochem 52:93.
26. Urieli-Shoval S, Gruenbaum Y, Sedat J, Razin A (1982). The absence of detectable methylated bases in Drosophila melanogaster DNA. FEBS Letts 146:148.
27. Hattman S, Henny C, Berger L, Pratt H (1978). Comparative study of DNA methylation in three unicellular eucaryotes. J Bacteriol 135:1156.
28. Antequera F, Tamame M, Villaneuva JR, Santos T (1984). DNA methylation in the fungi. J Biol Chem 259:8033.

29. Specht CA, Novotny CP, Ullrich RC (1984). Strain specific differences in ribosomal DNA from the fungus *Schizophyllum commune* rDNA contains 5-methylcytosine. Curr Genet 8:219.
30. Crain PF, McCloskey, JA (1983). Analysis of modified bases in DNA by stable isotope dilution gas chromatography-mass spectometry: 5-methylcytosine. Anal Biochem 132:124.
31. Rodland KD, Russell PJ (1983). Ribosomal genes of *Neurospora crassa*: constancy of gene number in the conidial and mycelial phases and homogeneity in length and restriction enzyme cleave sites within strains. Mol Gen Genet 192:285.
32. Laskin AI and Lechevalier HA, eds (1981). Handbook of Microbiology, Vol III. Boca Raton, FL: CRC Press, p 536.
33. Russell PJ, Wagner SW, Rodland KD, Feinbaum RL, Russell JP, Metzenberg RL (1984). Organization of the ribosomal ribonucleic acid genes in various wild-type strains and wild-collected strains of *Neurospora*. Mol Gen Genet 196:275.
34. Davies RW (1982). DNA sequencing. In Rickwood D, Hames BD (eds): "Gel Electrophoresis of Nucleic Acids: A Practical Approach," Oxford, England: IRL Press, p 117.
35. Southern EM (1975). Detection of specific sequences among DNA fragments separated by gel electrophoresis. J Mol Biol 98:503.
36. Rodland KD, Russell PJ (1983). Segregation of heterogeneous rDNA segments during demagnification of a *Neurospora crassa* strain possessing a double nucleolar organizer. Curr Genet 7:379.
37. Cedar H, Solage A, Glaser G, Razin A (1979). Direct detection of methylated cytosine in DNA by use of the restriction enzyme MspI. Nucleic Acids Res 6:2125.
38. Bird AP (1984). DNA methylation - how important in gene control? Nature 307:503.
39. Krumlauf R, Marzluf GA (1980). Genome organization and characterization of the repetitive and inverted repeat sequences in *Neurospora crassa*. J Biol Chem 255:1138.
40. Rodland KD, Russell PJ (1982). Regulation of ribosomal RNA cistron number in a strain of *Neurospora crassa* with a duplication of the nucleolus organizer region. Biochim Biophys Acta 697:162.

Molecular Genetics of Filamentous Fungi, pages 333–344

DNA METHYLATION IN COPRINUS CINEREUS[1]

Miriam E. Zolan† and Patricia J. Pukkila*

†Division of Biological Sciences, University of Michigan, Ann Arbor, Michigan 48109, and *Department of Biology and Curriculum in Genetics, University of North Carolina at Chapel Hill, Chapel Hill, North Carolina 27514

ABSTRACT Although modified bases occur in the DNAs of most organisms, analysis of the functions of eukaryotic DNA modification has been hampered by the paucity of such modifications in organisms that have been most amenable to genetic manipulation. Unlike most fungi, the basidiomycete Coprinus cinereus has a nuclear genome that is extensively methylated at the nucleotide doublet CpG. Long methylated tracts exist in the Coprinus genome, although there is apparently no sharp division of the DNA into methylated and unmethylated regions. We have begun a genetic characterization of cytosine methylation in Coprinus and have shown that methylated and unmethylated tracts in a single copy, centromere-linked sequence are inherited faithfully through meiosis.

INTRODUCTION

Extensive genetic and biochemical characterization of postreplicative methylation of cytosine and adenine in prokaryotes has shown that DNA methylation is a critical component of restriction-modification systems (1), mismatch repair mechanisms (2), and the expression of certain genes (3). Methylation is not, however, required for

[1]This work was supported by National Science Foundation grant PCM8215694 (to PJP) and National Research Service Award 5 F32 CA07395 (to MEZ).

viability, since *E. coli* mutants lacking all three of its methylation systems survive (4).

In eukaryotes, the principal methylated base is 5 methylcytosine (5meC). Although the function of 5meC is less well understood in eukaryotes than in prokaryotes, cytosine methylation has been postulated to have functional roles in both gene expression (5,6) and development (7). In animals (8) and in fungi (9), the major site of cytosine methylation is at the nucleotide doublet CpG; in plants methylation of the sequence CpXpG is also common (10). The methylation pattern, or distribution of 5meC along a particular DNA sequence, is conveniently studied using restriction endonucleases that contain CpG as part of their recognition sequences (11). In particular, enzyme pairs that cleave at the same sites, but whose activities are differentially affected by methylation, are powerful tools for the detection of 5meC in the genome. For example, HpaII will not cleave if the internal cytosine in its recognition sequence, CCGG, is methylated, whereas MspI, which recognizes the same sequence, will cleave DNA regardless of whether or not the internal cytosine is methylated (12). Therefore, if a sequence is methylated at a given HpaII/MspI recognition site, smaller fragments containing that sequence will be observed after digestion with MspI than after digestion with HpaII. Since eukaryotic DNA methylation occurs predominantly at the nucleotide doublet CpG, HpaII and MspI are a particularly useful enzyme pair for studying the methylation pattern of genomic sequences. It should be noted, however, that this enzyme pair does not detect all the 5meC in a given sequence, since it does not detect methylated CpG doublets in sequences other than its recognition sequence.

The maintenance of specific patterns of 5meC is believed to result from the activity of maintenance methylases, which act to methylate newly synthesized DNA, usually within minutes after DNA synthesis. In fact, it has been shown that mammalian methylases prefer as substrate hemimethylated DNA over unmethylated DNA (13). Sites such as CpG are symmetric, which allows semiconservative copying of methylation patterns. Biochemical studies, which have shown that most if not all new methyl groups are added to newly synthesized DNA (14), also support the model of semiconservative inheritance of DNA methylation.

Most studies of the maintenance of DNA methylation patterns have been performed in mammalian cells, using DNA-mediated cell transformation (15,16,17). In these experiments, DNA methylated *in vitro* to various extents was transformed into mammalian cells, and the pattern of methylation was followed for as many as 100 generations. These studies showed that methylation patterns, once set in a transformed cell, are clonally inherited for as many generations as have been followed, thus providing support for the concept of passive, semiconservative maintenance of methylation patterns in somatic cells.

There are very few good genetic systems for the study of DNA methylation. Neither *Drosophila* (18) nor most fungi (9) show appreciable amounts of methylation in their DNA, although small amounts of methylation, for example at a few specific sites, would not have been seen in the studies that have been done on these organisms. Nevertheless, the relative lack of methylation in lower eukaryotes, as compared to the greater levels observed for higher organisms, has led to the speculation that cytosine methylation may be a regulatory tool used primarily by higher eukaryotes (19).

Coprinus cinereus is a basidiomycete whose nuclear genome, unlike that of most lower eukaryotes, is extensively methylated at the nucleotide doublet CpG. It is also a good system for genetic analysis; tetrads of basidiospores can be easily recovered from gill segments (20), and over 70 loci have so far been assigned to 8 linkage groups (21). We have begun to characterize methylation in *Coprinus* by studying the meiotic inheritance of methylation patterns in a specific sequence of the nuclear genome.

RESULTS AND DISCUSSION

Methylation in *Coprinus cinereus*.

Digestion of total cellular DNA from *Coprinus cinereus* with HpaII or MspI reveals striking differences in the molecular weights of the digested DNAs. Figure 1 shows typical ethidium bromide staining patterns of total *Coprinus* DNA digested with either HpaII (A) or MspI (B). When digested with HpaII, there is a more or less continuous smear of DNA fragments from greater than 25 kb to less than 500 bp (figure 1A). The size distribution of DNA fragments after digestion with MspI is significantly smaller; the largest fragments are less than 3 kb (figure 1B). The prominent, bright bands that are identical after digestion of total DNA with either HpaII or MspI are from the C. cinereus mitochondrial genome (L. Casselton, personal communication).

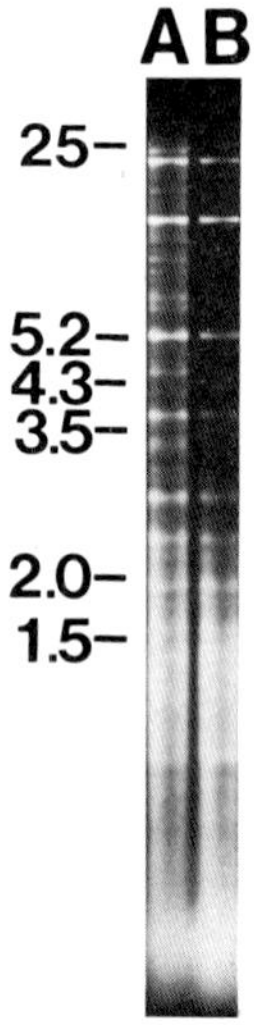

Figure 1. Methylation of Coprinus cinereus nuclear DNA. Total DNA was isolated (25) from one progenitor of the cross Okayama-7 x PJP52. 2 µg were digested with HpaII (A) or MspI (B), the fragments were separated on a 1% agarose gel, and the gel was stained with 0.5 µg/ml ethidium bromide. The prominent, bright bands that are identical in the two lanes are from the C. cinereus mitochondrial genome (L. Casselton, personal communication). The size scale on the left is in kb and was determined using restriction fragment markers generated by double digestion of λ DNA with HindIII and EcoRI.

The differences observed for Coprinus between HpaII- and MspI-digested nuclear DNAs are in contrast to the results found for most other fungi examined (9), for which no differences between HpaII and MspI digests are observed in the ethidium stain - i.e., there is little, or no, apparent methylation of total DNA. Antequera et al. (9) examined cytosine methylation in 20 species of fungi from six different classes. Only two of the species examined, a zygomycete, P. blakesleeanus, and a deuteromycete, S. dimorphosporum, showed any cytosine methylation, as judged by differences between HpaII and MspI digests.

It should be noted that even where no methylation is observed, by the criterion of identical HpaII and MspI digestion patterns, some amount of DNA methylation may exist. This is true for Neurospora, for which no difference is observable in total DNA, but which is methylated at 1-3% of its cytosines, and which shows considerable methylation at certain specific sequences (E. Selker, personal communciation, this volume). In addition, Specht et al. (22) have shown that the ribosomal RNA genes are in fact methylated in the same strain of Schizophyllum commune that Antequera et al. (9) examined and listed as "unmethylated". For Coprinus, although more of the DNA is cut with HpaII than remains high molecular weight, there is apparently no biphasic appearance to the molecular weight distribution. In contrast, the two methylated fungal species studied by Antequera et al. (9) exhibited a distinctly biphasic distribution of DNA fragments on agarose gels. In Coprinus, there is a continuum of lengths of methylated tracts, from less than 3 kb to greater than 25 kb. The extensive and easily assayed methylation in Coprinus should facilitate the recovery of mutants with altered levels of methylated cytosine in their nuclear genomes, enabling direct tests of hypotheses concerning the functions of eukaryotic DNA modification.

Methylation of Locus 16-1.

We have chosen to focus on the inheritance of the methylation pattern of a specific sequence of Coprinus nuclear DNA. A screen of wild geographic isolates of Coprinus with random λ clones of nuclear DNA (23) revealed that one strain, Okayama-7, carries a large insertion (relative to H9, the strain used to make the clones) homologous to clone λ 16. We have found that Okayama-7 is methylated in this region, whereas other strains, including strain PJP52, are not. However, the total amount of methylation in Okayama-7, as judged by ethidium bromide-stained agarose gels, is the same as that of PJP52 (data not shown). Analysis of subclones of λ 16 showed that both the insertion and the methylation are confined to a 6.8 kb SalI fragment, which is termed 16-1 and shown diagrammatically in figure 2. Further analysis showed that sequences homologous to the 2.4 kb SalI-SmaI sub-fragment of 16-1 are not methylated in the Okayama-7 genome. Further, the 0.4 kb SmaI fragment always hybridized to fragments larger than itself, and hybridized to a subset of the bands homologous to the 4.0 kb SalI-SmaI fragment. By this type of analysis it was found that one of the two SmaI sites present in H9 is missing in Okayama-7, and that both are missing from the genome of PJP52. To simplify our analysis, we used the 0.4 kb SmaI fragment, which we subcloned into pUC 9 (24), as a hybridization probe for most of our studies.

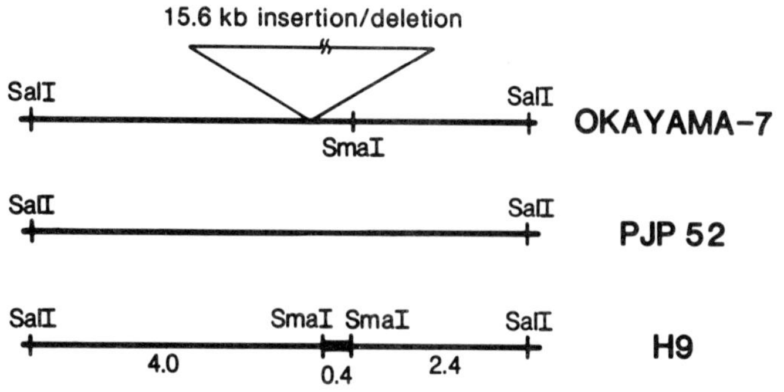

Figure 2. Map of locus 16-1. Only restriction sites relevant to this paper are shown. See ref. (23) for a more detailed restriction map of this locus. The exact position of the insertion/deletion has not been determined.

Hybridization of the 0.4 kb SmaI subclone to a genomic blot revealed that PJP52 is not methylated at locus 16-1 (figure 3). A 1.5 kb band is observed when total DNA from PJP52 is cut with either HpaII (figure 3, Panel A, lane c) or MspI (figure 3, Panel A, lane d). In contrast, the 0.4 kb SmaI clone hybridized to a 2.5 kb HpaII fragment (figure 3, Panel A, lane a), but a 1.0 kb MspI fragment (figure 3, Panel A, lane b) of Okayama-7 DNA. Partial digestion with MspI indicated that the difference between the HpaII and MspI patterns obtained for Okayama-7 represents the methylation of four potential HpaII/MspI sites. Experiments are currently in progress to determine the precise boundaries of the methylated region, and whether the methylation is confined to the insertion or the flanking DNA, or is present in both.

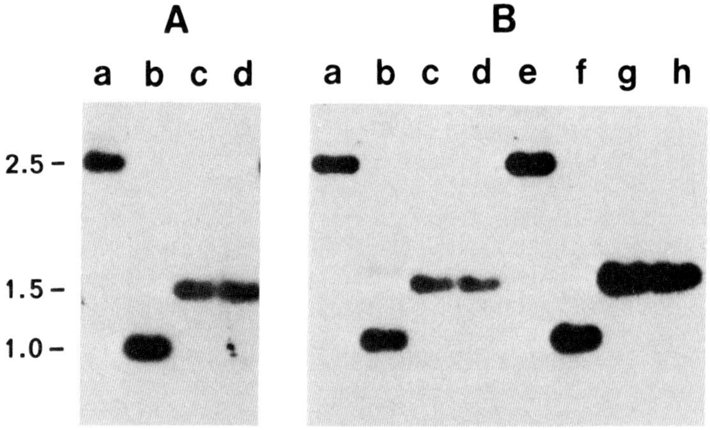

Figure 3. Inheritance of DNA methylation at locus 16-1. DNAs from Okayama-7 (Panel A, lanes a and b), PJP52 (Panel A, lanes c and d), and one progeny tetrad (Panel B) of a cross between Okayama-7 and PJP52 were isolated (25) and digested with HpaII (Panel A, lanes a and c; Panel B, lanes a,c,e, and g) or MspI (Panel A, lanes b and d; Panel B, lanes b,d,f, and h). The fragments were then separated on a 1% agarose gel, transferred to nitocellulose, and hybridized (28) to the 0.4 kb SmaI subclone that is shown in figure 2.

Inheritance of Methylation.

Since Okayama-7 and PJP52 differ both in the methylation state of this region, and at the sequence level, differently sized fragments homologous to clone 16-1 are generated by digestion with either HpaII or MspI (figure 3, Panel A). Therefore, progeny of a cross between these two strains could be analysed both for methylation and for the parental origin of the 16-1 locus. It was possible that, in a cross between Okayama-7 and PJP52, progeny would be found that were methylated at locus 16-1 in the PJP52 chromosome, perhaps due to the activity of a methylase that might segregate independently from locus 16-1, or to activity in _trans_ by a 16-1-linked methylase. Alternately, if methylated tracts are inherited semiconservatively through meiosis in _Coprinus_, progeny that receive the chromosome carrying locus 16-1 from Okayama-7 would be methylated, and those receiving their chromosome from PJP52 would be unmethylated, regardless of methylation at other loci.

Tetrad analysis of the progeny of a cross between Okayama-7 and PJP52 failed to localize locus 16-1 to any linkage groups marked in the cross (Table 1). The markers followed in the cross were trp-1, met-9, the mating type loci A and B, a restriction site polymorphism in the ribosomal RNA genes, and the restriction fragment length polymorphism in locus 16-1. The markers trp-1, met-9, A, and B are all within 10 map units of their centromeres (21), and the ribosomal RNA genes are also centromere linked (25). The low frequency of tetratype tetrads between locus 16-1 and all of these markers indicates that 16-1 is also located close to its centromere.

TABLE 1
TETRAD ANALYSIS OF PROGENY OF OKAYAMA-7 X PJP52

Markers	PD	NPD	T
trpl-16-1	5	6	1
met9-16-1	8	3	1
rDNA-16-1	7	3	2
A-16-1[a]	5	5	1
B-16-1[a]	5	4	2
trpl-met9[b]	4	7	1
trpl-rDNA[b]	4	7	1
met9-rDNA[b]	7	3	2
A-B[a]	10	0	1

[a]mating type data were not obtained for one of the twelve tetrads
[b]data from Cassidy et al. (25)

The restriction patterns of a typical tetrad (figure 3, Panel B) indicate that methylated and unmethylated tracts are inherited faithfully through meiosis in Coprinus cinereus. Two members of this tetrad are methylated, and have received chromosomes containing locus 16-1 from Okayama-7, the methylated parent (figure 3, Panel B, lanes a and b, e and f). Two members of the tetrad are unmethylated, and received chromosomes containing locus 16-1 from PJP52 (figure 3, Panel B, lanes c and d, g and h). In fact, in all twelve of the tetrads examined, two of the progeny were methylated, two were not, and the two that were methylated had received the chromosome carrying locus 16-1 from the methylated parent, Okayama-7. Thus, there was never any methylation of the PJP52 chromosome in this region. This indicates that the inheritance of methylation at this locus is probably the result of faithful replication of the parental methylation pattern, and not of the activity of an unlinked, segregating methylase. Although our data provide support for the idea of semiconservative maintenance of methylation through meiosis, it is still formally possible that methylation is altered in meiotic tissue. For example, it has been found that certain mouse satellite DNAs are hypomethylated in testes and sperm relative to somatic tissue (26). The material analyzed in our experiments was always monokaryotic mycelia, grown to approximately the same degree of confluence. Therefore, it is possible that methylation of this locus is increased or decreased in meiotic tissue. If this is the case, however, then the patterns are reformed precisely after spore germination.

Although methylated and unmethylated tracts are, in general, transmitted faithfully through meiosis, we did see some evidence for extension of tracts in a minority of meiotic progeny that had received the Okayama-7 chromosome carrying locus 16-1 (27). These hypermethylated progeny provided useful material for a cross in which the parents contained identical sequences at locus 16-1 which were methylated to different extents. One might argue that in our cross between Okayama-7 and PJP52, progeny receiving the 16-1 locus from PJP52 were never methylated because no potential methylatable sites were present. Locus 16-1 in the PJP52 chromosome does contain HpaII sites (data not shown). However, too little is known about the signals for eukaryotic DNA methylation for us to predict whether any of the HpaII sites in PJP52 are actually potential sites of methylation. Experiments in which both parents were methylated at locus 16-1, but to different extents, have confirmed that distinct methylation tracts are inherited faithfully through meiosis. These results, and others characterizing the rates of hypermethylation in mitotic and meiotic cells, will be reported elsewhere (27).

ACKNOWLEDGMENTS

We thank J. Cassidy, D. Moore, and S. Crowe for help with the tetrad analysis, D. Binninger for help with the Southern hybridizations, and J. Palmer for critical reading of this manuscript.

REFERENCES

1. Smith, JD, Arber, W, Kühnlein, U (1972). Host specificity of DNA produced by Escherichia coli XIV. The role of nucleotide methylation in in vivo B-specific modification. J Mol Biol 63:1-8.
2. Pukkila, PJ, Peterson, J, Herman, G, Modrich, P. Meselson, M (1983). Effects of high levels of DNA adenine methylation on methyl-directed mismatch repair in Escherichia coli. Genetics 104:571-582.
3. Kleckner, N, Morisato, D, Roberts, D, Bender, J (1984). Mechanism and regulation of Tn10 transposition. Cold Spring Harbor Symp Quant Biol 49:235-244.
4. Marinus, MG, Carraway, M, Frey, AZ, Brown, L, Arraj, JA (1983). Insertion mutations in the dam gene of Escherichia coli K-12. Molec Gen Genet 192:288-289.
5. Razin, A, Riggs, AD (1980). DNA methylation and gene function. Science 210:604-610.
6. Naveh-Many, T, Cedar, H (1981). Active gene sequences are undermethylated. Proc Natl Acad Sci USA 78:4246-4250.
7. Holliday, R, Pugh, JE (1975). DNA modification mechanisms and gene activity during development. Science 187:226-232.
8. Vanyushin, BF, Tkacheva, SG, Belozersky, AN (1970). Rare bases in animal DNA. Nature 225:948-949.
9. Antequera, F, Tamame, M, Villanueva, JR, Santos, T (1984). DNA methylation in the fungi. J Biol Chem 259:8033-8036.
10. Gruenbaum, Y, Naveh-Many, T, Cedar, H, Razin, A (1981). Sequence specificity of methylation in higher plant DNA. Nature 292:860-862.
11. Bird, AP, Southern, EM (1978). Use of restriction enzymes to study eukaryotic DNA methylation I. The methylation pattern in ribosomal DNA from Xenopus laevis. J Mol Biol 118:27-47.

12. Waalwijk, C, Flavell, RA (1978). MspI, an isoschizomer of HpaII which cleaves both unmethylated and methylated HpaII sites. Nucl Acids Res 5:3231-3236.
13. Taylor, SM, Jones, PA (1982). Mechanism of action of eukaryotic methyltransferase. Use of 5-azacytosine-containing DNA. J Mol Biol 162:677-692.
14. Grafstrom, RH, Hamilton, DL, Yuan, R (1984). DNA methylation: DNA replication and repair. In Razin, A, Cedar, H, Riggs, AD (eds): "DNA Methylation: Biochemistry and Biological Significance," New York: Springer-Verlag, p. 111.
15. Pollack, Y, Stein, R, Razin, A, Cedar, H (1980). Methylation of foreign DNA sequences in eukaryotic cells. Proc Natl Acad Sci USA 77:6463-6467.
16. Wigler, M, Levy, D, Perucho, M (1981). The somatic replication on DNA methylation. Cell 24:33-40.
17. Stein, R, Gruenbaum, Y, Pollack, Y, Razin, A, Cedar, H (1982). Clonal inheritance of the pattern of DNA methylation in mouse cells. Proc Natl Acad Sci USA 79:61-65.
18. Urieli-Shoval, S, Gruenbaum, Y, Sadat, J, Razin, A (1982). The absence of detectable methylated bases in Drosophila melanogaster DNA. FEBS Lett 146:148-152.
19. Razin, A, Cedar, H, Riggs, AD (eds): "DNA Methylation: Biochemistry and Biological Significance," New York: Springer-Verlag, Chapter 1.
20. Moore, D (1966). New method of isolating the tetrads of agarics. Nature 209:1157-1158.
21. Lewis, D, North, J (1974). Linkage maps of Coprinus lagopus. In Laskin AI, Lechaudlier, HA (eds): "CRC Handbook of Microbiology IV," Cleveland: CRC Press, Inc., p 691.
22. Specht, CA, Novotny, CP, Ullrich, RC (1984). Strain specific differences in ribosomal DNA from the fungus Schizophyllum commune. Current Genetics 8:219-222.
23. Wu, MMJ, Cassidy, JR, Pukkila, PJ (1983). Polymorphisms in DNA of Coprinus cinereus. Current Genetics 7:385-392.
24. Vieira, J, Messing, J (1982). The pUC plasmids, an M13mp7-derived system for insertion mutagenesis and sequencing with synthetic universal primers. Gene 19:259-268.
25. Cassidy, JR, Moore, D, Lu, BC, Pukkila, PJ (1985). Unusual organization and lack of recombination in the ribosomal RNA genes of Coprinus cinereus. Current Genetics 8:607-613.

26. Ponzetto-Zimmerman, C, Wolgemuth, DJ (1984). Methylation of satellite sequences in mouse spermatogenic and somatic DNAs. Nucl Acids Res 12:2809-2822.
27. Zolan, ME, Pukkila, PJ, manuscript in preparation.
28. Maniatis, T, Fritsch, EF, Sambrook, J (1982). "Molecular Cloning, A Laboratory Manual," New York: Cold Spring Harbor Laboratory, p. 387.

Molecular Genetics of Filamentous Fungi, pages 345–366

MOLDS, MANUFACTURING AND MOLECULAR GENETICS

J. W. Bennett

Department of Biology, Tulane University
New Orleans, Louisiana 70118

ABSTRACT Industrial mycology is a well established factor in the world's economy, valued in the tens of billions of dollars. Industrial processes involving molds include the production of low molecular weight primary and secondary metabolites; the production of enzymes; biotransformations; and the direct utilization of whole organisms, usually for food. Each of these applications is reviewed.

Virtually all industrial strain improvement to date has been effected by mutagenesis and screening. The success of this strategy, coupled with the gulf between academic genetics and industrial microbiology, has hampered the development of breeding programs. Molecular genetics offers new promises. Industrial enzymes, unique to fungi, can be isolated, cloned, and expressed in bacteria and yeast. In the future, it is hoped that molds will become the hosts for engineered fermentation processes, thus allowing the modification of multi-step pathways in situ. More basic research on industrially important mold species will be required before this promise can be met.

INTRODUCTION

Someone once defined technology as "science making money." Along with the space program and computer science, biotechnology is considered very high technology, very contemporary, and potentially very profitable. Unlike the space program and computer science, biotechnology has been a well established factor in the world's economy for a long time, with a history dating back to antiquity. Microorganisms ferment wine, leaven bread, and ripen cheeses.

Or as Arthur Koch of Indiana University once quipped, "Biotechnology is the second oldest profession."

The intense contemporary interest in biotechnology, particularly in industrial microbiology, has been fueled by developments in recombinant DNA technology. Several leading journals have devoted issues entirely to this topic: see, for example, the February 11, 1983 issue of "Science" entitled "Biotechnology" and the September, 1981 issue of "Scientific American" entitled "Industrial Microbiology."

In this paper, I will focus on the industrial applications of fungi, concentrating on molds. By necessity the discussion will be incomplete because many of the most interesting developments, like certain high yielding production strains, are proprietary property of the various manufacturers.

The multivolume series on "Economic Microbiology" edited by Anthony Rose (1,2,3,4,5) is a useful reference for historical developments and state-of-the-art in traditional fermentation industries, while Crueger and Crueger (6) provide a lucid introduction to the field. Other general overviews include the article by Demain (7), and the edited volumes by Smith and Berry (8), Smith et al (9) and Demain and Soloman (10). For a review of industrial fungi, with emphasis on the taxonomy of these organisms, see Onions et al (11).

ECONOMIC ASPECTS OF FUNGI

Rather than categorize the industrial fungi by applications (food, pharmaceutical, industrial) I have chosen a biological classification, and will discuss the following topics in turn: 1) primary metabolites, 2) secondary metabolites, 3) purified enzymes, 4) microbial transformations, and 5) the fungi themselves. I am arbitrarily excluding the negative impact of fungi in such areas as biodeterioration, plant and animal disease, and mycotoxin production.

Primary metabolites

Primary metabolites are cellular substances essential to life, consisting of metabolic products such as ethanol, organic acids, amino acids, polysaccharides, lipids, nucleotides, and nucleic acids, vitamins, and polyols. The major primary metabolites industrially produced by fungi are listed in Table 1.

TABLE 1
INDUSTRIALLY EXPLOITED PRIMARY METABOLITES FROM FUNGI[a]

Metabolites	Species in production
Alcohol:	
Ethanol	*Saccharomyces cerevisiae*
Organic acids:	
Itaconic acid	*Aspergillus terreus*
Gluconic acid	*Aspergillus niger*
Citric acid	*Aspergillus niger*, *Yarrowia lipolytica*
Vitamin:	
Riboflavin	*Eremothecium ashbyii*, *Ashbya gossypii*
Polysaccharides:	
Scleroglucan (Polytran)	*Sclerotium rolfsii*
Pullulan	*Aureobasidium pullulans*
Nucleotides:	
5'-IMP and 5'-GMP	hydrolosis of *S. cerevisiae* RNA

[a]after Bigelis (12) and Rose (2)

The major primary metabolite produced industrially by fungi is ethanol from yeast. It is widely used as a beverage, an industrial solvent, an antifreeze, a fuel, and in a number of other applications. For reviews see Rose (1) on beer, wine and spirit manufacturing; Spencer and Spencer (13) describe genetic improvement of industrial yeasts.

After ethanol, the most important primary metabolites from fungi are the organic acids, particularly citric acid. Citric acid is best known as an acidulant and flavor enhancer used extensively in the food and beverage industries. Industrial production involves fermentations with the mold *Aspergillus niger* or the yeast *Yarrowia lipolytica*. Fungal production of citric acid was first described in the late Nineteenth Century; historical landmarks are reviewed by Miall (14). Modern industrial strains are selected after repeated rounds of mutagenesis and screening, and are highly guarded secrets of the companies which utilize them. Although the patent literature reveals something of the fermentation conditions used, the details of the process are also guarded by the manufacturers, so we know little of exact conditions of the industrial process except to state that it is highly efficient.

An excellent comprehensive review of fungal primary metabolites is given by Bigelis (12). Organic acids, amino acids, vitamins, polysaccharides, lipids, and nucleotides are covered in chapters in Rose (2).

Secondary metabolites

Secondary metabolites are low molecular weight compounds usually produced during a restricted phase (idiophase) of the life cycle. Although secondary metabolites display enormous chemical diversity, they are all synthesized from a few simple precursors of primary metabolism. A comprehensive list of fungal secondary metabolites is given in Turner (15) and Turner and Aldridge (16), while aspects of biosynthesis, evolutionary function, regulatory control, and biological activity are addressed by Bu'Lock (17), Weinberg (18), Woodruff (19), and Bennett and Ciegler (20).

The most important secondary metabolites, in fact the most important of all the products made by mold fermentations, are the penicillins. With the other major group of β-lactam antibiotics, the cephalosporins, these compounds are the major drugs used to treat bacterial infections caused by gram-positive organisms. The production of natural and semi-synthetic β-lactams is a billion dollar business. The history of penicillin, and the story of the famous men involved -- Fleming, Florey, Chain and others -- make facinating reading (21,22,23).

Only two other important antibiotics are of fungal origin: griseofulvin and the fusidanes. In addition, a number of other fungal secondary metabolites are marketed for human or agricultural use (Table 2).

TABLE 2
INDUSTRIALLY EXPLOITED SECONDARY METABOLITES FROM FUNGI[a]

Compound	Activity	Production species
Antibiotics:		
penicillins	antibacterial: varies with specific form; in general, gram-positive bacteria and gram-negative cocci	Penicillium chrysogenum
cephalosporins	antibacterial: broad spectrum	Cephalosporium acremonium
griseofulvin	fungistatic	Penicillium griseofulvin P. patulum
fusidanes	antibacterial: Staphylococcus infections	Fusidium coccineum
Other drugs:		
ergot alkaloids	assuage postpartum hemorrhage; induce labor; treatment of migraine	Claviceps purpurea
Cyclosporin A	immunosuppressive; prevent rejection of organ transplants	Trichoderma polysporum
Agricultural:		
gibberellins	plant growth hormone	Fusarium moniliforme (Gibberella fujikurai)
zearalenone	growth promotant in cattle	Fusarium graminearum (Gibberella zeae)

[a]after Turner (26) and Rose (3)

Of the thousands of known secondary metabolites, relatively few have found economic application. For many compounds all that is known is the chemical structure and the name of the producing organism. Most of these compounds have never been screened for biological activity, or have only been screened for antimicrobial activity. Nature, not the laboratory, is the most important source of new products for commercial exploitation. The range of potentially useful pharmacological activity is vast: anaesthetic, coagulative, diuretic, emetic, hemolytic, hypolipidemic, sedative, vasodilatory, and more. See Demain (7,24) and Bu'Lock et al (25) for more detailed descriptions of the potential new applications of microbial secondary products.

Enzymes

The industrial use of microbial enzymes started with molds. Credit goes to Dr. Jokichi Takamine, a Japanese scientist born to a Samurai family, who lived much of his life in the United States. In 1894 (27) he patented a process for making diastatic enzymes from *Aspergillus oryzae*. This product is still sold under the trade name "Takadiastase." He established a company in Clifton, New Jersey which manufactured hydrolytic enzyme preparations from molds. Among the non-microbiological achievements of this "Japanese Pasteur" are the first commercial isolation of adrenalin, and the grove of Cherry Trees in Washington, D.C. for which he personally made the arrangements and the purchase (28).

In 1980, world-wide sales of enzymes were $300 million (29). Scientists and laymen unfamiliar with modern food processing are often surprised at the extent to which fungal enzymes are utilized in the mass production of food and beverages (Table 3). For example, more than half the white bread made in the United States is treated with enzymes derived from molds(30).

TABLE 3
EXAMPLES OF INDUSTRIAL ENZYMES FROM FUNGI[a]

Enzyme	Application	Production species
α-amylase	bread making	Aspergillus niger, A. oryzae, A. awamori
glucoamylase	malting	A. niger, A. foetidus, Rhizopus foetidus
lactase	feed additive, digestive aid, whey processing	A. oryzae, A. niger, Kluyveromyces lactis, K. fragilis, Torula cremoris
invertase	sucrose conversion (confectionery industry)	A. oryzae, A. niger, Saccharomyces cerevisiae
pectinases	fruit processing	Aspergillus niger, A. wentii, Rhizopus sp.
lipases	supplement for pancreatic lipase	Penicillium roqueforti, Rhizopus delmar
penicillin acylases	semi-synthetic penicillin	Penicillium chrysogenum, Aspergillus ochraceus, Trichophyton mentagrophytes others
acid proteases	breadmaking; chill-proofing bottled beer	Aspergillus saitoi, A. niger, A. oryzae, Trametes sanguineara, Mucor pusillus
microbial rennin	cheese making	Endothia parasitica, Mucor miehei, M. pusillus
cellulases	research	Trichoderma viride, Aspergillus niger

[a]after Crueger and Crueger (6)

Another specific example will illustrate the point. Rennet, the membrane that lines the fourth stomach of a calf, is used to curdle milk in the preparation of cheese. The purified enzyme is variously called rennin, chymosin, or chymase. During the 1960's, with the increasing world consumption of cheese, and a decline in the number of slaughtered calves, an alternative source of rennin became imperative. The first microbial substitute, from *Endothia parasitica* was placed on the market in 1967. Useful enzymes from *Mucor pusillus* and *M. miehei* soon followed (6). Although rennin ranks behind bacterial protease, glucamylase, alpha-amylase, and glucose isomerase in volume of production, it ranks close to first in dollar value of sales (20).

A nearly encyclopedic account of industrial food enzymes from fungi and other sources through 1966 is given by Reed (31). Details of enzyme preparation, manufacture and application are covered in deBeece (32), Blain (30), Rose (5) and Crueger and Crueger (6).

Microbial transformations

When microbes are harnessed to modify organic compounds the processes are described as "microbial transformations" or "bioconversions." The terms are often used to refer to processes in which microorganisms carry out individual steps in a long sequence of synthetic processes when the other steps are accomplished by non-biological processes. Bioconversions can employ immobilized enzymes, but it is cheaper to use whole organisms. Commercial applications of fungal bioconversions are developed for spores and mycelium, growing and non-growing cells, wet and dry cells, and for living and dead cells. The literature on microbial bioconversions is enormous. Useful access is provided by Rosazza (33), Rose (5), and Kieslich (34). Some of the industrially important conversions are listed in Table 4.

TABLE 4
EXAMPLES OF COMMERCIALLY USEFUL BIOCONVERSIONS EMPLOYING FUNGI[a]

Reaction	Species
Steroids	
Progesterone → 11- α-Hydroxyprogesterone	*Rhizopus nigricans*
"Compound S" → Cortisol	*Curvularia lunata*
Progesterone → 1,4-Androstadiene-3,17-dione	*Fusarium solani*
Progesterone → 11- α-Hydroxyprogesterone	*Aspergillus ochraceus*
Antibiotic	
Penicillin V → 6-amino-penicillanic acid (6-APA)	*Penicillium chrysogenum* *Cephalosporium* sp; *Emericellopsis minima*; *Trichophyton mentagrophytes*, *Fusarium avenaceum*, *F. conglutinans*, *F. semitectum*; *Pleurotus ostreatus*.

[a]After Crueger and Crueger (6) and Rose (5)

Historically, the most famous bioconversion story involves the manufacture of cortisone. The original synthesis involved 37 chemical steps. In the early 1950's it was found that a strain of Rhizopus orrhizus could introduce an oxygen at position 11 of the steroid molecule; this step was crucial in determining the biological activity of the molecule. Eventually the use of Rhizopus biotransformation reactions reduced the synthesis from 37 to 11 steps and the price of cortisone from $200./gm to $6./gm (35). By 1979, the cost was less than $1./gm (6). Details of this and other steroid bioconversions are covered in Charney and Tosa (36) and Kieslich (37).

Many modern antibiotics with improved effectiveness, absorption, and antimicrobial spectra are developed by modifying naturally occurring compounds. The starting material for most semi-synthetic penicillins is 6-amino-penicillanic acid (6-APA). Originally 6-APA was produced by direct fermentation. Low yields and expensive purifications soon rendered this approach obsolete when it became possible to modify major naturally occurring penicillins to 6-APA. Both chemical and microbial transformations are employed. Sebek (38) reports that in 1979 more than 30,000 kg of penicillin were converted to 6-APA and that chemists were capable of preparing more than 40,000 semisynthetic penicillins using 6-APA as the starting material. Microbial enzymes which hydrolyze penicillins into 6-APA and the side chain are called penicillin acylases (penicillin amidohydrolase EC 3.5.1.11). The phenoxymethypenicillin acylase is produced mostly by fungi, while the benzylpenicillin acylase is produced mostly by bacteria; see Vandamme (39) for an extensive listing of producing species for both types of major acylase.

Numerous other fungal bioconversions are known and have been catalogued (5) but do not currently find commercial applications. Some examples follow. Strains of Aspergillus fumigatus, Corticium practicola, and C. sasakii reduce the ergot alkaloid elymoclavine to agroclavine. Many species of basidiomycetes are capable of modifying morphine; e.g. Trametes sanguinea can convert thebaine into a variety of hydroxylated codeine derivatives. The schistosomicidal agent lucanthone is converted into the more active hycanthone by cultures of Aspergillus sclerotiorum, while the antihelminthic agent parbendazole is metabolized by Cunninghamella baineri to the same compounds found in the urine of farm animals exposed to parbendazole (5).

Direct Use of Fungi

Among the direct uses of entire fungi are the human consumption of edible macrofungi, mold-ripened cheeses, Oriental food fermentations, production of microbial biomass, and pesticides. With the exception of pesticides, these are all things we eat or feed to our animals.

Major examples of direct human food uses of fungi are listed in Table 5.

TABLE 5
DIRECT FOOD USES OF FUNGI[a]

Application	Species
Edible macrofungi	
Common edible mushroom	*Agaricus bisporus*
Shiitake	*Lentinus edodes*
Chinese or Straw mushroom	*Volvariella volvacea*
Winter mushroom	*Flammulina velutipes*
Oyster mushroom	*Pleurotus* sp.
Truffle	*Tuber melanospernum*
Cheeses	
Roquefort, Stilton, "Blue"	*Penicillium roqueforti*
Camembert, Brie, soft ripened cheeses	*Penicillium camembertii*
Oriental food fermentations	
Ang-kak	*Monascus purpurea*
Hamanatto	*Aspergillus oryzae*
Miso	*Aspergillus oryzae/sojae*
Ontjom	*Neurospora intermedia*
Shoyu (soy sauce)	*Aspergillus oryzae/sojae*
Tempeh	*Rhizopus oligosporus*

[a]After Gray (40), Hayes and Nair (41), Hayes and Wright (42) and Ayers et al (43), Perkins et al (76).

See Linsansky and Hall (44) for a review of fungal control of insects, and Rose (4) for several reviews on the commercial production of biomass, sometimes known as "single cell protein."

The practice of eating fungi is ancient. The Classical Romans held mushrooms in high esteem and Pliny is said to have praised Roquefort cheese in the first century A.D. (40). Currently, in Western culture, fungi largely serve as condiments or flavoring agents. Although the market for edible macrofungi is relatively small, it is not trivial, and is growing. Fungi are especially prized as "gourmet" food. The truffle, for example, is among the most valued of delicacies, currently selling for several hundred dollars per pound. These fleshy ascomycetes grow underground in association with oak trees. In traditional French agriculture, farmers use leashed female pigs to assist in the location of the truffles. The recent isolation of a boar sex hormone as a secondary metabolite of the truffle (45) may explain the particular interest of scws in this sport, if not the purported aphrodisiacal qualities of the ascocarp.

Commercial cultivation of Agaricus bisporus dates back to the reign of Louis XIV of France (1643-1715), where mushroom farms were established in underground caves in Paris on horse manure. Modern mushroom growing is reviewed by Hayes and Nair (41). Currently, mushroom cultivation is the only major process in biotechnology which sucessfully converts cellulosics into useful foods (46). Agaricus dominates in Western consumption of mushrooms, but in the Orient several other basidiomycetes are cultivated on a large scale. In addition, molds are used to process soybeans, rice, wheat, fish and other substrates. Soy sauce is the best known of these but other fermented foodstuffs are widely consumed in China, Japan, and South East Asia (Table 5). For more about mold fermented foods see Gray (40, 47) and Hesseltine (48).

GENETICS OF INDUSTRIAL FUNGI

Mutagenesis

Until quite recently, there was a wide gulf between basic fungal genetics and industrial mycology. With the exception of Saccharomyces cerevisiae, academicians and industrialists worked with different organisms. The best molds for genetic analysis were Aspergillus nidulans,

Neurospora crassa and *Podospora anserina*. None of these produced an industrially important product, while the organisms used in industry (see species listings in Table 1-5) were poorly characterized, if characterized at all, by geneticists. When Pontecorvo addressed the opening session of the meeting on Genetics of Industrial Microorganisms held in Sheffield, England, in 1974, he spoke to this point, saying, "One thing is clear ... the advances in the application of genetics to the improvement of strains of industrial microorganisms are trifles compared to the advances of the fundamental genetics of microorganisms. The main technique used is still a prehistoric one: mutation and selection" (49).

Although one might quibble with the accuracy of the adjective "prehistoric," in principle, Pontecorvo was right. Virtually all strain improvement of industrial fungi has been obtained by brute force mutagenesis, screening, and selection. This process has been slow, laborious, and expensive, but it has also been successful. The success has allowed industrialists to obtain their goal -- profitable, high-yielding strains -- without making an investment in the basic genetics of the producing species. Even the few attempts to employ breeding programs for strain improvement have had little success, although once a high-yielding parasexual diploid of *Penicillium chrysogenum* was briefly used as a production strain in a major penicillin fermentation (50).

In the traditonal, empirical approach to strain improvement, a suitable strain, usually the current high-yielding production strain, is mutagenized. Then a large number of the resultant isolates are randomly screened for higher titer of the given product or for some other desirable attribute. Sometimes a "rational" selection procedure is devised which enables a larger number of isolates to be screened more rapidly. Most of these "rational" screens are based on selection for some phenotype believed to be associated with the desired trait. Rational screens are discussed in more detail by Vournalkis and Elander (51) and by Rowlands (52). Rowlands has also provided an excellent review of mutagenesis as applied to industrial strain selection in fungi (53).

Breeding and Molecular Genetics

Until the advent of recombinant DNA technologies, hybridization was effected by the parasexual cycle, protoplast fusions, or, when available, conventional sexual crosses. There are two recent books on the genetics and breeding of industrial microorganisms (54, 55), and there are also recent works on fungal protoplasts (56), and overproduction of microbial products (57).

Recombinant DNA technology allows us to construct bacteria and yeasts that make proteins they never made in nature. Recombinant DNA technology holds the promise that we may be able to insert selected foreign genes into the existing genotype of our high-yielding strains of industrial molds, and the possibility of conducting gene transfer and cloning experiments in species which have no "normal" methods of gene transfer.

The popular imagination has been particularly captured by the molecular cloning of mammalian gene products (i.e. insulin, somatostatin, interferon) into bacteria. Magazine and newspaper articles describing the latest breakthrough are commonplace. So are scientific symposia that document, extol, and sometimes exaggerate the fruitfulness of this recent marriage of genetics and biotechnology (58, 59, 60, 61, 62).

In the euphoria surrounding the successes of recombinant DNA technology, it has become customary to conclude almost any discussion of industrial microbiology with a list of the potential benefits of genetic engineering to any and all processes. Lecturers blithely speak in broad generalizations about the soon-to-materialize profits that will be gleaned by engineering industrial microbes so that undesirable genes are eliminated, desirable ones are accentuated, and entirely new ones are inserted with ease. It is therefore well to interject a cautionary note. Molecular cloning will have an uneven impact on the solution of industrial problems.

Joseph Lein, founder of Panlabs, Inc., once formulated an "Industrial Microbiologist's Version of the Central Dogma."

$$\text{DNA} \rightarrow \text{RNA} \rightarrow \text{Enzymes} \rightarrow \text{Primary Metabolites}$$

$$\text{Primary Metabolites} + \text{Enzymes} \rightarrow \text{Secondary Metabolites}$$

This "Industrial's Central Dogma" highlights the levels of complexity we face in dealing with different products. Enzymes, and indirectly, bioconversions mediated by immobilized systems, are close to the gene. Primary metabolites are further away from the gene, and secondary metabolites further away still. Although it would be nice if recombinant DNA technology could suddenly allow us to breed exotic edible agarics or to improve the yield of French truffles, it is not readily apparent how this will come to pass.

For industrial enzymes, the future is now, and the benefits are already being reaped. For example, cDNA from calf prorennin has been cloned and expressed in E. coli, and although the process does not yet compete with enzymes derived from fungal fermentations, the story may yet come full cycle (6, 54). Most work, however, focuses on the isolation, cloning, and expression of unique fungal genes in bacteria or yeast. Workers at Cetus Corporation have cloned the genes for glucoamylase (GAI form) from Aspergillus awamori, exo-cellobiohydrolase from Trichoderma reesei, and pyranose-2-oxidase from Polyporus obtusus. All three genes contain introns (63). In further characterization of the glucoamylase gene, structural and regulatory features were modified because the Aspergillus promoter was not recognized, nor were the glucoamylase gene introns excised, in yeast. An intron-free glucoamylase gene was engineered and, in order to express the gene, the promoter and termination regions of the yeast enolase gene were used. Saccharomyces was able to process the preglucoamylase leader sequence, as well as glycosylate and secrete the enzyme. Thus, a strain of yeast which could grow on starch as the sole carbon source was constructed, demonstrating the feasibility of using S. cerevisiae as the host for the production, glycosylation, and secretion of industrial enzymes (64, 65). A general review of cloning fungal amylases and cellulases is presented by Montenecourt and Eveleigh (66).

Both primary and secondary metabolites are synthesized via multi-step pathways. Since it is not usually feasible to move entire pathways into bacteria or yeast, our objective is to move desired genes into the production strain. Before we can construct industrial molds to make gene products that are not part of their normal repertoire, we will have to go back and do some basic science. In the absence of a shuttle vector for filamentous fungi, it will be necessary to develop at least a rudimentary genetic system in the species of interest. Since Aspergillus niger and A.

oryzae are both on the Generally Regarded As Safe (GRAS) list of the FDA (67), these organisms are obvious targets for intensive research.

The marketable primary metabolites are largely commodity chemicals often manufactured on a massive scale. Even modest increases in energy efficiency are valuable when thousands of tons of a commodity are produced. Therefore, one obvious goal is to transfer genes involved in the breakdown of an inexpensive carbon source such as starch or cellulose into a production strain.

Another feat with enormous profit potential would be to identify the genes that make an over-producing strain over-produce. These, then, could be cloned, manipulated, and used in the construction of new production strains.

Antibiotics and other secondary metabolites hold the lion's share of the market for fermentation products. All antibiotics are secondary metabolites so the prospects for applying recombinant DNA technology to secondary metabolism deserve special scrutiny. Considerable progress has already been made for Streptomyces, the bacterial genus which produces nearly two-thirds of all known natural antibiotics (68, 69, 70, 71).

The immediate outlook for fungal secondary metabolites is not as promising. The biosynthetic pathways for most fungal secondary metabolites are not understood, nor are the genetic locations of the structural genes for the enzymes involved. Most producing species lack a sexual cycle. Penicillin is the best studied secondary metabolite, but the first enzyme for a penicillin pathway enzyme (isopenicillin N synthetase) was purified only in 1984, a full fifty-five years after the discovery of penicillin and after decades of industrial research (72).

Transformation systems have been reported for both Cephalosporium acremonium (74) and Penicillium chrysogenum (75), so it should not be long before more flexible genetic manipulations become feasible in these species. The economic incentives are high.

Cyclic peptides, which at first glance may seem obvious targets for genetic engineering, are also secondary metabolites. Fungal cyclic peptides such as cyclosporin A share properties with the peptide antibiotics of bacteria (73). They are not synthesized off a ribosome, but rather are made by a series of enzymatically catalyzed steps like other secondary metabolites. They frequently contain rare amino acids and D-amino acids not found in proteins.

In the short run, molecular cloning of enzymes used in biotransformations to manufacture semi-synthetic antibiotics and other drugs is a realizable objective. One obvious example would be to clone a fungal penicillin acylase gene, and develop a practical, competitive method for using this pure enzyme in industry.

SUMMARY

Any industrial fermentation must take into account certain economic criteria: the cost of starting materials, energy, and purification; the sale price of the final product; and the availability of competing substances produced by other means. If genetic engineering approaches do not provide economic advantages over traditional fermentations, they will not be implemented. Fungal enzymes can now be cloned and expressed in heterologous systems, but at least for the moment, these methodologies are not cost effective.

In the past, industrial fungi have been identified because they made a normal metabolic product useful in human commerce. Rounds of mutagenesis, screening, and selection have resulted in the isolation of "super" strains that "over-produce" the normal metabolic product. To put it another way, high-yielding strains have been selected, not constructed.

The biotechnological revolution tantalizes us with the prospect of making good production strains even better by direct manipulation of their genomes. However, before we can undertake genetic programming of industrial molds, we must learn a lot more about the basic genetics of these organisms. Perhaps, ultimately, the greatest benefit of all will be a new understanding of the fundamental biology of the simple molds that have served us so well for so long.

ACKNOWLEDGMENTS

Research in this laboratory on the genetics of secondary metabolism is supported by a Cooperative Agreement from the U.S. Department of Agriculture. I thank Amy Henderberg for help with the literature review, Irving LaValle for manuscript preparation, and Patricia Crickenberger for typing the final copy.

REFERENCES

1. Rose AH, ed (1977). "Economic Microbiology. Vol. 1. Alcoholic Beverages," London: Academic Press.
2. Rose AH, ed (1978). "Economic Microbiology. Vol. 2. Primary Products of Metabolism," London: Academic Press.
3. Rose AH, ed (1979). "Economic Microbiology. Vol. 3. Secondary Products of Metabolism," London: Academic Press.
4. Rose AH, ed (1979). "Economic Microbiology. Vol. 4. Microbial Biomass," London: Academic Press.
5. Rose AH, ed (1980). "Economic Microbiology. Vol. 5. Microbial Enzymes and Bioconversions," London: Academic Press.
6. Crueger W, Crueger A (1982). "Biotechnology: A Textbook of Industrial Microbiology," Sunderlands, MA: Sinauer Assoc.
7. Demain AL (1981). Industrial microbiology. Science 214:987.
8. Smith JE, Berry DR (1975). "The Filamentous Fungi. Vol. 1. Industrial Mycology," New York: John Wiley & Sons.
9. Smith JE, Berry DR, Kristiansen B (1983). "The Filamentous Fungi. Vol. 4. Fungal Technology." London: Edward Arnold.
10. Demain AL, Soloman N (1985). "Biology of Industrial Microorganisms," Menlo Park: Benjamin Cummings.
11. Onions AHS, Allsopp D, Eggins, HOW (1981) "Smith's Introduction to Industrial Mycology," Seventh Edition, London: Edward Arnold.
12. Bigelis R (1985). Primary metabolism and industrial fermentations. In Bennett JW, Lasure LL (eds): "Gene Manipulations in Fungi," New York: Academic Press, in press.
13. Spencer JFT, Spencer DM (1983). Genetic improvement of industrial yeasts. Ann Rev Microbiol 37:121.
14. Miall LM (1978). Organic acids. In Rose AH (ed): "Economic Microbiology. Vol. 3. Primary Products of Metabolism," London: Academic Press, p 47.
15. Turner WB (1971). "Fungal Metabolites." London: Academic Press.
16. Turner WB, Aldridge DC (1983). "Fungal Metabolites II." London: Academic Press.
17. Bu'Lock JD (1961). Intermediary metabolism and antibiotic synthesis. Ann Rev Appl Microbiol 3:293.

18. Weinberg ED (1971). Secondary metabolism: raison d' etre. Perspct Biol Med 14:703.
19. Woodruff HB (1980). Natural products from microorganisms. Science 208:1225.
20. Bennett JW, Ciegler A, eds. (1983). "Secondary Metabolism and Differentiation in Fungi." New York: Marcel Dekker.
21. Hare R (1970). "The Birth of Penicillin." London: George Allen and Unwin, Ltd.
22. Flynn ED, ed (1972). "Cephalosporins and Penicillins. Chemistry and Biology." New York: Academic Press.
23. Macfarlane G (1979). "Howard Florey." Oxford: Oxford University Press.
24. Demain AL (1983). New applications of microbial products. Science 219:709.
25. Bu'Lock JD, Nisbet LJ, Winstanley DJ, eds (1982). "Bioactive Microbial Products: Search and Discovery." London: Academic Press.
26. Turner WB (1975). Commercially important secondary metabolites. In Smith JE, Berry DR (eds): "The Filamentous Fungi. Vol. 1. Industrial Mycology," New York: John Wiley & Sons, p 122.
27. Takamine J (1894). United States Patents 525,820 and 525,823.
28. Kawakami KK (1928). "Jokichi Takamine. A Record of His American Achievements." New York: William Edwin Rudge.
29. Eveleigh DE (1981). The microbiological production of industrial chemicals. Sci Amer 245:154.
30. Blain JA (1975). Industrial enzyme production. In Smith JE, Berry DR (eds): "The Filamentous Fungi. Vol. 1. Industrial Mycology," New York: John Wiley & Sons, p 193.
31. Reed G (1966). "Enzymes in Food Processing," New York and London: Academic Press.
32. deBeeze GI (1970). Food enzymes. Critical Rev Food Technol 1:479.
33. Rosazza JP, ed (1982). "Microbial Transformations of Bioactive Compounds, Vol. 1 and 2." Boca Raton: CRC Press.
34. Kieslich K (1976). "Microbial Transformation of Non-Steroid Cyclic Compounds." Stuttgard: G. Thiem.
35. Aharonowitz Y, Cohen G (1981). The microbiological production of pharmaceuticals. Sci Amer 245:141.
36. Charney W, Tosa T (1980). "Microbial transformations of steroids," 2nd ed. New York: Academic Press.
37. Kieslich K (1980). Steroid conversions. In Rose AH (ed): "Economic Microbiology. Vol. 5. Microbial Enzymes and Bioconversions," London: Academic Press, p 396.

38. Sebek OK (1980). Microbial transformations of antibiotics. In Rose AH (ed): "Economic Microbiology. Vol. 5. Microbial Enzymes and Bioconversions," London: Academic Press, p 576.
39. Vandamme ED (1980). Penicillin acylases and beta-lactamases. In Rose AH (ed): "Economic Microbiology. Vol. 5. Microbial Enzymes and Bioconversions," London: Academic Press, p 468.
40. Gray WD (1970). The use of fungi as food and in food processing. CRC Critical Rev on Food Technology 1:225.
41. Hayes WA, Nair NG (1975). The cultivation of Agaricus bisporus and other edible mushrooms. In Smith JE, Berry DR (eds): "The Filamentous Fungi. Vol. 1. Industrial Mycology," New York: John Wiley & Sons.
42. Hayes WA, Wright SH (1979). Edible mushrooms. In Rose AH (ed): "Economic Microbiology. Vol. 4. Microbial Biomass." London, Academic Press, p 142.
43. Ayres JC, Mundt JO, Sandine WE (1980). "Microbiology of Foods." San Francisco: Freeman and Company.
44. Lisansky SG, Hall RA (1983). Fungal control of insects. In Smith JE, Berry DR, Kristiansen B (eds): "The Filamentous Fungi: Vol. 4. Fungal Technology," London: Edward Arnold.
45. Claus R, Hoppen HO, Karg H (1981). The secret of truffles: a steroidal pheromone? Experientia 37:1178.
46. Tautorus TE, Townsley PM (1984). Biotechnology in commercial mushroom fermentation. Biotechnology 2:696.
47. Gray WD (1981). Food technology and industrial mycology. In Cole GT, Kendrick B (eds): "Biology of Conidial Fungi. Vol. 2." New York: Academic Press.
48. Hesseltine CW (1983). Microbiology of oriental fermented foods. Ann Rev Microbiol 37:575.
49. Pontecorvo A (1976). Presidential address. In MacDonald KD (ed): "Second International Symposium on the Genetics of Industrial Microorganisms," London: Academic Press.
50. Malik VS (1979). Genetics of applied microbiology. Adv Genet 20:37.
51. Vournakis JN, Elander RP (1983). Genetic manipulation of antibiotic-producing microorganisms. Science 219:703.
52. Rowlands RT (1974). Industrial strain improvement: rational screens and genetic recombination techniques. Enzyme Microb Technol 6:290.

53. Rowlands RT (1983). Industrial fungal genetics and strain selection. In Smith JE, Berry DR, Kristiansen B (eds): "The Filamentous Fungi. Vol. 4. Fungal Technology," London: Edward Arnold.

54. Sakaguchi K, Okanishi M, eds (1980). "Molecular Breeding and Genetics of Applied Microorganisms." Tokyo: Kodansha Ltd.

55. Ball C, ed (1984). "Genetics and Breeding of Industrial Microorganisms." Boca Raton: CRC Press.

56. Peberdy JF, Ferenczy L, eds (1985). "Fungal Protoplasts. Applications in Biochemistry and Genetics." New York: Marcel Dekker.

57. Krumphanzl V, Sikyta B, Zanek Z (1982). "Overproduction of Microbial Products." London: Academic Press.

58. Paul JK, ed (1981). "Genetic Engineering Applications for Industry." Park Ridge, NJ: Noyes Publications.

59. Hollaender A, ed (1982). "Genetic Engineering of Microorganisms for Chemicals." New York: Plenum.

60. Whelan WJ, Black S (1982). "From Genetic Experimentation to Biotechnology -- The Critical Transition." Chichester: John Wiley & Sons.

61. Levin MA, Kidd GH, Zaugg RH, Swarz JR (1983). "Applied Genetic Engineering. Future Trends and Problems." Park Ridge, NJ: Noyes Publications.

62. Lurquin PF, Kleinhofs A, eds (1983). "Genetic Engineering in Eukaryotes." New York: Plenum.

63. White TJ, Meade JH, Shoemaker SP, Koths KE, Innis MA (1984). Enzyme cloning for the food fermentation industry. Food Technol, Feb 1984:90.

64. Nunberg JH, Meade JH, Cole G, Lawyer FC, McCabe P, Schweickart V, Tal R, Wittman VP, Flatgaard JE, Innis MA (1984). Molecular cloning and characterization of the glucoamylase gene of Aspergillus awamori. Mol Cell Biol 4:2306.

65. Innes MA, Holland MJ, McCabe PC, Cole GE, Wittman VP, Tal R, Watt WK, Gelfand DH, Holland JP, Meade JH (1985). Expression, glycosylation, and secretion of an Aspergillus glucoamylase by Saccharomyces cerevisiae. Science 228:21.

66. Montenecourt BS, Eveleigh DE (1985). Fungal carbohydrases: amylases and cellulases. In Bennett JW, Lasure LL (eds): "Gene Manipulations in Fungi," New York: Academic Press (in press).

67. Taylor MJ, Richardson T (1979). Applications of microbial enzymes in food systems and biotechnology. Adv Appl Microbiol 25:7.

68. Baltz RH (1982). Genetics and biochemistry of tylosin production: a model for genetic engineering in antibiotic-producing Streptomyces. In Hollaender A (ed): "Genetic Engineering of Microorganisms for Chemicals," New York: Plenum, p 431.
69. Burnett JP (1983). Commercial production of recombinant DNA-derived products. In Inouye M (ed): "Experimental Manipulation of Gene Expression." New York: Academic Press, p 259.
70. Malpartida V, Hopwood DA (1984). Molecular cloning of the whole biosynthetic pathway of a Streptomyces antibiotic and its expression in a heterologous host. Nature 309:462.
71. Martin JF, Gil JA (1984). Cloning and expression of antibiotic production genes. Biotechnology 2:63.
72. Hollander IJ, Sehn YQ, Heim J, Demain AL (1984). A pure enzyme catalyzing penicillin biosynthesis. Science 224:610.
73. Katz E, Demain AL (1977). The peptide antibiotics of Bacillus: chemistry, biogenesis and possible functions. Bacteriol Rev 41:449.
74. Queener SW, Ingolia TD, Skatrud PL, Chapman JL, Kaster KR (1984). Recombinant DNA studies in Cephalosporium acremonium. Abstr ASM Conf Genet Mol Biol Industrial Microorganisms, #30.
75. Solingen PVan, Muurling HD, Koekman BP (1985). Transformation of Penicillium chrysogenum. J Cellular Biochem, Suppl 9C. Abstract #1576, p 174.
76. Perkins DD, Turner BC, Barry EG (1976). Strains of Neurospora Collected from Nature. Evolution 30:281.

Molecular Genetics of Filamentous Fungi, pages 367–381

DEVELOPMENT OF THE GENETICS OF THE DIMORPHIC YEAST *YARROWIA LIPOLYTICA*[1]

Rod A. Wing and David M. Ogrydziak

Institute of Marine Resources, University of California
Davis, California 95616

ABSTRACT Two integrative transformation systems have been developed for *Y. lipolytica* in other laboratories. We have used one system, which consists of the *Y. lipolytica LEU2* gene inserted into pBR322 as the nonreplicating vector (pRW1) and a *Y. lipolytica leu2* strain as the host, in an attempt to isolate *Y. lipolytica* autonomously replicating sequences (*ARSs*). Two pools of plasmids containing random fragments of *Y. lipolytica* DNA (306 kb total) inserted into pRW1 were transformed into the *leu2* host. Slow growing, mitotically unstable transformants were obtained from one of the pools. When Southern blots of uncut DNA isolated from these transformants were hybridized with pBR322 a broad band with the mobility of chromosomal DNA and a sharper band of greater mobility (possibly plasmid DNA) were detected. Initial attempts to recover plasmids from the unstable transformants by transformation of *E. coli* were not successful. A plasmid indistinguishable from pRW1 was recovered from a stable transformant which had been transformed with pRW1. Other attempts to isolate *Y. lipolytica ARSs* will be reviewed.

INTRODUCTION

Yarrowia lipolytica (previously *Candida*, *Endomycopsis* or *Saccharomycopsis lipolytica*) is a dimorphic yeast. It is

[1]This work was supported by Central Research Division, Pfizer Inc., Groton, Conn. and by Agriculture Experiment Station fund CA-D-FST-3490-H.

quite different from *Saccharomyces cerevisiae*. In The 1960's, before the oil crisis, *Y. lipolytica* was extensively studied as a source of single cell protein and fermentation chemicals because of its ability to grow on hydrocarbons. Patents have been issued for its use in the production of erythritol and mannitol and it has been used industrially to produce citric acid. *Y. lipolytica* has been studied as a source of lysine (1) and as a possible genetically tractable model system for studying the yeast to mycelial transformation (2). We have been using *Y. lipolytica* as a model system for the study of secretion by eucaryotes. It secretes high levels of an alkaline extracellular protease (about 2% of total protein synthesized) and more modest levels of a ribonuclease, several acid proteases and a lipase. *Y. lipolytica* may have potential as a host for the synthesis and secretion of foreign proteins.

Sexuality of *Y. lipolytica* was first reported in 1969 by Wickerham, Kurtzman, and Herman (3). They established that *Y. lipolytica* was heterothallic and they published procedures for mating, sporulation, and stabilization of haploids and diploids. Inbreeding programs in R. Mortimer's and H. Heslot's laboratories resulted in improvements in the percentage of four-spored asci and in spore viability which made tetrad analysis feasible (4, 5). Most of the classical genetic techniques which have been developed for *S. cerevisiae* are now possible with *Y. lipolytica*. A primitive genetic map consisting of five linkage fragments on which a total of twenty six genes have been mapped has also been developed (6). Several studies involving fusion of *Y. lipolytica* protoplasts have been reported and a chromosome loss method for mapping has been developed (J. DeZeeuw, personal communication).

Two integrative transformation systems have been developed for *Y. lipolytica*. The system developed at Pfizer Inc. uses the *Y. lipolytica LEU2* gene inserted into pBR322 for the vector, a *Y. lipolytica leu2* host, and the lithium acetate transformation procedure. Transformation frequencies as high as several hundred transformants per microgram of uncut DNA and 10^4 per microgram of DNA linearized with restriction enzymes have been obtained. Another system developed in H. Heslot's laboratory uses *Y. lipolytica* DNA inserted into YIp333 (which carries the *S. cerevisiae LYS2* gene) for the vector (7), a *Y. lipolytica lys2* host, and the spheroplast transformation procedure. Transformation frequencies of 1-10 transformants per microgram of uncut DNA have been obtained.

No *Y. lipolytica* *ARS* has yet been isolated. In this manuscript our attempts and attempts by others to isolate a *Y. lipolytica* *ARS* will be described.

METHODS

Bacteria, Yeast and DNA.

Y. lipolytica CX161-1B *ade1 A* (4) was the source of the DNA screened for *ARS* activity. *Y. lipolytica* PC30827 *leu2-35 ura3-11 A* (obtained from J. DeZeeuw) was the host in transformation experiments. *Escherichia coli* strains HB101 (8) and MC1066 (9) were used. The plasmid pLD25 (obtained from L. Davidow) contains the *Y. lipolytica* *LEU2* gene on a 6.7 kb partial *Sau*3A fragment inserted into the *Bam*H1 site of pBR322.

Media.

YM, D, and YEPD were used as rich medium (4,10). YLT used as defined leucine-free medium contained 20 g/l glycerol, 6.7 g/l Yeast Nitrogen Base without amino acids plus adenine, uracil and amino acid supplements (except for leucine) in 50 mM citrate buffer at pH 5.5. YLT+leu contained L-leucine at 300 mg/l final concentration. LB and M9 supplemented with tryptophan and uracil were used as rich and defined media, respectively, for *E. coli* (11). Ampicillin was used at 50 ug/ml.

Enzymes.

T4 DNA ligase and all restriction enzymes were purchased from New England Biolabs and used according to manufacturers specification. Calf intestinal alkaline phosphatase was purchased from Boehringer-Mannheim.

DNA Isolation.

Two methods were used for extracting DNA from yeast. DNA for the plasmid library was isolated using a method which involves extraction over several days with SDS and toluene (12). In subsequent DNA isolations with this method the cells were not lyophilized. The other method involves lysing spheroplasts (13) which were made essentially as described by Stahl (14). Large scale plasmid preparations from E. coli were done using a cleared lysate method (15) followed by two CsCl/ethidium bromide centrifugations (11). Small scale plasmid preparations were done using the miniscreen procedure (15).

Construction of Hybrid DNA Molecules.

pRW1. The plasmids pLD25 and pBR322 were digested with SalI, ligated, transformed into E. coli MC1066 (which contains a leuB mutation) and ampicillin resistant transformants selected. Transformants were replica plated onto M9 medium containing tryptophan and uracil, Leu^+ transformants picked and those containing the pLD25 SalI fragment in pBR322 were identified by the miniscreen procedure.

Pools I and II. Y. lipolytica DNA (200 ug) extracted by the SDS/toluene procedure was partially digested with Sau3A and separated on a sucrose gradient (11). Fractions containing 8 to 12 kb fragments were pooled and ligated into BamH1 digested, dephosphorylated pRW1 and the ligation mixture transformed into E. coli HB101. Ampicillin resistant transformants were screened for inserts by agarose gel electrophoresis of plasmid miniscreen preparations. Forty six transformants were divided into two pools of 23 each and patched onto LB plates containing ampicillin.

After overnight growth the cells were washed from the plates, resuspended in LB plus ampicillin medium and grown for two cell divisions. DNA was extracted by the cleared lysate method and purified by CsCl/ethidium bromide gradient centrifugation.

DNA Transformation.

Y. lipolytica. The transformation procedure is a modification of the lithium acetate procedure developed by Ito et al. for S. cerevisiae (16). Late log phase cells of PC30827 (about 1.5×10^9 cells/ml) grown in YEPD were washed, resuspended at 12-fold concentration in 10 mM Tris-HCl, pH 7.5, 1 mM EDTA (TE), and treated with 0.1 M lithuim acetate (final concentration) for 1 hour at 28°C. Sample DNA and 25 μg of sonicated E. coli DNA were added to 0.5 ml aliquots of the competent cells. The cells were incubated for 0.5 h at 28°C and then 3.5 ml of 40% polyethylene glycol in TE was added and the mixture incubated for another hour at 28°C. The cells were heat shocked at 37°C for 10 minutes, harvested, resuspended in 0.8 ml of TE and plated on four plates of YLT. Plates were incubated at 28°C for one to two days.

E. coli. E. coli cells were transformed using either the calcium chloride (11) or the Hanahan (17) procedures.

Southern Blot Analysis

Agarose gel electrophoresis was done using Tris-acetate buffer (11). DNA transferred to nitrocellulose (18) was hybridized to ^{32}P-labelled pBR322 using the BRL nick translation protocol followed by Bio-gel P60 gel chromatography.

RESULTS

The method chosen for attempting to isolate ARSs from Y. lipolytica is similar to the method used by Stinchomb et al. (19) for isolating ARSs using S. cerevisiae. Random Y. lipolytica DNA fragments were inserted into a plasmid which does not replicate in Y. lipolytica and can only transform by integration. Plasmids containing an ARS would be detected by an increased transformation frequency and/or slow growth and plasmid instability of transformants. Specifically, random Y. lipolytica DNA fragments were inserted into the BamH1 site of the plasmid pRW1 which contains the Y. lipolytica LEU2 gene inserted into the SalI site of pBR322. A background level of 10 to several 100 transformants per μg of DNA are obtained by integrative transformation with pRW1. For a given experiment the transformation frequency is about 100-fold higher if pRW1 is first linearized by cutting with a restriction enzyme within the Y. lipolytica DNA insert in the plasmid.

Forty six plasmids with inserts were isolated and divided into two pools of 23 plasmids each. Pool I and Pool II contained 144 kb and 162 kb of insert DNA, respectively. DNA from each pool was purified by CsCl/ethidium bromide gradient centrifugation and used to transform Y. lipolytica. The transformation frequencies (transformants per μg DNA) were 12 for closed circular pRW1, 8 for Pool II, 34 for Pool I and 1100 for pRW1 linearized with BglII. Of the 295 transformants from Pool I, 26 formed small colonies and were restreaked on YLT medium. Six grew as rapidly as pRW1 transformants and 20 grew significantly more slowly. Of the 71 transformants from Pool II, 5 which formed small colonies were restreaked on YLT medium and all grew as rapidly as pRW1 transformants.

The stability of the plasmids was tested by streaking the transformants on rich medium (YM) and replica plating the colonies on YLT medium. For the pRW1 transformants and the faster growing transformants, all the replicated colonies grew on YLT. For the 20 slow growing transformants from Pool I, in each case, less than 10% of the replicated colonies grew. The somewhat higher transformation frequency with Pool I DNA and the instability of some of the transformants suggested that at least one of the plasmid inserts in Pool I contained a Y. lipolytica ARS.

When an unstable transformant was streaked on YLT, there were colonies as large as those for the stable transformants and there were distinctly smaller, slow growing colonies. Single cells from the large and small colonies of the unstable transformants were micromanipulated to defined positions on rich (D) and defined (YLT) medium. From the large colonies, 17 of 20 dissected cells formed visible colonies on D medium and 14 of 20 on YLT. From the small colonies, 21 of 32 dissected cells formed visible colonies on D medium and 0 of 60 on YLT. These results suggested that the plasmid had stably integrated in the cells from the large colonies.

The mitotic stability of the plasmid was investigated in more detail by comparing cell counts on YLT and YLT plus leucine plates for a stable transformant (RW1) transformed with linearized pRW1 and for an unstable transformant (ARS25). Cells were grown on YLT plates. Several small colonies were resuspended in sterile water; this suspension was plated on YLT and YLT +leu plates and inoculated into rich (YM) and defined (YLT) liquid medium. For these initial suspensions the percentage stability was 100% for RW1 and only 4.5% for ARS25 (TABLE 1). The Leu^+ phenotype was stable for RW1 after over 9 generations of growth. For ARS25 in nonselective medium, the percentage of cells containing the plasmid decreased to 0.07% which corresponds to a mitotic rate of loss of 42% per generation. For ARS25 in selective medium the percentage of cells containing the plasmid increased to 22% and then decreased to 12%. The increase in selective medium was probably due to the formation of stable integrants which grow faster than the unstable transformants and become a significant proportion of the population. The doubling time for RW1 in YLT was about 2.4 h whereas the doubling times for the unstable transformants were greater than 4 h.

TABLE 1
STABILITY OF TRANSFORMANTS

Culture	Growth Medium	Number of Generations	Percentage Stability[a]
RW1	YLT	1[b]	100
RW1	YLT	9.8	100
RW1	YM	9.5	97[c]
ARS25	YLT	1[d]	4.5
ARS25	YLT	7.1	22
ARS25	YLT	9.5	12
ARS25	YM	7.6	0.07

[a]Plate counts on YLT/(YLT + leu) x 100%.
[b]Inoculum - cell density of 2.2×10^4 cells/ml.
[c]Not significantly different from 100%.
[d]Inoculum - cell density of 7.5×10^4 cells/ml.

Four of the unstable transformants (13, 16, 17 and 25) were grown in YLT medium and DNA was extracted using the spheroplast procedure. The DNA was separated by agarose gel electrophoresis, transferred to nitrocellulose, and hybridized with nick-translated pBR322. The pBR322 probe hybridized to a broad band of high molecular weight DNA which suggests that the plasmid had integrated into the Y. lipolytica chromosomal DNA (Figure 1). The probe also hybridized to sharper bands of greater mobility which possibly consist of plasmid DNA. The mobility of these bands was similar to that of the open circular form of pRW1, but the form of the DNA was not known. The plasmid in ARS25 appeared to be somewhat larger than the plasmid in ARS17.

A large scale DNA preparation from ARS17 (500 ml culture) was prepared using the SDS/toluene extraction

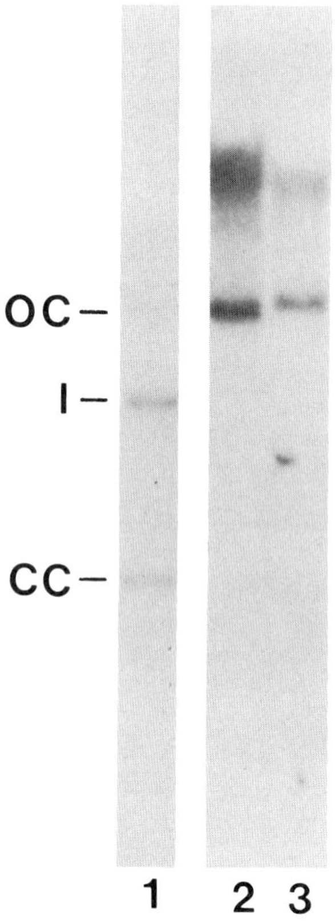

FIGURE 1. Southern blot of uncut DNA from unstable transformants hybridized with ^{32}P-labelled pBR322. Lane 1, pRW1; oc indicates open circle; ℓ indicates linear; and cc indicates closed circle. Lane 2, ARS17 DNA. Lane 3, ARS25 DNA.

procedure followed by CsCl/ethidium bromide gradient centrifugation. The gradient was fractionated and the fractions were analyzed on Southern blots by hybridization to ^{32}P-labeled pBR322. Both chromosomal and plasmid DNA were detected in the same fractions. This indicated that the plasmid was either in the open circle or linear form.

If the plasmid in ARS17 was in the open circular form, then based on the results in Figure 1 it was similar in size to pRW1 and this would suggest that it contained little or no insert DNA.

Initial attempts to isolate ARS17 plasmid DNA by transformation in E. coli and selection for ampicillin resistance have been unsuccesful. Digestion of ARS17 DNA with BglII or BamH1 (which cut pRW1 only once) followed by gel electrophoresis, Southern blotting and hybridization to 32P-labeled pBR322 revealed a very complicated situation. If the free plasmid in ARS17 was identical to pRW1, then a single band of 9.7 kb would be present in both digestions (Figure 2). For BamH1 two intense bands at 21 and 14.4 kb and a light band at 3 kb are seen. The BglII digest is quite different giving two intense bands of 14.4 kb and 7.3 kb, a medium intensity band of 8.8 kb and a very light band of 4.1 kb. Digests of chromosomal DNA with these two enzymes and Southern blot analysis showed that the LEU2 gene lies on a 11-12 kb BamH1 fragment and on two BglII fragments of 5.2 and 2.8 kb (unpublished results). Using this information, if it was assumed that ARS17 contained an insert with or without BamH1 and/or BglII sites and that the ARS17 DNA preparation contained free plasmid and plasmid DNA integrated (either single or multiple integrants) into the LEU2 region and/or into the region homologous to the insert then it was not possible to develop a consistent model to explain this data.

Multiple forms of RW1 including covalently closed circle, linear and open circle are recovered from E. coli. The presumed plasmids in the unstable transformants have only been detected in a single form. Based on the CsCl/ethidium bromide gradient results this form is probably linear or open circle. If the linear form resulted from random cutting of the plasmid, then after digestion with a restriction enzyme a broad smear of greater mobility than the plasmid probably should have been detected and it was not (Figure 2) which suggests that the plasmid is in the open circle form.

In an experiment in which stable transformants were examined by Southern blotting an unexpected result was obtained. In one of the transformants (RW1-8) the pBR322 probe hybridized to the chromosomal DNA and to what appeared to be multiple forms of the plasmid. The DNA was extracted by the SDS/toluene procedure and transformed into E. coli HB101. Four transformants were obtained, and the plasmids were recovered using the miniscreen procedure. Based on their EcoR1 and SalI restriction patterns, the plasmids were identical to pRW1. Since there could be important differences in the plasmid which would not be detectable

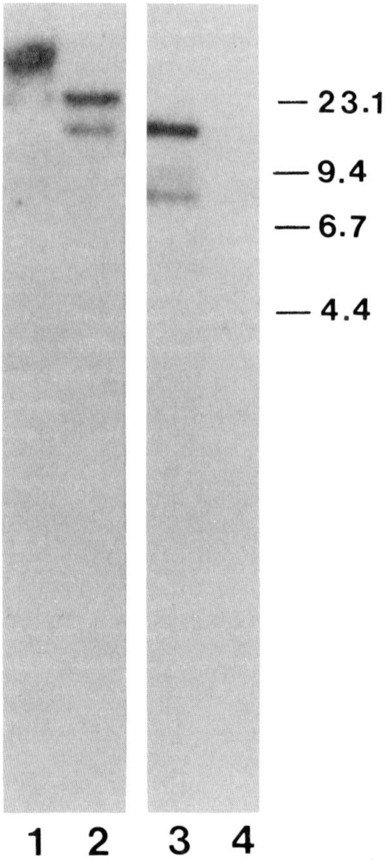

FIGURE 2. Southern blot ofDNA from unstable transformant ARS17 hybridized with ^{32}P-labelled pBR322. Lane 1, uncut DNA. Lane 2, DNA digested with BamH1. Lane 3, DNA digested with BglII. Lane 4, DNA digested with λ HindIII.

by differences in the sizes of restriction fragments, a large scale DNA preparation was made from RW1-8 and used to transform Y. lipolytica. The transformation frequency was less than twice that for uncut pRW1, no slow growing transformants were seen, and the Leu$^+$ phenotype was stable.

DNA isolated from transformant RW1-8 and from three stable transformants for which hybridization was only detected in the chromosomal DNA region was digested with *HindIII*, blotted and probed with pBR322. For all four transformants the probe hybridized to a band of high molecular weight material. This was expected since Southern blots of total genomic DNA indicated that the *LEU2* gene is on a very large *HindIII* fragment (>23 kb). For transformant RW1-8 there was also an intense band of 9.7 kb as would be expected for a free plasmid.

Perhaps plasmids have not yet been recovered from unstable transformants for technical reasons. Use of more gentle extraction procedures such as the one developed for isolating ring chromosome III from *S. cerevisiae* (20) in conjunction with transformation frequencies of 10^7-10^8 per μg of DNA which are now possible in *E. coli* may make recovery possible.

We have also attempted to find a *Y. lipolytica* *ARS* by screening in *S. cerevisiae*. Random *Y. lipolytica* DNA fragments were cloned into the nonreplicating plasmid YIp5 (19) and a fragment which had *ARS* activity in *S. cerevisiae* was isolated. Most of the insert DNA was subcloned into pRW1 and transformed into *Y. lipolytica*, but none of these fragments had *ARS* activity in *Y. lipolytica*.

DISCUSSION

The method of inserting random DNA fragments into a nonreplicating plasmid to find *ARSs* has been used successfully in *S. cerevisiae* (19,21), *Schizosaccharomyces pombe* (22), and *Kluveromyces lactis* (23).

In *S. cerevisiae* there is an average of one *ARS* per 32 kbp of DNA (21) and in *S. pombe* *ARSs* occur even more often. If *ARSs* occur in *Y. lipolytica* at an at all comparable frequency, then several should have been found in the over 300 kbp which were examined. Gaillardin screened an even larger plasmid library constructed by inserting *EcoR1* cut *Y. lipolytica* DNA into a nonreplicating plasmid, and he found no potential *Y. lipolytica* *ARS* (personal communication). This plasmid library was also screened for DNA fragments which functioned as *ARSs* in *S. cerevisiae*. Several were identified and recovered by transformation of *E. coli*, but as was true for our results none of these fragments behaved like an *ARS* in *Y. lipolytica*. Based on results with other yeast there was reason to hope that this approach might work. About half of the *S. pombe* fragments which are *ARSs* in *S. cerevisiae* are also *ARSs* in *S. pombe*.

The ARS KARS2 isolated in K. lactis also functions in Kluveromyces fragilis and in S. cerevisiae (24).

One explanation for the difficulty in isolating ARSs from filamentous fungi involves cytoplasmic mixing within the hypha and the resulting lack of direct selection on individual nuclei (25). We do not believe this is the major reason fo the difficulty in finding ARSs in Y. lipolytica since during growth of unstable transformants most of the cells are not filamentous

The rate of plasmid loss in nonselective medium is quite high and the percentage of cells carrying the plasmid under selective conditions is low (4.5%). Based on the micromanipulation results and results from experiments in which the percentage of cells containing the plasmid was determined for individual small colonies, the 4.5% value may be an overestimate. Since many colonies were picked to get sufficient cells for the inoculum for the stability study, some of these colonies may have contained cells in which the plasmid had already integrated. The increase in stability for ARS25 to 22% (Table 1) might be explained by a takeover by the faster growing cells containing integrated plasmid but the later decrease to 12% is then difficult to explain. At low cell densities the doubling times of unstable transformants in selective medium are about 4 to 6 h, but as the culture grows at some point the doubling time decreases rapidly to about 2.5 h. When this change occurs varies from culture to culture even for cultures for which the inocula are identical. Probably the timing of this increase in growth rate is related to when the integration event(s) occurred.

Southern blots which appear to have closed circle, open circle and linear forms of the plasmid have only been obtained for the stable transformant RW1-8. Several results suggest that an open circular form is being detected in the unstable transformants. Whether or not this is the state of the plasmid in vivo is unclear. Perhaps the extraction process nicks and linearizes the plasmid and the linearized form is degraded.

The Southern blots of the restriction digests (Figure 2) are not easy to explain. They do not seem to be due to cotransformation and integration of plasmids with different size inserts as was found by Gaillardin (personal communication). Perhaps rearrangements have occurred.

One possible explanation for this data could be that the unstable transformants contain plasmids with ARSs which segregate at high frequency. Perhaps a low average copy number, rearrangement of the plasmid, and/or technical problems with plasmid recovery from *Y. lipolytica* could explain why plasmids have not yet been recovered from unstable transformants. Another explanation would be that the plasmids do not replicate but they do integrate, and the plasmids detected on the Southern blots are pop outs. The high level of instability argues against the model. Nonreplicating plasmids have been recovered by transformation of *E. coli* from *S. pombe* and *Aspergillus* (25). The fact that slow growing, unstable transformants were not obtained after transformation with pRW1 or with Pool II argues that something different came out of Pool I and that such results would not be obtained with all nonreplicating plasmids. If the plasmid does contain an ARS then once the plasmid is recovered it should be possible to stabilize it with a *Y. lipolytica* centromere sequence.

Y. lipolytica genes have been cloned in *Y. lipolytica* using the integrative transformation system (L. Davidow, personal communication). Such cloning is more difficult than cloning with a replicating plasmid, and this is one reason to keep searching for a *Y. lipolytica* ARS. Another reason is that maybe the difficulties in cloning a *Y. lipolytica* ARS reflect some basic differences between *Y. lipolytica* and the other yeast where ARSs have been found. Perhaps these differences are relevant to filamentous fungi.

ACKNOWLEDGMENTS

We thank L. Davidow for pLD25 and J. DeZeeuw for PC30827 and W. Timberlake for helpful discussion. We thank L. Davidow, D. Apostolakos, M. O'Donnell, A. Proctor, I. Stasko, J. DeZeeuw, C. Gaillardin, A. Ribet, and H. Heslot for communication of unpublished results.

REFERENCES

1. Gaillardin CM, Fournier P, Sylvestre G, Heslot H (1976). Mutants of *Saccharomycopsis lipolytica* defective in lysine catabolism. J Bacteriol 125:48.
2. Rodriguez C, Dominguez A (1984). The growth characteristics of *Saccharomycopsis lipolytica*: morphology and induction of mycelium formation. Can J Microbiol 30:605.
3. Wickerham LJ, Kurtzman CP, Herman AI (1969). "Recent

trends in yeast research." Chazy: Miner Institute, p 81.

4. Ogrydziak D, Bassel J, Contopoulou R, Mortimer R (1978). Development of genetic techniques and the genetic map of the yeast Saccharomycopsis lipolytica. Mol Gen Genet 163:229.
5. Gaillardin CM, Charoy V, Heslot H (1973). A study of copulation, sporulation and mitotic segregation in Candida lipolytica. Arch Microbiol 92:69.
6. Ogrydziak DM, Bassel J, Mortimer RK (1982). Development of the genetic map of the yeast Saccharomycopsis lipolytica. Mol Gen Genet 188:179.
7. Eibel H, Philippsen P (1983). Identification of the cloned Saccharomycopsis cerevisiae LYS2 gene by an integrative transformation approach. Mol Gen Genet 191:66.
8. Boyer HW, Roulland D (1969). A complementation analysis of the restriction and modification of DNA in Escherichia coli. J Mol Biol 41:459.
9. Casadaban MJ, Martinez-Arias A, Shapira SK, Chou J (1983). β-Galactosidase gene fusions for analyzing gene expression. In Wu R, Grossman L, Moldave K (eds): "Methods of Enzymology, Recombinant DNA Part B", 100:293 New York: Academic Press, Inc.
10. Sherman P, Fink GR, Lawrence CW (1979). "Methods in Yeast Genetics." Cold Spring Harbor: Cold Spring Harbor Laboratory.
11. Maniatis T, Fritsch EF, Sambrook J (1982). "Molecular Cloning (A Laboratory Manual)." Cold Spring Harbor: Cold Spring Harbor Laboratory.
12. Specht CA, DiRusso CC, Novotny CP, Ullrich RC (1982). A method for extraction of high-molecular-weight deoxyribonucleic acid from fungi. Anal Biochem 119:158.
13. MacKay VL (1983). In Wu R, Grossman L, Moldave K (eds): Cloning of yeast STE genes in 2 μm Vectors. "Methods of Enzymology, Recombinant DNA Part C," 101:325 New York: Academic Press Inc.
14. Stahl U (1978). Zygote formation and recombination between like mating types in the yeast Saccharomycopsis lipolytica by protoplast fusion. Mol Gen Genet 160:111.
15. Rodriguez RL, Tait RC (1983). "Recombinant DNA Techniques: an Introduction." Reading: Addison-Wesley, p 41.
16. Ito H, Fukuda Y, Murata K, Kimura A (1983).

Transformation of intact yeast cells treated with alkali cations. J Bacteriol 153:163.
17. Hanahan D (183). Studies on transformation of Escherichia coli with plasmids. J Mol Biol 166:557.
18. Southern EM (1975). Detection of specific sequences among DNA fragments separated by gel electrophoresis. J Mol Biol 98:503.
19. Stinchomb DT, Thomas M, Kelly J, Selker E, Davis RW (1980). Eukaryotic DNA segments capable of autonomous replication in yeast. Proc Natl Acad Sci USA. 77:4559.
20. DeVenish RJ, Newlon CS (1982). Isolation and characterization of yeast ring chromosome III by a method applicable to other circular DNAs. Gene 18:277.
21. Chan CSM, Tye B-K (1980). Autonomously replicating sequences in Saccharomyces cerevisiae. Proc Natl Acad Sci USA 77:6329.
22. Beach D, Nurse P (1981). High-frequency transformation of the fission yeast Schizosaccharomyces pombe. Nature 290:140.
23. Das S, Hollenberg CP (1982). A high-frequency transformation system for the yeast Kluyveromyces lactis. Curr Genet 6:123.
24. Das S, Kellermann E, Hollenberg CP (1984). Transformation Kluyveromyces fragilis. J Bacteriol 158:1165.
25. Buxton FP, Radford A (1984). The transformation of mycelial spheroplasts of Neurospora crassa and the attempted isolation of an autonomous replicator. Mol Gen Genet 196:339.
26. Turner G, Ballance DJ (1985). Cloning and transformation in Aspergillus. In Bennett JW, Lasure L (eds): "Gene Manipulations in Fungi," New York:Academic Press, Inc. (in press).

Molecular Genetics of Filamentous Fungi, pages 383–403

MOLECULAR ANALYSIS OF THE PLANT-FUNGUS INTERACTION[1]

O. C. Yoder and B. G. Turgeon

Department of Plant Pathology, Cornell University
Ithaca, New York 14853

INTRODUCTION

The purposes of this chapter are to: 1) introduce briefly the four subsequent chapters in this section by summarizing the definitive information now available on the molecular nature of the plant-fungus interaction, and 2) outline the technology needed for molecular dissection of the pathogenic process. Portions of this discussion are taken from a previous article (1).

Plant pathogenic fungi confront two fundamental problems in their encounters with plants. First, the protective outer surface of the plant must be breached. Then, once within the plant, the fungus must determine whether or not the plant will support its growth and reproduction.

Fungi employ a number of strategies to deal with such problems. These include the elaboration of infection structures and enzymes needed for penetration of the plant epidermis, and the production of toxins, enzymes, and perhaps even hormones, which act as chemical messages to help determine whether or not the plant that has been attacked is an appropriate host.

Although many fungal structures and metabolites have been proposed as pathogenicity factors, few have been evaluated rigorously for roles in disease. Certain host-parasite systems however, have features that facilitate experimental analysis and in these cases the roles in disease of particular fungal products have been described.

[1]This work was supported by grants from NSF, USDA/CRGO, and Pioneer Hibred International.

FUNGAL PRODUCTS INVOLVED IN PATHOGENESIS

Infection Structures.

Particular morphological changes appear necessary for fungal penetration into plants, although extracellular enzymatic action is also involved, at least in some cases. Rust fungi provide favorable systems for studying prepenetration activities because their infection structures are distinct, elaborate, and inducible. Recent advances in understanding rust differentiation have been reviewed by Staples and Macko (2). The emphasis in this section, following a brief description of the phenomenology, will be on the prospect for genetically dissecting the infection process and determining how the genes control pathogenicity.

Spores of the bean rust fungus *Uromyces appendiculatus* (also known as *U. phaseoli*) germinate on the leaf surface and produce a germ tube which quickly attaches itself snugly to the surface of the leaf. This attachment is a critical step. If it fails, for example if the spore germinates on the surface of a waxless leaf to which it cannot stick, none of the subsequent developmental changes occur (3). As the germ tube grows across the surface of the leaf, it orients its direction of growth perpendicular to the surface ridges of the leaf. The result of this orientation is that germ tubes grow in straight lines, rather than randomly, so that the probability of finding a stomate, the site of penetration, is maximized (3). Upon contacting a stomate, growth of the germ tube stops, the tip of the germ tube differentiates into a structure known as an appressorium, which covers the stomate, and growth resumes from the appressorium downward via a peg through the pore of the stomate into the substomatal cavity. There a substomatal vesicle forms. This gives rise to infection hyphae which in turn produce haustorial mother cells. Upon contact with a leaf mesophyll cell, haustoria form that are capable of breaching the wall. At this point, the fungus must determine whether or not the plant is a suitable host. If so, it will establish a nutritional relationship with the plant, colonize the tissues extensively, and eventually sporulate. If not, the fungus grows no further, fails to reproduce, and the plant recovers from the attack.

It is important to note that most if not all fungi that attack foliage perform on plant surfaces in a manner similar to that described above, although there are many variations on the theme. Some fungi for example, penetrate directly through epidermal walls rather than through stomates, and in many cases the infection structures are not as distinct as they are in rust fungi. Conversely, saprophytic fungi undergo none of the differentiation phases described for parasitic ones. Saprophytes grow profusely on leaf surfaces but do not attach themselves to the epidermal layer, do not orient their growth with respect to the topography of the leaf surface, do not elaborate infection structures, and do not attempt to penetrate, even when they contact an opening such as a stomate by chance. This observation alone constitutes strong circumstantial evidence that production of infection structures is required for fungal pathogenicity to foliage.

Differentiation of infection structures is an inducible process. Differentiation does not occur on smooth surfaces. For U. appendiculatus the stimulus appears to be physical only; no chemicals are needed. If positive leaf replicas are made of chemically inert polystyrene and inoculated with spores, the fungus differentiates normally as though it were on a leaf surface (4). However, on a smooth surface in the absence of physical stimuli certain chemicals or other treatments will induce differentiation. All or part of the infection structures will be produced on a noninducing surface in the presence of potassium ions (5) or cyclic AMP (6), or after heat shock or treatment with ultrasound (7). The most effective inducers are physical stimuli such as a stomate or a scratch on an artificial membrane (8).

Several biochemical changes occur in the fungal cell as it starts to differentiate. DNA synthesis begins, nuclei divide (9), several new proteins appear (10), and the elements of the cytoskeleton take on a new orientation (2). It seems likely that at least some of the new proteins are components of the cytoskeleton (see below). The basic molecular question then is how the fungus recognizes a signal from the leaf surface and transmits that signal to the nucleus, where several genes are activated presumably to start the differentiation process.

There is indirect evidence that the cytoskeleton of U. appendiculatus (11) plays an important role in differentiation. First, one of the proteins whose synthesis is greatly stimulated after induction is calmodulin, a component of the cytoskeleton (R. C. Staples, personal communication). Second, the greatest concentration of cytoskeletal filaments is near the interface between the fungus and the substrate. Third, chemicals such as colcemid, griseofulvin, and nocodazole, which depolymerize microtubules, or podophyllotoxin and the cytochalasins, which depolymerize microfilaments, induce the start of differentiation on a smooth surface where differentiation would not normally occur (7). Heat shock and ultrasound also depolymerize the cytoskeleton. Further, chemicals that stabilize microtubules (deuterium oxide) and microfilaments (phallotoxin or taxol) inhibit differentiation under conditions where it would normally occur.

It is possible to construct a model to explain differentiation of infection structures in terms of the cytoskeleton. In this model the cytoskeleton could be involved in any or all of three processes. First, the actual sensing of the shape of the substrate could be mediated by proteins associated with the cytoskeleton, since there is a high concentration of microfilaments at the growing tip of the germ tube (11) and proteins in the extracellular matrix of the germ tube are necessary for sensing to occur (L. Epstein and R. C. Staples, personal communication). These extracellular proteins may be connected to the cytoskeleton. Second, cytoskeletal cables could transmit the signal from the germ tube tip to the nucleus, where differentiation-specific genes are activated. Third, the cytoskeleton could determine the particular shapes of the various infection structures. To test this model it will be necessary to isolate genes that control production of the differentiation-specific proteins, as well as other genes that are activated during differentiation, and determine their metabolic functions as well as the mechanisms by which they are regulated in response to tactile stimuli.

Cutinase.

A long-standing issue regarding fungal penetration of plants is whether the process is strictly mechanical or involves enzymatic digestion of the plant cell surface. Although both mechanisms are probably involved, direct evidence in support of either of them has been difficult to obtain. Recently, Kolattukudy and his colleagues evaluated the role of cutinase in penetration and concluded that in certain cases at least this enzyme contributes to successful entry of the fungus into the plant. This topic is treated fully in the following chapter by Kolattukudy and will not be discussed further here.

Pisatin Demethylase.

Nectria haematococca is a fungal pathogen of peas. When under attack, the pea plant responds by activating a biosynthetic pathway which gives rise to an antibiotic called pisatin. This metabolite is inhibitory to the growth in culture of most fungi, including some strains of Nectria.

Nectria isolates collected from nature exhibit a range of virulences toward pea. All of the strains with high virulence are tolerant of pisatin whereas strains with low virulence range from tolerant to sensitive to pisatin in culture (12).

Tolerance of pisatin is associated with the production by the fungus of an enzyme called pisatin demethylase (PDA), which is a cytochrome P-450 monooxygenase that detoxifies pisatin by removing a methyl group (13). In pisatin-tolerant isolates PDA is inducible but the level of enzyme produced varies among isolates (14). Chemicals that will induce PDA include only pisatin itself and certain closely related compounds (15). Isolates that are sensitive to pisatin produce no PDA at all.

Genes controlling production of PDA and the involvement of PDA in the virulence of the fungus have been investigated by genetic analysis (16, 17), which has resulted in several conclusions. First, several genes control PDA production. A single gene, Pda1, specifies high PDA production and another gene, Pda2, determines low PDA production. A third gene, Pda3, also controls production of low PDA (18). It is not yet known whether all three loci control production of the same enzyme. Second, high PDA production is required for high virulence of the fungus toward peas, i.e., highly virulent isolates always produce high levels of PDA. Third, weakly virulent isolates may have either no detectable PDA, low PDA, or even high PDA; the latter class demonstrates that more than PDA alone is required for high virulence to peas. Fourth, only Pda$^-$ strains, which lack PDA completely, are sensitive to pisatin in culture.

Toxins.

Toxic metabolites are commonly produced by plant pathogenic fungi. Many of these compounds are generally toxic to living cells (19, 20) whereas some of them are specific to only certain genotypes of higher plants (21, 22, 23, 24). In a number of cases, the toxin-sensitive plant is also the host of the toxin-producing fungus. Such fungal metabolites are called host specific toxins because they have high activity only against fungal host plants.

None of the nonspecific toxins produced by pathogenic fungi have been critically evaluated for possible roles in plant disease, but several of the host-specific ones have undergone extensive analysis (24). For purposes of this discussion, only host-specific toxins will be considered because their unusual specificity for host plants alone implicates them as pathogenicity molecules. Further, as described below, some of them have been subjected to genetic analysis, which has demonstrated that they are causally involved in pathogenesis.

Chemically, the host-specific toxins are a diverse group of low molecular weight secondary metabolites (25). Two of them, HC-toxin and AM-toxin, are cyclic tetrapeptides of MW 436 and 445, respectively. HS-toxin is a sesquiterpene galactofuranoside (MW 884), T-toxin is a linear polyketol (MW 768), AK-toxin is an ester of epoxydecatrienoic acid (MW 413), and AAL-toxin is a dimethylheptadecapentol ester of propanetricarboxylic acid (MW 508). Several of these toxins are found in culture filtrates as families of isomers. The chemistry and biosynthesis of many nonspecific toxins have been investigated (20).

The genetics of toxin production by many fungal pathogens has not been analyzed because most of them have no known sexual cycle. An exception to this is the genus Cochliobolus, an Ascomycete with a meiotic cycle much like that of Neurospora (26). Several members of the genus produce host-specific toxins, e.g. HV-toxin, HC-toxin,

T-toxin, and HS-toxin. In each case, (except for HS-toxin, which has not been genetically analyzed), both Tox^+ and Tox^- strains are found to occur naturally. Tox^+ strains are always pathogenic (or highly virulent) whereas Tox^- strains are nonpathogenic (or weakly virulent). When a cross is made between a pathogenic Tox^+ strain and a nonpathogenic Tox^- strain there is 1:1 segregation for toxin production, indicating a single gene involved in each case. All toxin-producing progeny are pathogenic whereas all toxin-less progeny are nonpathogenic. Furthermore, only parental types and no recombinants are found among progeny. These correlations between pathogenicity and ability to produce toxin provide convincing evidence that each of the toxins examined in this way is casually involved in pathogenesis (24).

The level of involvement in pathogenesis varies from toxin to toxin. HV-toxin and HC-toxin appear to be required by the respective toxin-producing fungi in order to cause any disease at all; these toxins are called pathogenicity factors (24). T-toxin on the other hand is not required for pathogenicity because Tox^- strains can cause low levels of disease. The toxin appears to regulate the level of virulence of the fungus since Tox^+ strains simply cause more disease than Tox^-; this toxin is known as a virulence factor.

The mechanisms by which toxins interact with plant cells to cause damage are generally unknown. Three cellular components have been proposed as sites of toxin action: the plasma membrane, the mitochondrion, and a specific enzyme (1). There is direct evidence supporting only the enzyme site; this is discussed in detail in the following chapter by Gilchrist.

Double-stranded RNA.

There is only one well-documented case of double-stranded RNA playing a causal role in virulence of a fungus toward its host. The disease is Chestnut Blight caused by the fungus *Endothia parasitica*. There is now substantial evidence that if dsRNA is present in *E. parasitica*, it reduces the virulence of the fungus toward chestnuts. Highly virulent strains lack dsRNA, although they become weakly virulent if they acquire dsRNA from a strain that carries it. This system is described fully in the following chapter by Van Alfen.

CLONING AND ANALYSIS OF PATHOLOGICALLY-IMPORTANT GENES FROM FUNGI

Choosing a System.

Before serious molecular biological work can be done with fungal plant pathogens, they must be sufficiently domesticated to permit efficient laboratory manipulation. Pathogens, as they are recovered from nature, are too variable and unpredictable for careful experimental work. It is therefore necessary to breed and select strains that perform well under laboratory conditions. But first an appropriate pathogen should be chosen. Criteria that are useful in deciding upon an experimental subject include culturability, an easily manipulated sexual cycle, rapid vegetative growth, short life and disease cycles, and a host plant that also satisfies these criteria. In addition to these items, it is important that a fungus is amenable to mutation and selection of useful mutants such as auxotrophs, and more importantly, mutations that affect pathogenicity. Moreover, protocols must be developed for the production of large numbers of protoplasts which will regenerate at high frequency and methods for isolation of DNA that is of large size and digestable with restriction endonucleases must be available (27).

Another important consideration is the ability of the fungus to withstand long term storage. Many important pathogenic strains have been lost because of improper storage. In addition, it is clear that fungi can change rapidly after being isolated from nature and cultured (28). Thus, it is extremely important that all strains be metabolically immobilized in permanent storage immediately after acquisition and recovered directly from such storage for each experiment. In the past, convenient and reliable long-term storage methods have not been available. But with the development of inexpensive low temperature (-80 C or -135 C) freezers, it is now possible to store large numbers of strains in cryovials in glycerol (29). This method is advantageous because preparation of microbes for storage and their recovery from storage are simple rapid procedures and because the storage medium is aqueous, which obviates the need to maintain dry storage compartments. An alternative to low temperature glycerol storage is the silica gel method (30), which is known to preserve Neurospora for at least 20 years (31).

Vector Construction

Once the basic microbiological tools are in hand, it is necessary to construct a suitable transformation vector or set of vectors. To date only a few transformation systems for fungi have been described (32). The purpose of the discussion here is to emphasize the points that may be particularly significant in construction of vectors for undeveloped organisms.

At this time, high frequency transformation has not been achieved for plant pathogenic fungi. Among the filamentous fungi, only Aspergillus and Neurospora have efficient transformation systems. The modest number of successes so far reflects the unexpected difficulties that have been encountered by those who are developing transformation systems for fungi. From the information now available, it is apparent that the use of homologous sequences in vector construction is more likely to be successful than dependence on heterologous components. Thus, it is important to obtain selectable genes, promoters, and origins of DNA replication from the organism to be transformed.

Selectable genes are usually found in nuclear DNA and can sometimes be isolated by complementation of heterologous organisms (eg. E. coli or yeast) with libraries of DNA constructed in a vector. Promoters and DNA origins of replication can be found on chromosomes or on native plasmids. Those on plasmids may be easier to isolate and more highly expressed than those from chromosomes. Indeed, one only need consider the achievements made possible in yeast by the resident 2μ circle to recognize the potential importance of native plasmids in vector construction.

An essential part of a vector is a selectable gene. Most selectable genes that have been used for transformation of eukaryotic microbes are those that correct nutritional deficiencies For example, in yeast the most frequently employed selectable genes are URA3, LEU2, TRP1, and HIS3; in Neurospora the most popular selectable genes are qa2 and am; in Aspergillus, trpC and argB are used. However, in order to use such genes the corresponding mutations must be available, yet particular mutants are often difficult to obtain in filamentous fungi.

An exceptional nutritional gene is amdS from Aspergillus nidulans, which codes for acetamidase, an enzyme that converts acetamide to acetate and ammonia. Wild type A. nidulans is amdS+ and can use acetamide as a sole source of nitrogen, whereas amdS- strains cannot. Hynes et al. (33) have cloned the amdS gene, which has been used as a selectable marker for transformation of both A. nidulans (34) and A. niger (35). An advantage of using amdS is that a particular mutation is not required in recipient cells; the only requirement is that cells lack (or have low expression of) acetamidase. This makes any genetically undeveloped fungus (which includes most of the economically important ones) a potential candidate for transformation by amdS, assuming it does not already produce high levels of acetamidase. Indeed, we have recently shown that amdS (kindly provided to us by Dr. M. Hynes) can serve as a selectable marker for the transformation of the plant pathogen Cochliobolus (B. G. Turgeon, R. C. Garber, and O. C. Yoder, unpublished).

An alternative to nutritional genes are genes for drug-resistance. Such genes are preferable because they are selectable in all wild type sensitive strains without the need for induced mutations. The most commonly-used drug-resistance gene for eukaryote transformation is the kan (kanamycin-resistance) gene of E. coli. This gene codes for an enzyme (aminoglycoside phosphotransferase) which detoxifies not only kanamycin but the experimental drug G418 as well. G418 inhibits growth of both prokaryotic and eukaryotic cells. The kan gene has been shown to confer resistance to G418 in cells of yeast (36); it is also expressed in cells of higher plants (37), and mammals (38) if fused to an appropriate eukaryotic promoter. However, the kan gene has not been readily useable in filamentous fungi and only recently was reported to be expressed in Neurospora (39). One problem is that high concentrations (0.5-1.0 mg/ml) of G418 are needed to adequately inhibit growth of many fungi.

A more toxic drug, hygromycin, has been reported for use against both prokaryotes and eukaryotes. Hygromycin is effective against fungi at approximately 10-fold lower concentration than is G418, and there is now a cloned gene, from Streptomyces, for resistance to it. This gene, hyh, confers hygromycin-resistance in yeast provided any of several yeast promoters are fused to it (40, 41).

Thus, it is a likely prospect for selectability in filamentous fungi if coupled to a homologous promoter.

One desirable feature of vectors is ability to replicate autonomously in the eukaryotic system. To do this the vector needs an appropriate origin of DNA replication. Sources of origins include native plasmids, the best example of which in eukaryotes is the yeast 2μ circle, and chromosomes. There is a convenient method for testing ability of DNA sequences to replicate autonomously in yeast (42). The sequence to be tested is inserted into a vector such as YIp5, which carries a gene selectable in yeast (URA3) but no yeast origin of DNA replication, and transformed into yeast cells. Sequences that promote autonomous replication in yeast are called ARSs (autonomously replicating sequences) rather than origins, since they are not known to be authentic origins of DNA replication in the homologous systems.

We have isolated ARSs from *Cochliobolus* using the Stinchcomb et al. (42) procedure. Approximately 750 yeast transformants were recovered from an EcoRI library of *Cochliobolus* DNA in YIp5. Forty transformants were examined further and found to contain plasmids with *Cochliobolus* inserts; each plasmid promoted high frequency transformation of yeast (compared to a low frequency integrating plasmid) and was mitotically unstable (compared to a relatively stable plasmid carrying the yeast 2μ circle replicon). Genomic locations and frequencies were established for ten of these ARSs by gel transfer analysis. Two of the ten were a 1.05 kb fragment of the mitochondrial chromosome, three were the 5.7 kb EcoRI fragment of the 9 kb tandemly repeated unit that carries the 5.8S, 17S, and 25S ribosomal RNA genes (R. C. Garber, B. G. Turgeon, E. U. Selker, and O. C. Yoder, unpublished), and the remaining five were from nuclear DNA that had no homology to the tandemly repeated rDNA. Of the latter five, ARS15 was recovered three times from the group of ten, yet it hybridized to a single EcoRI fragment of genomic DNA, which hints that ARS15 is on a tandemly repeated sequence. ARS20 was found once among the ten and it hybridized to multiple EcoRI fragments in genomic DNA, suggesting a dispersed sequence.

The value of ARSs from most eukaryotes, especially those that have been isolated by complementation of *S. cerevisiae*, has not yet been demonstrated. In general ARSs returned to the organism of origin have not been found to function there as true replicons. This may be due in part to the lack of suitable transformation vectors for most of the organisms from which ARSs have been isolated. But in the case of *N. crassa* ARS8, which was isolated by complementation of *S. cerevisiae*, there was no detectable replicon function in *N. crassa* (43). It is better to isolate ARSs by complementing the homologous system, which has been done for *K. lactis* (44, 45) and *S. pombe* (46, 47).

Centromeres may also become important components of vectors for filamentous fungi although only yeast centromeres have been cloned so far. Chromosomal sequences from both *S. cerevisiae* (48) and *S. pombe* (49) have been identified as functional centromeres. The effect of a chromosomal centromere in an ARS-containing plasmid is to maintain the copy number at about one per cell and to cause the plasmid to segregate stably in mitosis and meiosis as though it were a normal chromosome.

Stable vectors with low copy number are important for two reasons. First, some gene products at high concentration are toxic to cells. If such products are over-produced as the result of being encoded by a gene on a multicopy plasmid, cells containing the gene of interest can be killed. The low copy number of centromere-containing plasmids reduces the possibility of missing a gene because its product is toxic. Second, the stability provided by a centromere permits growth of cells containing an autonomously replicating plasmid on non-selective medium. This could be important for plant pathogenic fungi. For example, some toxins are produced in large amounts by fungi on a rich medium and little or not at all on a nutritionally-exacting selective one. Thus, analysis of a gene controlling toxin production would be facilitated by a plasmid containing a centromere.

Centromeres have been isolated from yeasts either by "walking" from a cloned centromere-linked gene or by selecting for ability of DNA fragments to mitotically stabilize an otherwise unstable plasmid (48). For plant pathogenic fungi, centromere-linked genes are generally unknown, so the alternative approach is more plausible. However, autonomously-replicating vectors are so-far not available, making heterologous systems necessary for centromere isolation. In choosing such a system it may be important to consider what is known about conservation of centromere sequences. S. cerevisiae centromeres appear to be highly conserved among chromosomes within the strains examined and are very small (∿ 200 bp). They are quite different from centromeres of S. pombe, which are several kb long (49) and similar in size to centromeres from higher eukaryotes. Thus, it may be found that isolation of centromeres by selecting for mitotic plasmid stability in a heterologous system will depend on choosing a system which recognizes the relevant centromere sequences.

Approaches to Isolation of Pathogenicity Genes.

Isolation of genes controlling pathogenicity in fungi can be approached by a variety of methods. Functional complementation is most generally useful because it permits recovery of any gene for which there is a mutation that renders the gene nonfunctional. The main requirement for this approach is an efficient transformation system and the strategy used to isolate a pathogenicity gene (PAT^+) is the following. A library of DNA fragments from a PAT^+ strain is constructed in a vector carrying a gene (SEL^+) selectable in the pathogen of interest. The PAT^+ and SEL^+ alleles must be dominant. Protoplasts of a sel^-; pat^- strain are transformed with the library and transformants are selected for Sel^+. They are then screened for the presence of the PAT^+ allele. This may be done by an in vitro assay (50) if the metabolic product of PAT^+ is known (eg. if it is a toxin or an enzyme). If no product is known the transformants would be screened in vivo for ability to cause disease on a plant susceptible to the PAT^+ wild type strain. When a transformant with the desired pathological phenotype is found, the transforming DNA must be recovered. If the vector is autonomously-replicating, plasmid DNA can be isolated directly and amplified in E. coli. If an integrating cosmid vector is used, the transforming DNA can be recovered by packaging chromosomal sequences containing the cos site in the bacteriophage lambda, and transfecting E. coli (51, 52).

There are methods for isolating genes for which no mutant alleles are available. If the gene is inducible or developmentally-regulated it can be isolated *via* an abundant mRNA. For example, a large number of developmentally-regulated conidiation genes have been isolated from *A. nidulans* by virtue of their differential expression in conidiating vs. mycelial cultures (53). This approach requires the removal by hybridization of all mRNAs common to both expressing and nonexpressing cultures. If the protein product of the gene is known and can be purified, antibodies to the purified protein can be prepared and used to screen a library of DNA fragments from a wild type strain carried in an *E. coli* expression vector such as λgt11 (54). Alternatively, the amino acid sequence of the protein can be determined and short synthetic oligonucleotides can be prepared which correspond to regions of the genetic code that are least degenerate. These synthetic probes can be used to screen a library constructed in *E. coli*. For example, the *Neurospora am* gene for glutamate dehydrogenase was isolated by this method (55).

Uses for Cloned Pathogenicity Genes.

Analyze roles of metabolites in pathogenesis. Physically defined genes provide the most powerful tools available to analyze whether or not a metabolite plays a role in pathogenesis, and if so, the nature and extent of the role. In the past, metabolites of fungal pathogens have been evaluated as factors in disease using experimental designs that do not exclude uncontrolled variables (24). If a gene controlling production of a fungal metabolite is cloned, experiments can be performed in which there is a single variable, i.e. the cloned gene itself. To do such an experiment, protoplasts of a strain lacking the metabolite in question would be transformed with a plasmid containing the gene that determines production of the metabolite. Control protoplasts would receive the same plasmid but with an inactive gene for production of the metabolite, thus defining the single variable in the experiment. Colonies derived from the transformed protoplasts would be purified and tested both for the production of the metabolite and for any change in pathogenicity or virulence toward plants. If such a change is detectable, that change indicates the role of the metabolite in disease development. A complication in this experiment could arise if the gene had pleiotropic effects. This possibility could be reduced if the cloned sequence were known to be a structural gene, if there were no other apparent changes in the life-cycle of the fungus, and if the metabolic effects of the gene and its product were fully understood.

Determine distribution of homologous sequences. A cloned gene can be used as a probe for the isolation of other alleles at its locus, to determine the number of copies per genome, and to assess the phylogenetic boundaries of the gene's range. Isolation of homologous alleles from the same organism would lead to an understanding of whether the difference between mutant and wild type strains is determined at the level of transcription or translation. It would also be of interest to probe genomic DNAs of a variety of organisms, both pathogens and nonpathogens, with a cloned pathogenicity gene to see which organisms have it. This information would bear on the issue of whether or not pathogens have unique genes for pathogenicity, and therefore unique metabolites for pathogenicity. Alternatively both pathogens and nonpathogens may have the same sets of structural genes but different mechanisms of regulation. The latter view would be supported if it were found that a gene proven to be necessary for pathogenicity and first isolated from a pathogen, also resided and functioned in the genomes of nonpathogens.

Determination of gene function and regulation. The nucleotide sequence of a cloned pathogenicity gene would reveal the coding and the regulatory regions. By comparison with sequences of other genes, it may be possible to find clues to the gene's function and to identify the elements involved in regulation. Suggestions about function may be found by scrutinizing the sequence (with computer assistance) and by analysis of its polypeptide product. Elements involved in regulation can be identified and studied by the construction of hybrid regulatory regions (56) and assessment of their effects on expression of a gene in the homologous or heterologous genome. For example, when the regulatory region of the yeast CYC1 gene (which is regulated by catabolite repression) is placed upstream of the LEU2 gene (which is not normally regulated by catabolite repression), expression of the LEU2 gene falls under control by catabolite repression (57). It should be possible to identify the regulatory regions of pathogenicity genes by a similar approach.

Construction of agents for biological pest control. Pathogenic fungi are potential biological control agents for pests such as weed plants and insects (58), yet few fungi have been used successfully for biological pest control. There are a number of reasons for this, but one of the most important is that in general fungal pathogens, while being highly specific for their hosts, are not virulent enough to effectively reduce the host population under field conditions. It may be possible to remedy this deficiency by genetic engineering techniques once a supply of cloned pathogenicity genes is available.

Genes useful in construction of biocontrol agents can be classified according to their functions. One class of genes is involved in the production of specialized structures that are needed by the fungus to penetrate the host surface. These genes are thought to control general attacking mechanisms, since virtually all pathogenic fungi form some type of infection structure, whereas nonpathogens do not. Examples of attacking mechanisms include the enzyme cutinase and components of the cytoskeleton.

A second class of genes is invoked after the fungus has penetrated into the plant and must decide whether the plant tissue is a congenial or hostile environment in which to complete its life cycle. Genes involved in this decision control molecular mechanisms that determine host range as well as levels of virulence. The types of metabolic gene products involved here include toxins that condition the host tissue for colonization and enzymes that detoxify toxic compounds produced by the host.

Yet another group of genes is needed by the pathogen for fitness in nature. These genes are probably not confined to pathogens and may confer qualities such as resistance to cold, heat, or dessication, the quality and quantity of spores, and for facultative organisms, the ability to compete saprophytically. Survivability in the field may be altered by genes that affect fitness. For example, a gene that changes the shape of the conidium may facilitate aerial dispersal. Conversely, if the life span of the fungus were to be limited, specific genes for senescence may be added, or genes for cold-sensitive ribosomes (59) may be used to prevent overwintering of inoculum.

There are currently projects underway in various laboratories which aim to isolate genes representing each of the three classes. When such genes are available, realistic attempts to design fungal biocontrol agents will be possible. The strategy will be to take advantage of the natural host-specificity of fungal pathogens to target the control to a particular pest or group of pests, and then to genetically engineer increased virulence by using genes of the various types described above.

REFERENCES

1. Yoder OC, Turgeon BG (1985). Molecular bases of fungal pathogenicity to plants. In Bennett J, Lasure L (eds): "Gene Manipulations in Fungi", New York: Academic Press, in press.
2. Staples RC, Macko V (1984). Germination of urediospores and differentiation of infection structures. In Bushnell WR, Roelphs P (eds): "The Cereal Rusts. Vol. 1, Origins, Specificity, Structure, and Physiology," Academic Press, Inc. p 255.
3. Wynn WK, Staples RC (1981). Tropisms of fungi in host recognition. In Staples RC, Toenniessen GH (eds): "Plant Disease Control: Resistance and Susceptibility" New York: John Wiley, p 45.
4. Wynn WK (1976). Appressorium formation over stomates by the bean rust fungus: response to a surface contact stimulus. Phytopathology 66:136.
5. Staples RC, Grambow H-J, Hoch HC (1983). Potassium ion induces rust fungi to develop infection structures. Exp Mycol 7:40.
6. Hoch HC, Staples RC (1984). Evidence that cAMP initiates nuclear division and infection structure formation in the bean rust fungus, Uromyces phaseoli. Exp Mycol 8:37.
7. Staples RC, Hoch HC (1982). A possible role for microtubules and microfilaments in the induction of nuclear division in bean rust uredospore germlings. Exp Mycol 6:293.
8. Staples RC, Grambow H-J, Hoch HC, Wynn WK (1983). Contact with membrane grooves induces wheat stem rust uredospore germlings to differentiate appressoria but not vesicles. Phytopathology 73:1436.

9. Staples RC, App AA, Ricci P (1975). DNA synthesis and nuclear division during formation of infection structures by bean rust uredospore germlings. Arch Microbiol 104:123.
10. Huang B-F, Staples RC (1982). Synthesis of proteins during differentiation of the bean rust fungus. Exp Mycol 6:7.
11. Hoch HC, Staples RC (1983). Visualization of actin _in situ_ by rhodamine-conjugated phalloin in the fungus _Uromyces phaseoli_. Eur J Cell Biol 32:52.
12. VanEtten HD, Matthews PS, Tegtmeier KJ, Dietert MF, Stein JI (1980). The association of pisatin tolerance and demethylation with virulence on pea in _Nectria haematococca_. Physiol Plant Pathol 16:257.
13. Matthews DE, VanEtten HD (1983). Detoxification of the phytoalexin pisatin by a fungal cytochrome P-450. Arch Biochem Biophy 224:494.
14. VanEtten HD, Matthews PS (1984). Naturally occurring variation in the inducibility of pisatin demethylating activity in _Nectria haematococca_ mating population VI. Physiol Plant Pathol 25:149.
15. VanEtten HD, Barz W (1981). Expression of pisatin demethylating ability in _Nectria haematococca_. Arch Microbiol 129:56.
16. Kistler HC, VanEtten HD (1984). Regulation of pisatin demethylation in _Nectria haematococca_ and its influence on pisatin tolerance and virulence. J Gen Microbiol 130:2605.
17. Tegtmeier KJ, VanEtten HD (1982). The role of pisatin tolerance and degradation in the virulence of _Nectria haematococca_ on peas: a genetic analysis. Phytopathology 72:608.
18. Kistler HC, VanEtten HD (1984). Three non-allelic genes for pisatin demethylation in the fungus _Nectria haematococca_. J Gen Microbiol 130:2595.
19. Rudolph K (1976). Non-specific toxins. In Heitefuss R, Williams PH (eds): "Encyclopedia of Plant Physiology, New Series, Vol. 4, Physiological Plant Pathology," New York: Springer-Verlag, p 270.
20. Stoessl A (1981). Structure and biogenetic relations: fungal nonhost-specific. In Durbin RD (ed): "Toxins in Plant Disease," New York: Academic Press, p 110.
21. Daly JM, Knoche HW (1982). The chemistry and biology of pathotoxins exhibiting host selectivity. In Ingram DS, Williams PH (eds): "Advances in Plant Pathology," Academic Press, vol. 1.

22. Nishimura S, Kohmoto K (1983). Host-specific toxins and chemical structures from Alternaria species. Ann Rev Phytopathol 21:87.
23. Scheffer RP, Livingston RS (1984). Host-selective toxins and their role in plant disease. Science 223:17.
24. Yoder O C (1980). Toxins in pathogenesis. Ann Rev Phytopathol 18:103.
25. Macko V (1983). Structural aspects of toxins. In Daly JM, Deverall BJ (eds): "Toxins and Plant Pathogenesis," Australia: Academic Press, p 41.
26. Guzman D, Garber RC, Yoder OC (1982). Cytology of meiosis I and chromosome number of Cochliobolus heterostrophus (Ascomycetes). Can J Bot 60:1138.
27. Garber RC, Yoder OC (1983). Isolation of DNA from filamentous fungi and separation into nuclear, mitochondria, ribosomal, and plasmid components. Anal Biochem 135:416.
28. Griffiths AJF, Bertrand H (1984). Unstable cytoplasms in Hawaiian strains of Neurospora intermedia. Cur Genet 8:387.
29. Maniatis T, Fritsch EF, Sambrook J (1982). Molecular Cloning: A Laboratory Manual. Cold Spring Harbor Laboratory.
30. Perkins DD (1977). Details for preparing silica gel stocks. Neurospora Newsl 24:16.
31. Catcheside DEA, Catcheside DG (1979). Survival of Neurospora conidia on silica gel. Neurospora Newsl 26:24.
32. Bennett J, Lasure L (1985). "Gene Manipulations in Fungi," New York: Academic Press, in press.
33. Hynes MJ, Corrick CM, King JA (1983). Isolation of genomic clones containing the amdS gene of Aspergillus nidulans and their use in the analysis of strucutral and regulatory mutations. Molec Cell Biol 3:1430.
34. Tilburn J, Scazzocchio C, Taylor GG, Zabicky-Zissman JH, Lockington RA, Davis RW (1983). Transformation by integration in Aspergillus nidulans. Gene 26:205.
35. Kelly JM, Hynes MJ (1985). Transformation of Aspergillus niger by the amdS gene of Aspergillus nidulans. EMBO J 4:475.
36. Jimenez A, Davis J (1980). Expression of a transposable antibiotic resistance element in Saccharomyces. Nature 287:869.

37. Fraley RT, Rogers SG, Horsch RB, Sanders PR, Flick JS, Adams SP, Bittner ML, Brand LA, Fink CL, Fry JS, Galluppi GR, Goldberg SB, Hoffman NL, Woo SC (1983). Expression of bacterial genes in plant cells. Proc Natl Acad Sci 80:4803.
38. Gorman C, Padmanabhan R, Howard BH (1983). High efficiency DNA-mediated transformation of primate cells. Science 221:551.
39. Bull JH., Wootton JC (1984). Heavily methylated amplified DNA in transformants of *Neurospora crassa*. Nature 310:701.
40. Gritz L, Davies J (1983). Plasmid-encoded hygromycin B resistance: the sequence of hygromycin B phosphotransferase gene and its expression in *Escherichia coli* and *Saccharomyces cerevisiae*. Gene 25:179.
41. Kaster KR, Burgett SG, Ingolia TD (1984). Hygromycin B resistance as dominant selectable marker in yeast. Cur Genet 8:353.
42. Stinchcomb DT, Thomas M, Kelly J, Selker E, Davis RW (1980). Eukaryotic DNA segments capable of autonomous replication in yeast. Proc Natl Acad Sci 77:4559.
43. Suzci A, Radford A (1983). *ARS8* sequences in the *Neurospora* genome. Neurospora Newsl 30:13.
44. Sreekrishna K, Webster TD, Dickson RC (1984). Transformation of *Kluyveromyces lactis* with the kanamycin (G418) resistance gene of Tn903. Gene 28:73.
45. Das S, Hollenberg CP (1982). A high-frequency transformation system for the yeast *Kluyveromyces lactis*. Cur Genet 6:123.
46. Beach D, Nurse P (1981). High-frequency transformation of the fission yeast *Schizosacharomyces pombe*. Nature 290:140.
47. Losson R, Lacroute F (1983). Plasmids carrying the yeast OMP decarboxylase structural and regulatory genes: Transcription regulation in a foreign environment. Cell 32:371.
48. Carbon J (1984). Yeast centromeres: structure and function. Cell 37:351.
49. Carbon J, Clarke L (1984). Yeast centromeres and minichromosomes. Molec Basis Plant Dis Conf, Univ. Calif., Davis, abstr.

50. Yoder OC (1981). Assay. In Durbin RD (ed): "Toxins in Plant Disease," New York: Academic Press, p 45.
51. Lund T, Grosveld FG, Flavell RA (1982). Isolation of transforming DNA by cosmid rescue. Proc Natl Acad Sci 79:520.
52. Yelton MM, Timberlake WE, van den Hondel CAMJJ (1985). A cosmid for selecting genes by complementation in Aspergillus nidulans: selection of the developmentally regulated yA locus. Proc Natl Acad Sci 82:834.
53. Zimmermann CR, Orr WC, Leclerc RF, Barnard EC, Timberlake WE (1980). Molecular cloning and selection of genes regulated in Aspergillus development. Cell 21:709.
54. Young RA, Davis RW (1983). Efficient isolation of genes by using antibody probes. Proc Natl Acad Sci 80:1194.
55. Kinnaird JH, Keighren MA, Kinsey JA, Eaton M, Fincham JRS (1982). Cloning of the am (glutamate dehydrogenase) gene of Neurospora crassa through the use of a synthetic DNA probe. Gene 20:387.
56. Guarente L, Yocum R, Gifford P (1982). A GAL10-CYC1 hybrid yeast promoter identifies the GAL4 regulatory region as an upstream site. Proc Natl Acad Sci 79:7410.
57. Guarente L, Lalonde B, Gifford P, Alani E (1984). Distinctly regulated tandem upstream activation sites mediate catabolite repression of the CYC1 gene of S. cerevisiae. Cell 36:503.
58. Yoder OC (1983). Use of pathogen-produced toxins in genetic engineering of plants and pathogens. In Kosuge T, Meredith C, Hollaender A (eds): "Basic Life Sciences, Vol. 26, Genetic Engineering of Plants: An Agricultural Perspective," New York: Plenum Publ. Corp., p. 335.
59. Russell PJ, Granville RR, Tublitz N (1980). A cold-sensitive mutant of Neurospora crassa obtained using tritium-suicide enrichment that is conditionally defective in the biosynthesis of cytoplasmic ribosomes. Exp Mycol 4:23.

Molecular Genetics of Filamentous Fungi, pages 405–420

PHYTOTOXINS AS MOLECULAR DETERMINANTS OF PATHOGENICITY AND VIRULENCE[1]

D. G. Gilchrist, S. D. Clouse[2],
B. L. McFarland[3], and A. N. Martensen

Department of Plant Pathology
University of California
Davis, California 95616

ABSTRACT Extracellular phytotoxins produced by plant pathogenic fungi are known to be causally involved in many plant diseases. Their role in disease extends from being uniquely required to incite disease (pathogenicity factor) to situations where infection can occur in the absence of the toxin but the toxin contributes to the severity of disease by influencing the relative virulence of the pathogen (virulence factor). Studies with both levels of toxin involvement have potential economic and scientific value. Disease resistant plant genotypes potentially may be identified from whole plants or cell cultures using toxins as screening tools. Toxins functioning as pathogenicity factors would be expected to select for high levels of resistance whereas an intermediate level would likely emerge from cells with differential sensitivity to a virulence determining toxin. Phytotoxins historically have been viewed as key chemical probes of mechanisms of disease induced plant stress as well as the molecular basis of genetically controlled disease avoidance.

[1]Supported in part by Grant No. 59-2063-10406 CRGO, USDA, SEA, and NSF Grant No. PCM 80-1173.

[2]Present address: Plant Molecular and Cellular Biology Lab, The Salk Institute, P.O. Box 85800, San Diego, CA 92138.

[3]Present address: Chevron Biotechnology Group, Chevron Chemical Co., 940 Hensley, Richmond, CA 94804.

Fungi in all classes have been shown to produce phytotoxins among which several species of *Helminthosporium* and *Alternaria* are particularly noteworthy. Evidence obtained with AAL-toxins produced by *Alternaria alternata* f. sp. *lycopersici* specifically pathogenic on tomato indicates that the toxins are produced both in culture and in infected host tissue. The host reaction to both the pathogen and the toxin is controlled by the same gene. Only isolates of *A. alternata* which produce AAL-toxins are pathogenic on tomato and the toxins reproduce the typical disease symptoms. Current *in vitro* results suggest that the AAL-toxins interact differentially with aspartate carbamoyl transferase (ACTase) in the respective host genotypes to effect a AAL-toxin:UMP synergistic inhibition. Alteration in regulation around the key ACTase step of pyrimidine biosynthesis with potential integrated metabolic disruption through interlocking pathway loops indicates a number of testable mechanisms to explain cell death due to AAL-toxins.

INTRODUCTION

Disease of higher plants, the result of unrelieved pathogen induced physiological cell stress, is recognized by a variety of symptoms including abnormal growth or morphogenetic development, wilting, chlorosis, and premature death of massive cell members. Host-plant resistance to potential pathogens is characterized by absence or a significant reduction in symptom expression and may be controlled by one, few, or many genes (1). The presence or absence of symptoms expressed in either compatible or incompatible host-parasite interactions is generally regarded as the culmination of a series of undoubtedly discrete, likely complex, molecular events but which lack a universal mechanism (2). Whether or not this prognosis is correct, neither the molecular events controlling the development of the disease state nor the biochemical basis of physiological stress have been completely resolved for any single host-parasite interaction.

Compatible host-parasite interactions require expression of specific key attributes of pathogenicity by the pathogen coupled with the failure of the host to possess or express an adequate defensive response. Incompatibility implies the converse, regardless of whether the determinative event involves the presence or absence of active gene products. Among the attributes required for successful pathogenesis by fungi are the ability (a) to adhere to plant surfaces, (b) penetrate into or among subsurface cells, (c) invade and/or metabolize within the host, (d) reproduce in or on the host, and (e) incite the symptomatic expression of disease stress (2). Each of these attributes involves the coordinant expression of biochemical interactions, one or more of which may play a determinative role in the ultimate outcome of the host-parasite encounter.

This discussion will deal with biochemical factors of pathogen origin which are involved in inciting physiological stress and host sites which may determine the extent of the stress impact. The focus will be on small molecular weight extracellular phytotoxins produced by specific plant pathogenic fungi which, along with macromolecules involved in vascular plugging, enzymes, capable of degrading polymeric host components, and phytohormones, are recognized as biochemical factors causally involved in pathogenesis (3).

Fungal plant pathogens are known to produce an extensive array of phytotoxic compounds when grown in pure culture but there is general agreement that solid evidence for a causal role in disease is lacking for the vast majority (4). Among the toxins for which evidence supporting a causal role is perhaps the strongest are those which are specifically toxic to the natural hosts of the pathogens which produce them but are not toxic to non-hosts. Such toxins are classified as host-specific (or selective) (5) and are known to be produced by several fungal genera which include four species of Bipolaris (Helminthosporium), six species of Alternaria, along with one species each of Phyllosticta, Periconia, and Corynespora (6, 7). Structural characterization has been completed on the toxins produced by all but three of the 13 species. Each species appears to produce a unique toxin which suggests a different initial site of action in each case and therefore a different functional controlling element at the genetic level.

However, since the toxin and pathogen host ranges are equivalent it is reasonable to expect that the host genetic elements controlling each disease would be functionally related to the site or mode of toxin interaction in each case. Genetic studies of complimentary host-parasite and host-toxin interactions have confirmed that a single genetic locus in the host controls both interactions in those cases where such studies have been completed. Thus, structurally characterized host-specific toxins interacting with defined single genes in the host which regulate the disease process provide chemical probes to study mechanisms of disease induced physiological stress as well as the molecular basis of genetically controlled disease avoidance (8).

In addition, the biological role of the various host-specific toxins extends from being uniquely required to incite disease (pathogenicity factor) to situations where infection can occur in the absence of the toxin but the toxin contributes to the severity of disease by influencing the relative virulence of the producing organism (virulence factor) (9). Supporting data exists for the role of certain host-specific toxins as pathogenicity factors. Dunkle (2) recently summarized evidence that toxins produced by Bipolaris (Helminthosporium) victoriae (HV-toxin), Periconia circinata (PC-toxin), B. carbonum (HC-toxin), and Alternaria kikuchiana (AK-toxin) are pathogenicity factors. Isolates or mutants of each that lack or have lost the ability to produce their respective toxins are non-pathogenic on their individual hosts. In contrast B. maydis occurs as two races, race T which produces HmT toxin and race 0 which does not produce the toxin. Race T is more virulent than race O on maize genotypes with Texas male-sterile cytoplasm. However, both races are equally virulent on most genotypes with normal (fertile) cytoplasm. Thus, in the case of HmT toxin the production of toxin determines a quantitative trait, virulence, as apposed to the qualitative expression of pathogenicity by HV-, PC-, HC-, and AK-toxins. PM-toxin produced by Phyllosticta maydis is structurally similar to HmT toxin and also appears to be a virulence factor.

Studies with both levels of toxin involvement have potential economic and scientific value. Disease resistant plant genotypes potentially may be identified from whole plants or cell cultures using toxins as screening tools. Toxins functioning as pathogenicity factors would be expected to select for high levels of resistance whereas an intermediate level would likely emerge from cells with differential sensitivity to a virulence determining toxin (9).

An additional example of a host-specific toxin which is proposed to be a pathogenicity factor is produced by Alternaria alternata f. sp. lycopersici causal agent of the Alternaria stem canker disease of tomato. This disease first appeared in San Diego County, California in the early 1960's and the etiology was confirmed in 1975 (10). The disease is characterized by dark brown to black cankers with concencentric zonation located on the stem at the base of the plant. Associated with canker formation are foliar symptoms which include epinasty of the petiole with angular interveinal necrotic areas on one or both sides of the leaf midrib. Interestingly, the pathogen remained confined to the stem canker and could not be recovered from leaf tissue which suggested that the foliar symptoms were due to a translocatable toxin.

To further investigate the role of a possible phytotoxin in this disease we grew a variety of A. alternata isolates in liquid culture and treated excised tomato leaflets with the cell-free culture filtrates. Leaflets from cultivars of tomato susceptible to A. alternata showed symptoms characteristic of leaves on naturally infected plants when exposed to culture filtrates from pathogenic isolates but not A. alternata isolates which were nonpathogenic. Furthermore, susceptible genotypes were 1,000-fold more sensitive to toxic culture filtrates than were resistant cultivars, (determined by dilution end point in a detached leaflet bioassay). The host range of the pathogen and the host-specificity of toxic cell-free culture filtrate were examined using 38 lines of tomato, eight solanaceous species and two representatives each from the Chenopodiaceae, Cucurbitaceae, Compositae, Convolvulaceae, Cruciferae, Gramineae, Leguminosae, and Umbelliferae. Only certain lines of tomato were susceptible to canker formation and only susceptible lines of tomato were sensitive to the toxin at the dilution end point (11).

A preliminary genetic analysis using F_1 and F_2 progeny from a cross between homozygous resistant and susceptible cultivars showed the host reaction to both toxin and pathogen were governed by a single gene with two alleles (termed the R locus). Resistance to the pathogen exhibited complete dominance (Rr was resistant and a 3:1 ratio of resistant to susceptible plants occurred in the F_2) while sensitivity to the toxin showed incomplete dominance (Rr was intermediate in toxin sensitivity and a 1:2:1 ratio occurred in the F_2). Finally, a study of A. alternata isolates form a variety of sources including both pathogenic and nonpathogenic forms showed complete correspondence between pathogenicity of the isolate and toxin production in liquid culture (11).

Thus, the toxin produced by A. alternata f. sp. lycopersici conformed to the currently accepted definition of a host-specific toxin (5). In order to more critically evaluate the role of the toxic fraction as a determinant of disease in this system the toxin was purified and structurally characterized.

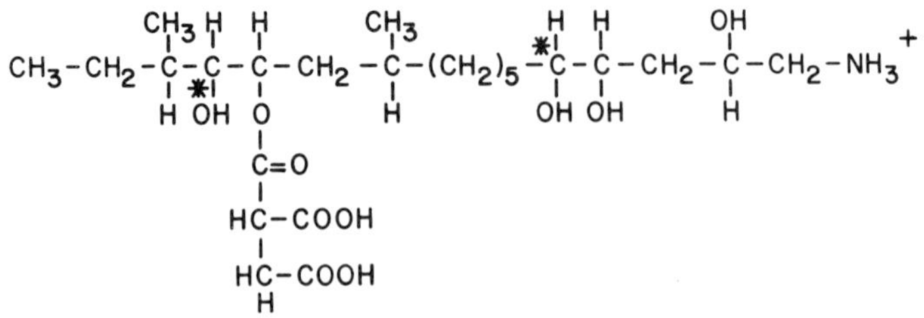

Figure 1. AAL-toxin TA_1 produced by Alternaria alternata f. sp. lycopersici. Esterification of the tricarballyate moiety at $*C_{14}$ of the aminopentol is designated as TA_2. Absence of the hydroxyl group at $*C_5$ produces the alternative pair of TB isomers TB_1 and TB_2. The ratio of TA to TB produced in culture is 3:1.

The AAL-toxins occur as two closely related esters of 1,3,3-propanetricarboxylic acid and 1-amino-11,15-dimethylheptadeca-2,4,5,13,15-pentol each occurring in culture as two isomers (TA and TB) (12). Esterification sites are the terminal carboxyl of the acid and C_{13} (TA_1) or C_{14} (TA_2) of the amino pentol (Figure 1). The TB form also consists of two components with the same carbon skeleton as TA_1 and TA_2 minus the C_5 hydroxyl. All four AAL-toxin components have been separated quantitatively from culture filtrates by high performance liquid chromatography (HPLC) (13) and recovered from infected tomato leaves (14). Phytotoxic fractions isolated from extracts of necrotic leaves of tomato plants infected with A. alternata f. sp. lycopersici were indistinguishable from AAL-toxins obtained from culture filtrates of the pathogen based on chromatographic mobility in both thin-layer and HPLC systems. Extracts from both sources possessed equal specific activity and were equally host-specific. Quantitative estimates indicated that toxin concentration in diseased leaves was similar to the concentration of AAL-toxin required to induce necrosis in detached leaves of the same cultivar (10 ng/ml).

Contrary to nearly all of the previously studied host-specific toxins, AAL-toxins did not cause ion leakage in tomato prior to the onset of visible necrosis (15, 16) nor early disruption of cell membranes (17). It appeared, therefore, that the primary site of toxin action was intracellular rather than at the membrane surface. The presence of an L-aspartate-like moiety (tricarballyate) in the toxin structure prompted initial studies on the mode of toxin action. L-aspartate (at 34 mM), but not D-aspartate, when incubated with leaves prior to AAL-toxin bioassay reduced the dilution endpoint by 100-fold (15). The protective effect of orotate, an intermediate unique to the *de novo* pathway of pyrimidine biosynthesis (18), was even more pronounced. These findings suggested that aspartate carbamoyl transferase (ACTase, E.C.2.1.2.3), a key regulatory enzyme in the biosynthetic pathway to pyrimidines in higher plants, might be the initial site of AAL-toxin action in tomato (Figure 2).

Higher plant ACTase has been partially characterized from etiolated lettuce (19) and mung bean (20) seedlings and from wheat germ (21). In all cases studied, the enzyme has been regulated *in vitro* by the pathway endproduct uridine-5'-monophosphate (UMP). ACTase from green tomato seedlings was highly labile, however, and kinetic characterization would not have been possible without the discovery that 1 mM UMP stabilized the enzyme *in vitro*. A comparative kinetic analysis of ACTase (partially purified by manganese sulfate and ammonium sulfate precipitation and Sephadex G-150 gel filtration) from RR and rr cultivars was then undertaken (15).

Convincing biological arguments for toxin induced necrosis using *in vitro* enzyme inhibition data thus require enzyme preparations that are both stable and responsive to known *in vivo* allosteric effectors throughout the course of the experiments. Also, since the enzyme encounters the toxin in the presence of the allosteric effectors *in planta* cell-free preparations used for enzyme:toxin inhibition studies must be regulated *in vitro* and assessed in the presence of the effectors and the toxins.

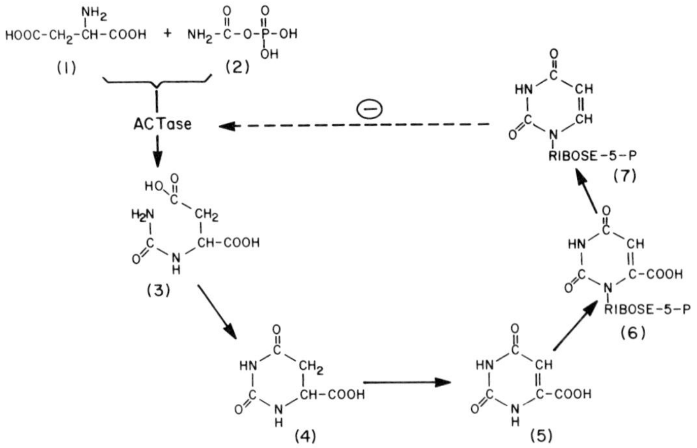

Figure 2. Orotic acid pathway of pyrimidine biosynthesis in higher plants. Condensation of aspartate (1) and carbamoyl phosphate (2), catalyzed by aspartate carbamoyl transferase (ACTase) is the first committed step leading to the synthesis of pyrimidines with orotate (5) as the unique pathway intermediate. UMP (7) is the only pathway product reported to feedback inhibit the pathway at the ACTase step (---) in higher plants.

METHODS

Genetic Analysis of Pathogen and Toxin Interactions.

The starting material for near-isogenic line development was the F_1 hybrid (Rr) from an ACE 55 VFN (RR) x Earlypak 7 (rr) cross. The F_1 was selfed to produce a segregating F_2 population and rr genotypes were identified and elminated by inoculating with A. alternata f. sp. lycopersici. The heterozygous resistant (Rr) plants were distinguished from the homozygous resistant (RR) by the intermediate AAL-toxin sensitivity of Rr genotypes in the detacted leaflet bioassay. Ten Rr individuals were selected and selfed to produce the segregating F_3. The process was continued through the F_6 generation at which time a single selfed fruit was taken from an F_6 Rr plant and 35 seeds were grown out in individual 30 cm pots. The F_7 plants were inoculated with A. alternata and selected resistant and susceptible plants were selfed for progeny testing. Plants identified as RR and rr by progeny testing were selfed and intercrossed to obtain RR, Rr and rr individuals in an F_8 background (22).

Interaction of ACTase with Substrates, UMP, and AAL-toxins.

AAL-toxins were purified as described previously (13). ACTase was isolated from RR and rr cultivars (Po generation table 1), purified 110 fold in the presence of UMP (1 mM), and assayed by method two of Prescott and Jones (23). The partially purified enzyme from both genotypes exhibited no metal cofactor or sulfhydryl group requirement, had an estimated molecular weight of 79,000, and had a pH optimum 9.9 (15). Prior to assays to test the sensitivity of ACTase to AAL-toxins and substrate or effector interactions various kinetic parameters were evaluated for the enzyme for both RR and rr genotypes. Only preparations which displayed sigmoid kinetics for both substrates carbamoyl phosphate (CAP) and aspartate (ASP), were inhibited by the dicarboxylic substrate analog succinate (competitively with respect to ASP and uncompetitively for CAP), and also were inhibited by UMP were used to test the AAL-toxin interaction.

RESULTS

Genetic Analysis of F_9 Paired Lines

A minimum of 28 F_9 individuals from each of 109 selfed F_8's were inoculated with A. alternata f. sp. lycopersici. Plants giving 100% susceptible progeny were considered rr, those giving zero out of 28 were RR, and those giving at least one canker (but less than 100%) were Rr. A 22RR:66Rr:21rr ratio was obtained which fit the proposed model at the 0.05 probability level. The pooled F_8 Rr plants produced a segregating population of 2,064 individuals in which 511 cankers were recorded (3.04:1 ratio of resistant to susceptible plants). The data fit the model of a single dominant gene with two alleles for resistance to Alternaria stem canker at the 0.975 probability level.

To investigate possible independent assortment of the genetic control of host reaction to the pathogen and host sensitivity to AAL-toxins, a cross between F_7-85-4 (rr) and F_7-85-12 (RR) was made. The F_8 heterozygote was selfed and 390 F_9 progeny were grown for four weeks under greenhouse conditions. Each plant was individually tagged and two leaflets from each plant were removed for detached leaflet bioassay. The plants were then inoculated with _A. alternata_ as described above and rated for canker development. Of the 99 plants which developed cankers, 100% were sensitive (toxin severity index of 3 or above) to AAL-toxin (TA) at 2.6 µg/ml. Of the remaining plants (291) which did not develop cankers, none were as sensitive (toxin severity index of 2 or less) to 2.6 µg/ml TA in the detached leaflet bioassay as the canker-susceptible plants.

To further correlate toxin sensitivity with the allelic state of the R locus, 35 plants from each of the parents (ACE and Earlypak 7), the F_1 progeny (Rr), the near-isogenic F_8 homozygotes and their crossed progeny (Rr), were tested for canker and toxin sensitivity as above (22). An analysis of variance of the results followed by Duncan's multiple range test, indicated three significantly different levels of toxin sensitivity which corresponded to the genotype at the R locus (Table 1).

Kinetic Parameters in the Absence of AAL-toxins.

ACTase from RR and rr genotypes displayed positive homotropic cooeratively for both ASP and CAP with Hill values for ASP of 1.64 (RR) and 1.74 (rr) compared to values for CAP of 1.32 (RR) and 1.23 (rr). Succinate inhibited ACTase from both genotypes competitively for ASP (Kiapp = 32 mM) and uncompetitively for CAP ($I_{0.5}$ = 167 mM). UMP inhibited the enzyme competitively for CAP (Kiapp = 0.33 mM) and uncompetitively for ASP ($I_{0.5}$ = 2.4 mM) with no significant genotype-specific differences among the two genotypes.

Effect of AAL-toxins on ACTase.

ACTase from both genotypes was inhibited by AAL-toxins in the absence of UMP, with the apparent dissociation constants (Kiapp) of 2.4 and 4.2 mM for _rr_ and _RR_ respectively. Assay of ACTase at 2.0 mM AAL-toxin decreased the Hill value for CAP to 0.72 (_RR_) and 0.64 (_rr_). Dixon plots (1/V vs. I) (24) were linear for CAP (2

TABLE 1

CORRELATION OF TOXIN SENSITIVITY WITH GENOTYPE

Genotype	Generation	Cankers[a]	Mean TSI[b]	Duncan Grouping[c]
rr	P_0	35	3.97	A
rr	F_9	35	3.97	A
Rr	F_1	0	2.20	B
Rr	F_8	0	2.20	B
RR	P_0	0	0.68	C
RR	F_9	0	0.81	C

a) 35 plants of each genotype were inoculated.
b) TSI = Toxin Severity Index, 0 to 4. 70 leaflets (2/plant) of each were bioassayed.
c) Means with the same letter are not significantly different. Alpha = 0.05.

to 16 mM) at 0, 0.5, 1.0, and 2.0 mM toxin and indicated uncompetitive binding. Hyperbolic velocity and linear Dixon plots suggested either single, or multiple noncooperative, AAL-site(s) on the enzyme. Comparable experiments with variable ASP at saturating CAP (10 mM) revealed noncompetitive toxin binding with respect to ASP for both genotypes. This type of binding differed from that of succinate noted earlier. The noncompetitive pattern of toxin vs ASP is taken to indicate that the enzyme binds AAL-toxin regardless of whether ASP is present and the substrate saturation alone will not relieve the inhibition.

Effect of UMP and AAL-toxins on ACTase.

In the presence of UMP, which itself decreases carbamoyl phosphate binding (competitive inhibition), AAL-toxin interferes with the binding of carbamoyl phosphate in a competitive manner for ACTase from both rr and RR tomato genotypes. Moreover, an interesting synergism occurred between toxin and UMP for the rr ACTase and was twice as great as that for the ACTase from tolerant tissue (RR) at ASP saturation. When assayed at variable ASP concentration the AAL-toxins appeared to be noncompetitive inhibitors with respct to aspartate, and, in the presence of UMP (uncompetitive with respect to aspartate), act synergistically to produce a double, dead-end inhibitor complex with the net result that ACTase activity is decreased to zero by concentrations of UMP that are not normally (i.e., in the absence of the AAL-toxins) inhibitory. Additionally, this synergism occurred in a genotype-specific manner; the synergism was forty-fold greater for the rr genotype at subsaturating aspartate (2.5 mM).

The interaction of UMP and AAL-toxins on ACTase from RR and rr tissues can be summarized through an inhibitor potency index rating similar to that used by Wedler et al to compare regulation of two glutamine synthetase isozymes from Bacillus caldolyticus (25). At saturating substrates (ASP and CAP) the effect of adding increasing amounts of

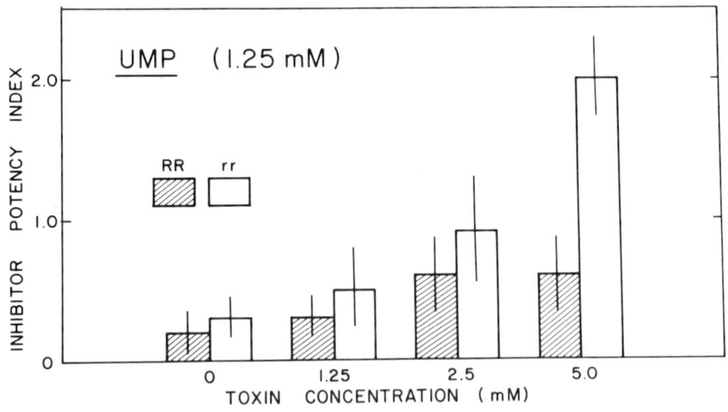

Figure 3. Inhibitor potency index ratings for UMP and AAL-toxins at saturating substrate concentrations. Values for the inhibitor potency index (25), $i_{max}/I_{0.5}$, represent means and standard deviations of the respective kinetic parameters obtained from plots of 1/i vs 1/[I] in two separate replicated experiments. Inhibitor combinations of major significance have high i_{max} and low $I_{0.5}$ values.

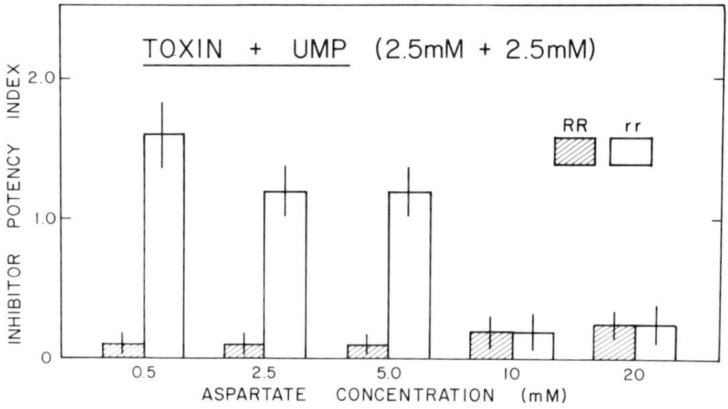

Figure 4. Inhibitor potency index ratings for UMP and AAL-toxins, each at 2.5 mM, were evaluated at various aspartate concentrations and saturating cabamoyl phosphate. Values for the inhibitor potency index (25), $i_{max}/I_{0.5}$, represent means and standard deviations of the respective kinetic parameters obtained from plots of 1/i vs 1/[I] in two separate replicated experiments. Inhibitor combinations of major significance have high i_{max} and low $I_{0.5}$ values.

AAL-toxins to ACTase from RR and rr tissues in the presence of 1.25 mM UMP revealed a genotype-specific differential sensitivity to the combination of the two inhibitors (figure 3). The combination of UMP and AAL-toxins, both at 2.5 mM, revealed that the degree of synergistic interaction of the two was dependent on the concentration of aspartate in the reaction mixture (figure 4).

DISCUSSION

Evidence presented in this paper suggests the existance of an allosteric site on ACTase from tomato which binds the AAL-toxins leading to a proposed conformational change with an enhanced sensitivity to the natural feedback inhibitor UMP. Based on kinetic data the metabolic consequence of the synergistic interaction between the AAL-toxins and UMP is the ability of this combination to drive the reaction velocity to zero, an effect not observed with either compound alone. The further observation that the pathogen susceptible and toxin sensitive genotype (rr) is up to 40-fold more sensitive to this synergistic interaction than the resistant genotype (RR) is consistent with a genotype-specific explanation of the physiological basis of AAL-toxin induced cell stress.

Precisely how the inhibition of ACTase by such a mechanism leads to cell stress and eventually cell death is unknown. However, consideration of certain alternative explanations may be relevant to other host-toxin interactions and provide a general format for stress induced in other plants by structurally unrelated pathotoxins. In the present case, inhibition of ACTase through AAL-toxin - UMP synergism could lead eventually to a limitation in essential pyrimidines causing metabolic stress. However, recycling of existing pyrimidines in the cell may forestall an immediate loss in cell integrity. Alternatively, interlocking metabolic pathways could be affected through feedback loops eventually leading to inhibition of specific processes for which no short-term compensatory mechanisms exist (26).

Recent development of F_8 paired lines, near-isogenic for the R-locus, coupled with a procedure to purify ACTase from tomato to homogeneity will permit physical and kinetic comparison of the enzyme from F_8 RR, Rr, and rr plants. Studies also are in progress to clone the ACTase gene from the respective near-isogenic lines to develop additional information on the role of ACTase in regulating this host-parasite interaction (8).

REFERENCES

1. Day PR (1974). "Genetics of the host-parasite interaction." San Francisco: W. H. Freeman, p .
2. Dunkle LD (1984). Factors in pathogenesis. In Kosuge T, Nester EW (eds): "Plant-Microbe Interactions: Molecular and Genetic Perspectives," New York: MacMillan, Inc., p 19.
3. Callow JA (1983). "Biochemical Plant Pathology." Chichester: John Wiley, 484 p.
4. Wheeler H (1981). Role in pathogenesis. In Durbin RD (ed): "Toxins in Plant Disease," New York: Academic Press, p 477.
5. Scheffer RP, Briggs SP (1981). A perspective of toxin studies in Plant Pathology. In Durbin RD (ed): "Toxins in Plant Disease." New York: Academic Press, p 1.

6. Scheffer RP (1983). Toxins as chemical determinants of plant disease. In Daly JM, Deverall BJ (eds): "Toxins and Plant Pathogenesis," Australia: Academic Press, p 1.
7. Macko V (1983). Structural aspects of toxins. In Daly JM, Deverall BJ (eds): "Toxins and Plant Pathogenesis," Australia: Academic Press, p 41.
8. Gilchrist DG, Yoder OC (1984). Genetics of host-parasite systems: A prospectus for molecular biology. In Kosuge T, Nester E (eds): "Plant-Microbe Interactions, Molecular and Genetic Perspectives," New York: MacMillan, p 69.
9. Yoder OC (1983). Use of pathogen produced toxins in genetic engineering of plant and pathogens. In Kosuge T, Meredith C, Hollaender A (eds): "Genetic Engineering of Plants," New York: Plenum Press, p 335.
10. Grogan RG, Kimble KA, Misaghi I (1975). A stem canker disease of tomato caused by Alternaria alternata f. sp. lycopersici. Phytopathology 65:880.
11. Gilchrist DG, Grogan RG (1976). Production and nature of a host-specific toxin from Alternaria alternata f. sp. lycopersici. Phytopathology 66(2):165.
12. Bottini AT, Bowen JR, Gilchrist DG (1981). Phytotoxins II. Characterization of a phytotoxic fraction from Alternaria alternata f. sp. lycopersici. Tetrahedron Lett 22(29):2723.
13. Siler DJ, Gilchrist DG (1982). Determination of host-selective phytotoxins from Alternaria alternata f. sp. lycopersici as their maleyl derivatives by high-performance liquid chromatography. J Chromatography 238:167.
14. Siler DJ, Gilchrist DG (1983). Properties of host specific toxins produced by Alternaria alternata f. sp. lycopersici in culture and in tomato plants. Physiol Plant Pathol 23:265.
15. McFarland BL (1984). Studies on the interaction of tomato and Alternaria alternata f. sp. lycopersici host-specific toxins. Davis CA: University of CA: Ph.D. dissertation, 265 p.
16. Kohmoto K, Verma VS, Nishimura S, Tagami M, Scheffer RP (1982). New outbreak of Alternaria stem canker of tomato in Japan and production of host-selective toxins by the causal fungi. J Fac Agric, Tottori Univ 17:1.

17. Park P, Nishimura S, Kohomoto K, and Otani H (1981). Comparative effects of host-specific toxins from four pathotypes of Alternaria alternata on the ultrastructure of host cells. Ann Phytopathol Soc Japan 47:488.
18. Ross C (1981). In Stumpf P, Conn E (eds in-chief): "The Biochemistry of Plants: A Comprehensive Treatise," V 6 Marcus A (ed), New York: Academic Press, p 169.
19. Neuman J, Jones ME (1962). ACTase from lettuce seedlings: Case of end product inhibition. Nature 195:709.
20. Ong BL, Jackson JJ (1972). Pyrimidine nucleotide biosynthesis in Phaseolus aureus: Enzyme aspects of control of CAP synthesis and utilization. Biochem J 129:583.
21. Yon R (1972). Chromatography of lipophylic proteins on absorbants containing mixed hydrophobic and ionic groups. Biochem J 126:765.
22. Clouse SD (1985). Genetic and biochemical studies on Alternaria stem canker of tomato. Davis, CA: University of california: Ph.D. dissertation, 265 p.
23. Prescott LM, Jone ME (1969). Modified methods for the determination of carbamyl aspartate. Anal Biochem 32:408.
24. Segel IH (1975). "Enzyme Kinetics: Analysis of Rapid Equilibrium and Steady-State Enzyme Systems." New York: Wiley and Sons, p. 475.
25. Wedler FC, Shreve DS, Fisher KE, Merkler DJ (1981). Complimentarity of regulation for the two glutamine synthetases, E.C.-6.3.1.2, from Bacillus caldolyticus, an exteme thermophile. Arch Biochem Biophys 211:276.
26. Gilchrist DG (1983). Molecular modes of action. In Daly JM, Deverall B (eds): "Toxins and Plant Pathogenesis," Australia: Academic Press, p 81.

Molecular Genetics of Filamentous Fungi, pages 421–438

MOLECULAR BIOLOGY OF THE EARLY EVENTS IN THE FUNGAL PENETRATION INTO PLANTS[1]

P.E. Kolattukudy, C.L. Soliday, C.P. Woloshuk and M. Crawford

Institute of Biological Chemistry
Biochemistry/Biophysics Program
Washington State University
Pullman, Washington 99164-6340

ABSTRACT Penetration of pathogenic fungi into plants involves breaching of the plant cuticle and cell wall. The major structural component of the cuticle is a biopolyester called cutin which is broken by the fungal extracellular enzyme, cutinase. Inhibition of this enzyme prevents fungal penetration into the plant and thus prevents fungal infection. The ability of fungal isolates to produce cutinase can determine virulence. The cDNA for this enzyme has been cloned and sequenced. The genomic DNA containing cutinase gene was also cloned and sequenced. This gene contains one 51 bp intron which shows typical junctions and splicing signals that are homologous to those found in other fungi. Evidence is presented that a fungal spore senses the contact with the plant surface using the low constitutive level of cutinase which releases cutin monomers from the host cuticle. Low levels of monomers that are unique to cutin were shown to induce cutinase production by the spores. This induction involved transcriptional control and the new cutinase transcripts became detectable by dot blot analysis within 15 minutes after the spore came into contact with cutin. The induction of this enzyme is followed by induction of polygalacturonase which is presumably used to degrade the pectin barrier which lies under the cuticle.

[1]Scientific Paper #7182, Project 2001, College of Agriculture Research Center, Washington State University, Pullman, WA 99164. This work was support in part by grant DMB-8306835 from the National Science Foundation.

INTRODUCTION

Aerial parts of plants are protected by the cuticle which is composed of cutin, the structural component, and associated waxes (1,2). Cutin, the insoluble polymer, constitutes the major physical barrier to penetration by pathogenic organisms. This polymer is a biopolyester composed of hydroxy and hydroxyepoxy fatty acids derived from the common cellular fatty acids. Dihydroxy C_{16} acid, 18-hydroxy,9,10-epoxy C_{18} acid and 9,10,18-trihydroxy C_{18} acid are usually the major components of cutin (Figure 1). These monomers are interesterified via the primary hydroxy groups. About one-half of the mid-chain hydroxy groups are also involved in ester linkages representing branching and/or cross-linking. How pathogenic fungi breach the cuticular barrier has been debated for nearly a century (3,4). The view that the penetration process is entirely mediated by the physical force of growth of the germinating spore has been refuted by suggestions that an extracellular enzyme that catalyzes degradation of cutin is involved. Ultrastructural examination of the penetration areas has given hints that the process might involve degradative enzymes. However, in the absence of more direct evidence, the controversy continued.

MAJOR CUTIN ACIDS

$$CH_2(CH_2)_x \; CH(CH_2)_y \; COOH$$
$$\;\;|\quad\quad\quad\;\;|$$
$$OH\quad\quad\quad OH$$

$(y = 8, 7, 6, \text{or } 5 \quad x + y = 13)$

$$CH_2(CH_2)_7 \; CH-CH(CH_2)_7 \; COOH$$
$$\;\;|\quad\quad\quad\;\;\backslash\;\;/$$
$$OH\quad\quad\quad\quad O$$

$$CH_2(CH_2)_7 \; CH-CH(CH_2)_7 COOH$$
$$\;\;|\quad\quad\quad\;\;|\quad\;|$$
$$OH\quad\quad\quad OH\;\;OH$$

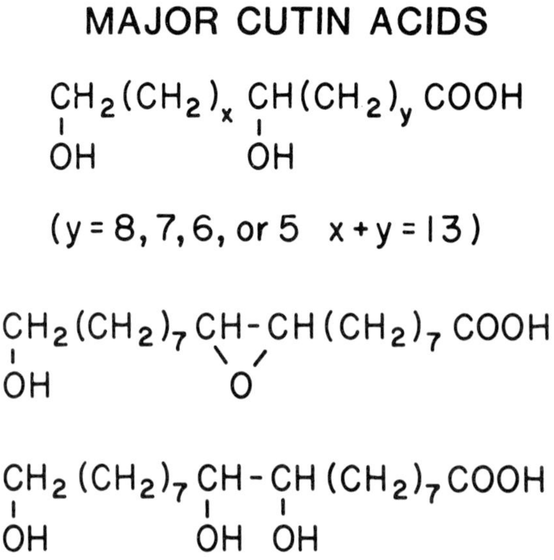

FIGURE 1. Structure of the major monomers of cutin.

ISOLATION AND CHARACTERIZATION OF CUTINASE

Cutinase has been isolated from the extracellular fluid of pathogenic fungi grown on cutin as the sole source of carbon (5,6). The most widely applicable purification method involves an acetone precipitation, decolorization by passage through QAE-sepharose, and octyl sepharose chromatography followed by SP-sepharose cation exchange chromatography (7). Such a procedure has been used to purify cutinase produced by a variety of pathogenic fungi including several isolates of *Fusarium solani* f. sp. *pisi*, *F. roseum* f. sp. *culmorum*, *F. roseum* f. sp. *sambucinum*, *Ulocladium consortiale*, *Helminthosporum sativum*, *Streptomyces scabies*, *Colletotrichum gloeosporioides*, *Colletotrichum capsici*, *Colletotrichum graminicola*, *Phytophthora cactorum and Botrytis cinerea* (4,7). The properties of cutinase obtained from the various fungal sources appear to be quite similar. The molecular weight is near 25,000 and the amino acid composition appears to be quite similar. Fungal cutinases are glycoproteins containing a few percent O-glycosidically attached carbohydrates. The enzyme shows optimal activity at a pH around 9.0 and shows specificity for hydrolysis of primary alcohol esters. The enzyme activity can be measured using a spectrophotometric assay which depends on the fact that the enzyme catalyzes hydrolysis of C_2 - C_{18} fatty acid esters of *p*-nitrophenol. The amino group of the N-terminal glycine is attached to glucuronic acid by amide linkage (8,9). Cutinase utilizes the classical catalytic triad involving active serine, histidine and a carboxy group to hydrolyze ester bonds (10). Thus, the enzyme is inhibited severely by organic phosphates, alkyl and aryl boronic acids and other inhibitors of the classical serine hydrolases.

ROLE OF CUTINASE IN PATHOGENESIS

With the availability of pure cutinase, immunological methods could be used to test whether this enzyme is secreted by germinating spores involved in a real infection. Suspensions of *Fusarium solani* f. sp. *pisi* spores were placed on pea stem segments and the progress of germination and penetration was followed using scanning electron microscopy (Figure 2). At the time when penetration appeared to be happening, this area was treated

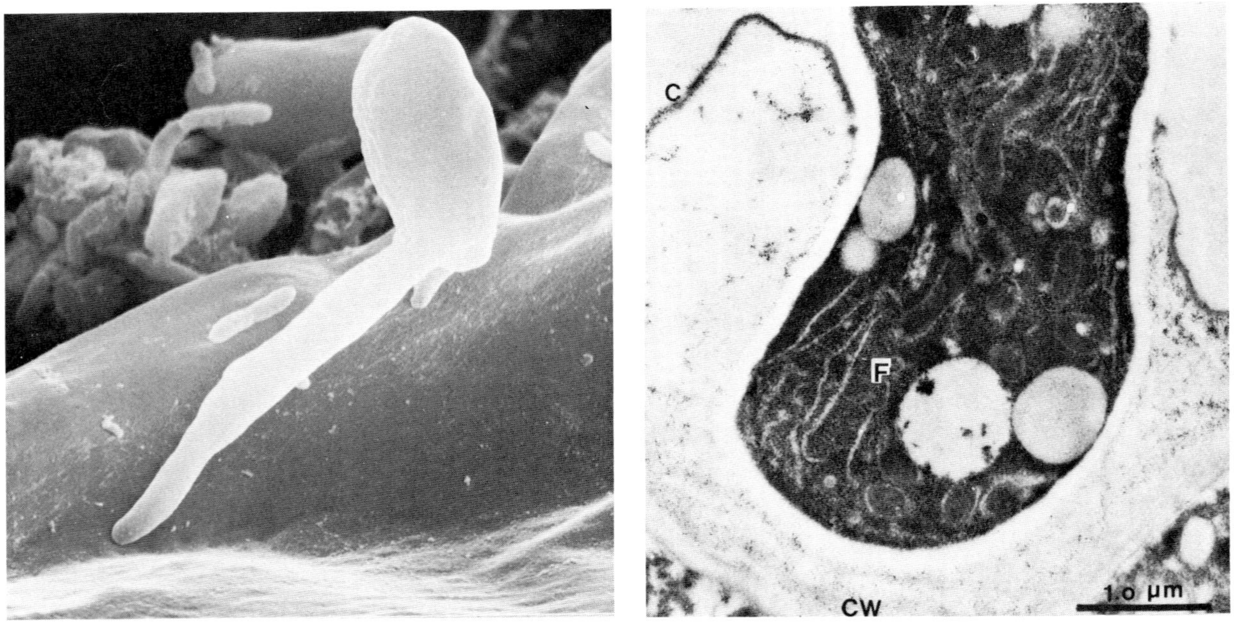

FIGURE 2. Scanning (left) and transmission (right) electron micrographs of the penetration of pea stem cuticle by mycelia of *F. solani* f. sp. *pisi*. Both figures represent the state of events 12 hr after placing the spores on the stem of the pea seedling. The infection area was treated with ferritin-conjugated, anticutinase IgG. Ferritin granules can be seen at the site of penetration in the transmission electron micrograph. C, cuticle; CW, cell wall; F, fungus.

with ferritin-labeled antibodies prepared against cutinase. Electron microscopic examination of the infection area revealed penetrating fungi. Ferritin granules representing cutinase-antibody complexes were associated with the broken cuticle showing that the penetrating *Fusarium* did, in fact, secrete cutinase. Similar experiments were done with *Colletotrichum gloeosporioides* spores placed on papaya fruits and using ferritin-conjugated antibodies prepared against cutinase purified from this organism (M. Dickman, S.S. Patil, C.L. Soliday and P.E. Kolattukudy, unpublished results). The results clearly showed that this organism produced cutinase during the actual infection process.

If the cutinase produced by the germinating spores is, in fact, needed for penetration, specific inhibition of the enzyme should prevent penetration and consequently the infection process. Incorporation of antibodies or specific cutinase inhibitors into these spore suspension droplets prevented infection of pea stem sections by *Fusarium solani* f. sp. *pisi* (11) and papaya fruits by *C. gloeosporioides* (12). More recently, bioassays were conducted with apple seedlings in which case *Venturia inequalis* spore suspension was placed on the leaves of apple seedlings with or without the inclusion of cutinase inhibitors (R. Chacko and P.E. Kolattukudy, manuscript in preparation). Similar experiments were also done by placing a spore suspension of *Colletotrichum graminicola* on the leaves of corn seedlings. In both cases the presence of cutinase inhibitors completely prevented infection of the seedlings. These results clearly showed that inhibition of cutinase can prevent infection, thus providing strong evidence that cutinase is involved in the penetration of the cuticular barrier by these fungi.

Cutinaseless mutants of *Colletotrichum gloeosporioides* were isolated and these mutants could not produce cutinase as measured by the level of enzyme activity and the level of immunologically detectable proteins (13). These mutants could not infect papaya fruits. That the mutations did not affect the other processes involved in infection was shown by the fact that these cutinaseless mutants infected papaya fruits when the cuticle was mechanically breached or when the spore suspension was mixed with exogenous cutinase. Another example of the need for the ability to produce cutinase for setting up infection was obtained when a wound pathogen of papaya, namely a *Mycosphaerella* sp., was placed on papaya fruits with or without the inclusion of cutinase. Without added cutinase, this organism could not infect papaya fruits but the exogenous cutinase allowed it to infect the fruits (12).

A series of isolates of *F. solani* f. sp. *pisi* originally isolated at Cornell University by Professor Hans Van Etten and his colleagues were tested for pathogenicity on pea stem sections with or without intact cuticular barrier. Two strains called T-8 and T-30 were found to be highly virulent on stem sections with broken cuticular barrier whereas when the intact cuticular barrier was present, T-30 was avirulent but T-8 was highly virulent (14). To test whether the lack of virulence of the T-30 strain on intact stem resulted from the inability of this isolate to produce cutinase, the spores of the isolates were tested for cutinase activity using a highly radioactive cutin preparation as the substrate. The spores of the isolate T-8 showed a high level of cutinase activity whereas T-30 showed little or no cutinase activity. All of the above results strongly suggested that virulence can be strongly influenced by the ability of the pathogen to produce cutinase and therefore the biosynthesis of the enzyme was thought to be highly relevant to pathogenesis.

CLONING AND SEQUENCING OF CUTINASE cDNA

When *F. solani* f. sp. *pisi* was grown on glucose as the carbon source, little or no cutinase activity could be detected in the extracellular fluid (15). When glucose in the medium was depleted, cutinase induction occurred if cutin or cutin hydrolysate was added to the medium. These results suggested that the small amount of cutinase generated upon starvation produced small amounts of hydrolysate from the extracellular insoluble polymer and these monomers triggered the production of cutinase in the organism. This cutinase induction system provided a convenient material with which to study the regulation of cutinase gene expression. The *in vitro* translation of the mRNA isolated from induced cultures gave products which could be immunologically isolated. Such experiments clearly showed that the primary translation product of cutinase mRNA was 2,100 daltons larger than the mature protein suggesting that the primary translation product is a precursor form of cutinase (16).

The mRNA preparation obtained from induced *Fusarium* cultures was used to prepare cDNA which was cloned using pBR322 as the vector (17). A group of 74 clones unique to the induced cultures was selected. By hybrid selected translation and colony hybridization, clones containing cutinase cDNA were selected. When the cDNA inserts were isolated from these clones, it was found that some of them were about 1,000bp long and Northern blot analysis showed that the mRNA contained 1,050 nucleotides. Thus, the cloned cDNA represented nearly full-length mRNA and therefore could be used for structure elucidation. The full-length cDNA and the restriction fragments subcloned from it were sequenced by a combination of the Maxam-Gilbert and the phage M13-dideoxy techniques. The complete primary structure of the enzyme, including that of the signal peptide region, could be deduced from the nucleotide sequence (Figure 3).

Three regions of cutinase from *F. solani* f. sp. *pisi* were isolated after proteolysis and the purified peptides were subjected to direct amino acid sequencing. These results completely agreed with the sequence deduced from the nucleotide sequence (shown by solid lines in Figure 3). The active serine involved in catalysis was identified by specific labeling of this serine with radioactive diisopropylfluorophosphate. The four cysteines were labeled by treatment with [^{14}C]iodoacetamide after reduction of the disulfide bridges (W. Ettinger and P.E. Kolattukudy, unpublished results). The histidine residue and the carboxy group involved in catalysis were also identified in the primary structure of cutinase. The three members of the catalytic triad were found to be located far apart in the primary structure. Obviously the three residues are held together in the proper juxtaposition for catalysis by the secondary and tertiary structures. The two disulfide bridges involving cys 125 with cys 47 and cys 187 with cys 194 (W. Ettinger and P.E. Kolattukudy, unpublished results) must be involved in maintaining the appropriate structure of the protein because breaking of these disulfide bridges results in inactivation of the enzyme (W. Köller and P.E. Kolattukudy, manuscript in preparation). The N-terminal glycine residue of the mature protein was the 32nd residue from the methionine start point. The 31 amino acid leader sequence showed characteristics similar to those observed for other leader sequences but its "post core region" appeared to be somewhat longer than the other leader sequences. How the precursor is processed by cleavage of the leader sequence and introduction of the glucuronic acid remains to be elucidated.

```
AACCACACATACCTTCACTTCATCAACATTCACTTCAACTTCTTCGCCTCTTCCTTTTCACTCTTTATCATCCTCACC

                                     10
MET LYS PHE PHE ALA LEU THR THR LEU LEU ALA ALA THR ALA SER ALA LEU PRO THR SER
ATG AAA TTC TTC GCT CTC ACC ACA CTT CTC GCC GCC ACG GCT TCG GCT CTG CCT ACT TCT
                        |-> C-57
                                     30   32
ASN PRO ALA GLN GLU LEU GLU ALA ARG GLN LEU GLY ARG THR THR ARG ASP ASP LEU ILE
AAC CCT GCC CAG GAG CTT GAG GCG CGC CAG CTT GGT AGA ACA ACT CGC GAC GAT CTG ATC
                            |-> C-4
                        47          50
ASN GLY ASN SER ALA SER CYS ARG ASP VAL ILE PHE ILE TYR ALA ARG GLY SER THR GLU
AAC GGC AAT AGC GCT TCC TGC CGC GAT GTC ATC TTC ATT TAT GCC CGA GGT TCA ACA GAG

                                     70
THR GLY ASN LEU GLY THR LEU GLY PRO SER ILE ALA SER ASN LEU GLU SER ALA PHE GLY
ACG GGC AAC TTG GGA ACT CTC GGT CCT AGC ATT GCC TCC AAC CTT GAG TCC GCC TTC GGC

                                     90                                  99
LYS ASP GLY VAL TRP ILE GLN GLY VAL GLY GLY ALA TYR ARG ALA THR LEU GLY ASP ASN
AAG GAC GGT GTC TGG ATT CAG GGC GTT GGC GGT GCC TAC CGA GCC ACT CTT GGA GAC AAT

                                     110
ALA LEU PRO ARG GLY THR SER SER ALA ALA ILE ARG GLU MET LEU GLY LEU PHE GLN GLN
GCT CTC CCT CGC GGA ACC TCT AGC GCC GCA ATC AGG GAG ATG CTC GGT CTC TTC CAG CAG

                    125              130                     136
ALA ASN THR LYS CYS PRO ASP ALA THR LEU ILE ALA GLY GLY TYR SER GLN GLY ALA ALA
GCC AAC ACC AAG TGC CCT GAC GCG ACT TTG ATC GCC GGT GGC TAC AGC CAG GGT GCT GCA

                                     150
LEU ALA ALA ALA SER ILE GLU ASP LEU ASP SER ALA ILE ARG ASP LYS ILE ALA GLY THR
CTT GCA GCC GCC TCC ATC GAG GAC CTC GAC TCG GCC ATT CGT GAC AAG ATC GCC GGA ACT

                                     170
VAL LEU PHE GLY TYR THR LYS ASN LEU GLN ASN ARG GLY ARG ILE PRO ASN TYR PRO ALA
GTT CTG TTC GGC TAC ACC AAG AAC CTA CAG AAC CGT GGC CGA ATC CCC AAC TAC CCT GCC

                        187          190           194
ASP ARG THR LYS VAL PHE CYS ASN THR GLY ASP LEU VAL CYS THR GLY SER LEU ILE VAL
GAC AGG ACC AAG GTC TTC TGC AAT ACA GGG GAT CTC GTT TGT ACT GGT AGC TTG ATC GTT

            204                      210
ALA ALA PRO HIS LEU ALA TYR GLY PRO ASP ALA ARG GLY PRO ALA PRO GLU PHE LEU ILE
GCT GCA CCT CAC TTG GCT TAT GGT CCT GAT GCT CGT GGC CCT GCC CCT GAG TTC CTC ATC

                                     230
GLU LYS VAL ARG ALA VAL ARG GLY SER ALA *** GGAGGATGAGAATTTTAGCAGGCGGGCCTGTTAAT
GAG AAG GTT CGG GCT GTC CGT GGT TCT GCT TGA

TATTGCGAGGTTTCAAGTTTTTCTTTTGGTGAATAGCCATGATAGATTGGTTCAACACTCAATGTACTACAATGCCTCCC

CCCCCCCCCCCCC
```

FIGURE 3. Nucleotide sequence of cloned cutinase cDNA and the amino acid sequence deduced from it. C-4 and C-57 indicate the beginning of the nucleotide sequence of two additional cDNA clones. The solid lines represent the regions for which the primary structure was confirmed by amino acid sequencing.

STRUCTURE AND EXPRESSION OF CUTINASE GENE

When the different isolates of *Fusarium solani* f. sp. *pisi* were examined for their ability to produce cutinase, it was found that they range from very high to very low (Table 1); the low producers were generally avirulent (C.

TABLE 1
CUTINASE PRODUCTION BY *F. solani* f. sp. *pisi*

Strain	PNB Activity (units/ml)
T-23	11.0
T-1	9.92
T-220	9.88
T-291	8.89
T-21	8.3
T-289	6.73
T-159	7.65
T-14	7.64
T-32	6.24
T-29	5.56
T-110	4.95
T-101	4.9
34-18	4.26
T-63	3.98
T-284	3.94
T-17	3.8
T-213	3.55
T-160	3.44
T-2	3.4
T-61	3.3
23-40	3.1
6-94	3.1
43-6	3.06
51-37	2.85
T-276	2.75
T-7	2.7
T-161	2.52
T-277	2.45
6-36	2.4
T-217	2.37
T-27	1.98
T-95	1.96
T-219	1.8
T-192	1.78
T-10	1.7
36-14	1.41
T-9	1.30
T-33	1.2
50-33	1.06
T-214	0.88
T-69	0.84
T-34	0.26

Isolates from Prof. Van Etten. Hydrolysis of *p*-nitrophenylbutyrate was used in this assay.

Woloshuk and P.E. Kolattukudy, unpublished results). To study how cutinase gene expression is regulated, the genomic DNA was subjected to restriction enzyme digestion, electrophoresis and hybridization with cutinase cDNA probes. Preliminary analysis suggests that the high producers have two copies of cutinase gene whereas the low producers have only one. Presumably the unique copy of the gene present in the high producers must be located in a position such that it is under the influence of a strong promoter. To elucidate the nature of the regulation of this gene, genomic clones were prepared in λ-phage and the library was screened for cutinase gene (C.L. Soliday and P.E. Kolattukudy, unpublished results). The cutinase gene containing fragments were isolated, subcloned and sequenced. Cutinase gene contains one intron which is 51bp long and has a typical intron junction (Figure 4). The

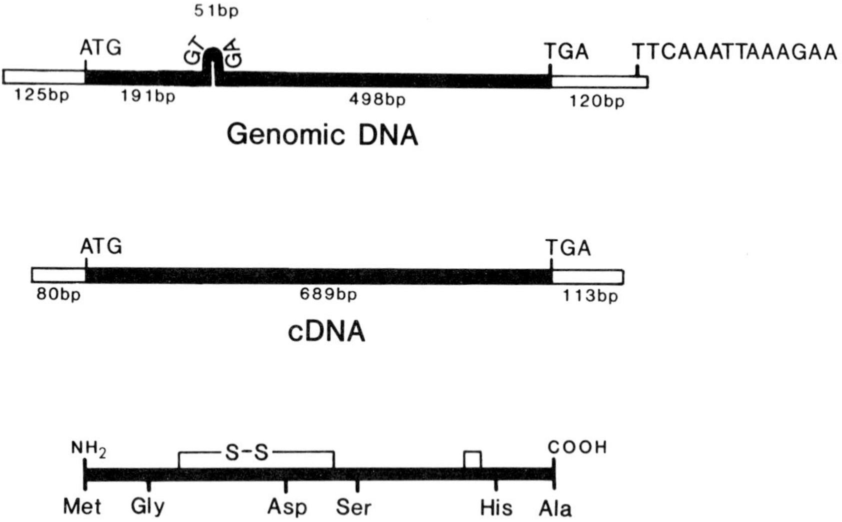

FIGURE 4. Schematic comparison of the cloned genomic DNA and cDNA for cutinase with the encoded protein; the catalytic triad, the disulfide bridges and the N-terminus of the mature protein (gly) are indicated. The intron and the polyadenylation signal are indicated on the genomic DNA.

polyadenylation site also appeared to be quite typical. Similarly, the intron structure also showed homology to intron structures observed in yeast and other filamentous fungi (Figure 5). The restriction map showing the 5'-region of the cutinase gene is shown in Figure 6. Sequences reminiscent of the CAT box and TATA box were found but whether they have any functional significance is not known.

Intron Sequences

5' ... 3'	
GTTCGT.............TCATGACA...........TAG	F. solani pisi cutinase
GTATGT...............TGCTGACT...........TAG GTATGT...............AGCTGACT..........CAG GTGTGT...............AGCTAACC...........CAG GTAAGT.............TACTAACA...........TAG	A. awamori glucoamylase[a]
GTACGT...............AGCTGACT...........CAG GTAAGT..............TGCTGACT...........TAG	N. crassa glutamate dehydrogenase[b]
GTAAGT..............TGCTAACG...........CAG GTAAGT...............TTGTAACA...........CAG GTACGT..............AACTAACA...........CAG	N. crassa histones[c]
GTATGT...............TACTAACA...........PyAG	yeast consensus[d]

FIGURE 5. Comparison of the intron structure in the cutinase gene with consensus sequences of introns in structural genes of other filamentous fungi and yeast. Intron sequences taken from [a]Nunberg et al. (1984) *Mol. Cell Biol.* 4:2306, [b]Kinnaird and Fincham (1983) *Gene* 26:253, [c]Woudt et al. (1983) *Nucl. Acids Res.* 11:5347, [d]Langford and Gallwitz (1983) *Cell* 33:519.

PLANT SURFACE CONTACT TURNS ON CUTINASE GENE IN THE FUNGAL SPORE

When fungal spores land on the plant surface, they are able to sense the presence of the plant and turn on the cutinase gene to gain access into the plant. The molecular mechanism by which the spores sense the contact with the

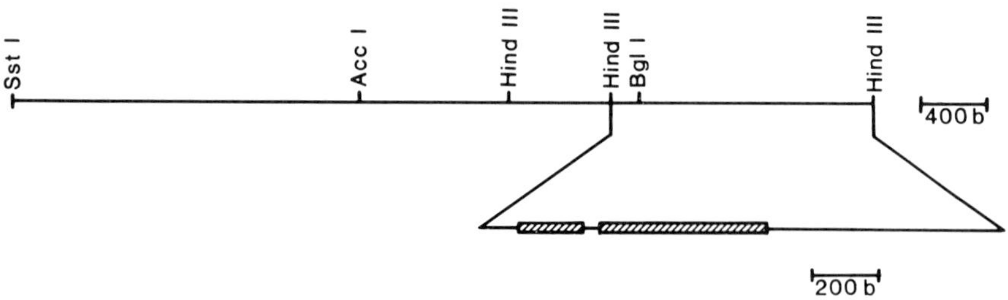

FIGURE 6. Restriction endonuclease map of *F. solani* f. sp. *pisi* genome surrounding the cutinase gene. The 5.5 kb fragment containing the cutinase-encoding regions (shaded line) was isolated form a Charon 35 clone containing a 21 kb genomic insert.

plant surface was not known and how the induction of the enzyme occurs in the spore had also not been examined. It was found that the spores produced cutinase only when cutin was present in the medium (C. Woloshuk and P.E. Kolattukudy, manuscript in preparation). The level of activity produced by the spores depended upon the amount of cutin present in the medium and the level increased with time. How the presence of an insoluble polymer triggers the induction process inside the spore was examined. It was thought that a small amount of cutinase that might be present in the spores could generate enough hydrolysate from the polymer and these hydrolysis products could enter the cells and trigger the induction of the enzyme. In fact, cutinase activity did appear in the medium when cutin hydrolysate was provided (Figure 7). Isolated cutin monomers and, in particular, dihydroxy C_{16} acid and trihydroxy C_{18} acid, two of the most unique components of plant cutin, induced the production of cutinase. That the appearance of the activity in the medium involved protein synthesis was shown by the fact that cycloheximide at 5 μg/ml completely inhibited the appearance of cutinase activity. In medium containing cutin, spores of *F. solani* f. sp. *pisi* incorporated [^{35}S]methionine mainly into cutinase (Figure 8), indicating that synthesis of this

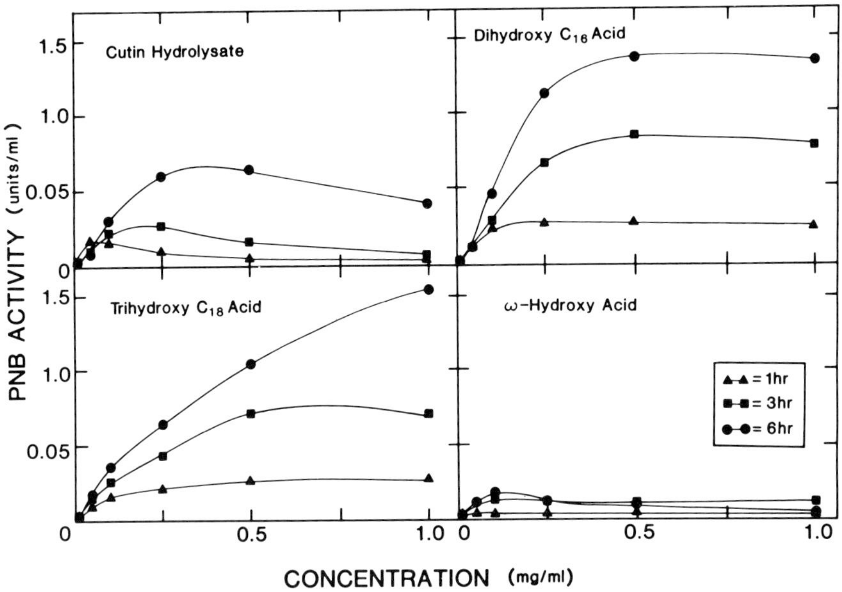

FIGURE 7. Time course of appearance of cutinase activity (*p*-nitrophenylbutyrate hydrolysis) in the extracellular fluid of spore suspensions of *F. solani* f. sp. *pisi* at various concentrations of cutin hydrolysate or isolated cutin monomers.

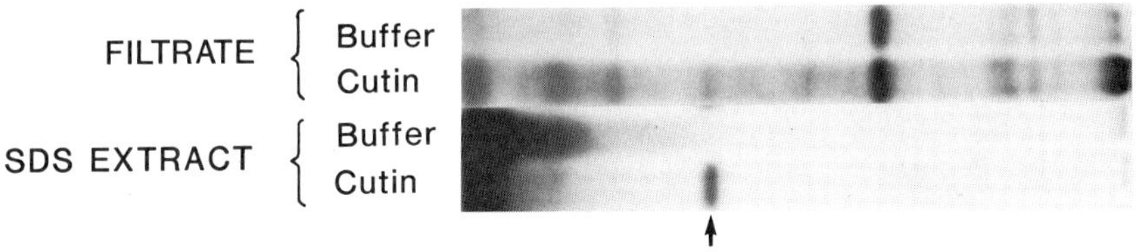

FIGURE 8. Fluorogram of an SDS-gel of electrophoretically separated total proteins generated from exogenous [^{35}S]methionine by spores of *F. solani* f. sp. *pisi* T-8. Spores were suspended at 1 mg dry wt/ml in buffer containing 1 μCi/ml of ^{35}S-Met with or without 1 mg/ml cutin. After 3 hr culture filtrate was collected by centrifugation and the spores were boiled in 4% SDS for 15 min and the proteins recovered in this extract are shown in the two lanes on the bottom. The arrow indicates the position of cutinase.

enzyme is a major activity during the very early stages of spore germination. Immunoblot analysis showed that the increase of activity in the medium represented a true increase in the amount of cutinase protein in the medium. Cutinase was detected in the extracellular fluid as early as 45 minutes after the addition of cutin or monomers of cutin into the medium. Whether this process involved increased transcriptional activity was tested by dot blot analysis of RNA using ^{32}P cDNA probe for cutinase. Within 15 minutes after the addition of cutin into the medium, mRNA transcripts were detectable and the level increased for several hours (Figure 9). If the cutinase induction

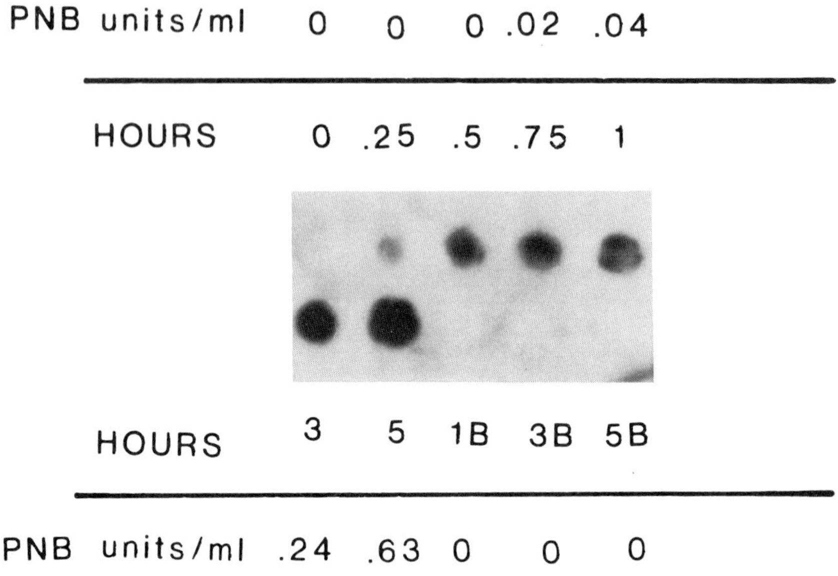

FIGURE 9. Dot blot analysis of cutinase mRNA content of total RNA from spores of *F. solani* f. sp. *pisi* T-8 exposed to cutin for various periods of time. Cutinase [^{32}P]cDNA was used as the probe. The duration (hr) of exposure of the spores to cutin is indicated by the numbers and B indicates control with no cutin. *p*-Nitrophenylbutyrate hydrolase activity (PNB) found in the extracellular fluid is also indicated.

was triggered by cutin monomers generated by the small amount of cutinase carried by the spores, exogenous cutinase should enhance the production of the cutinase transcripts induced by cutin. In fact, addition of cutinase with cutin enhanced the formation of cutinase transcripts.

ENZYMATIC PENETRATION OF PECTINACEOUS BARRIER

Once the cuticle is penetrated the fungus must then breach the barriers posed by the carbohydrate polymers. Recently it was found that spores produce a pectin hydrolase when the spores are in contact with pectin (M. Crawford and P.E. Kolattukudy, unpublished results). The maximal level of production of hydrolase coincided with the initiation of germination (Figure 10). Many hours later a pectate lyase was also produced. Both the pectin hydrolase and pectate lyase activity levels were much higher in the case of *Fusarium solani* f. sp. *pisi* isolate T-8 which is highly virulent when placed on pea stem with intact cuticle when compared to isolate T-30 which required mechanical breaching of the cuticle/wall barriers for infection. The lyase from T-8 was purified to homogeneity and antibodies were prepared and these antibodies inhibited penetration of *Fusarium* into pea stems and thus protected pea stem from infection. These preliminary studies suggest that the enzymes involved in the breaking of not only the cuticle but also the other wall barriers could be used as targets for interference with the penetration and therefore the infection process.

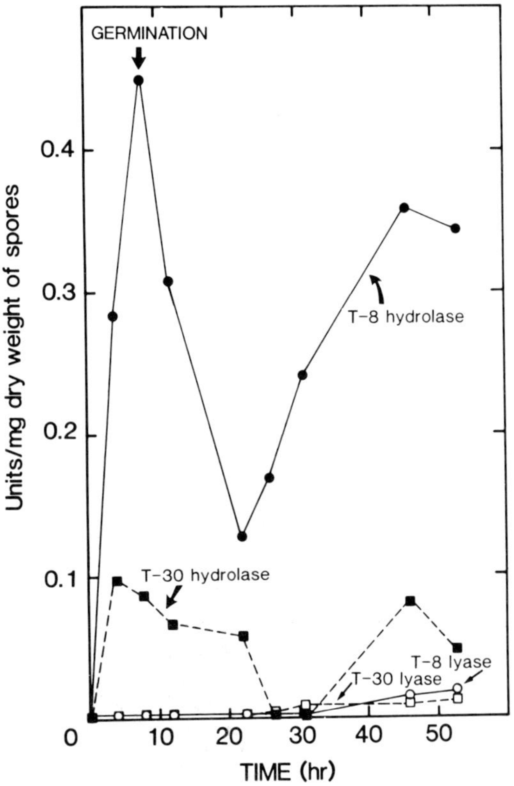

FIGURE 10. Pectin hydrolase and lyase activities released by the spores of *Fusarium solani* f. sp. *pisi* isolates T-8 and T-30. Spores were suspended at 100 μg (dry wt)/ml in a mineral medium containing 1 mg/ml pectin.

REFERENCES

1. Kolattukudy PE (1980). Biopolyester membranes of plants: Cutin and suberin. Science 208:990.
2. Kolattukudy PE (1981). Structure, biosynthesis, and biodegradation of cutin and suberin. Ann Rev Plant Physiol 32:539.
3. van den Ende G, Linskens HF (1974). Cutinolytic enzymes in relation to pathogenesis. Ann Rev Phytopathol 12:247.

4. Kolattukudy PE (1985). Enzymatic penetration of the plant cuticle by fungal pathogens. Ann Rev Phytopathol 23:223.
5. Purdy RE, Kolattukudy PE (1973). Depolymerization of a hydroxyfatty acid biopolymer, cutin, by an extracellular enzyme from *Fusarium solani* f. *pisi*: Isolation and some properties of the enzyme. Arch Biochem Biophys 159:61.
6. Purdy RE, Kolattukudy PE (1975). Hydrolysis of plant cuticle by plant pathogens. Purification, amino acid composition, and molecular weight of two isozymes of cutinase and a nonspecific esterase from *Fusarium solani* f. *pisi*. Biochemistry 14:2824.
7. Kolattukudy PE (1984). Cutinases from fungi and pollen. In Börgstrom B, Brockman H (eds): "Lipases," Amsterdam: Elsevier/North Holland, p 471.
8. Lin T-S, Kolattukudy PE (1977). Glucuronyl glycine, a novel N-terminus in a glycoprotein. Biochem Biophys Res Comm 75:87.
9. Lin T-S, Kolattukudy PE (1980). Structural studies on cutinase, a glycoprotein containing novel amino acids and glucuronic acid amide at the N-terminus. Eur J Biochem 106:341.
10. Köller W, Kolattukudy PE (1982). Mechanism of action of cutinase: Chemical modification of the catalytic triad characteristic for serine hydrolases. Biochemistry 21:3083.
11. Maiti IB, Kolattukudy PE (1979). Prevention of fungal infection of plants by specific inhibition of cutinase. Science 205:507.
12. Dickman MB, Patil SS, Kolattukudy PE (1982). Purification, characterization and role in infection of an extracellular cutinolytic enzyme from *Colletotrichum gloeosporioides* Penz. on *Carica papaya* L. Physiol Plant Pathol 20:333.
13. Dickman MB, Patil SS (1984). Genetic and molecular analysis of a virulence determinant, cutinase from *Colletotrichum gloeosporioides*. Molec Basis Plant Dis Conf, Univ Calif Davis, Poster Abstr, Aug. 19-23, 1984, Abstr 41, p 31.
14. Köller W, Allan CR, Kolattukudy PE (1982). Role of cutinase and cell wall degrading enzymes in infection of *Pisum sativum* by *Fusarium solani* f. sp. *pisi*. Phys Plant Path 20:47.
15. Lin T-S, Kolattukudy PE (1978). Induction of a biopolyester hydrolase (cutinase) by low levels of

cutin monomers in *Fusarium solani* f. sp. *pisi*. J Bacteriol 133:942.

16. Flurkey WH, Kolattukudy PE (1981). *In vitro* translation of cutinase in mRNA: Evidence for a precursor form of an extracellular fungal enzyme. Arch Biochem Biophys 212:154.
17. Soliday CL, Flurkey WH, Okita TW, Kolattukudy PE (1984). Cloning and structure determination of cDNA for cutinase, an enzyme involved in fungal penetration of plants. Proc Natl Acad Sci USA 81:3939.

Molecular Genetics of Filamentous Fungi, pages 439–445

ANALYSIS OF CLONES FROM A FUNGAL WILT PATHOGEN LIBRARY (PHOMA TRACHEIPHILA) FOR USE IN A SPECIFIC DETECTION ASSAY

Franco Rollo[1]
Isabella Di Silvestro[2], and William V. Zucker[2]

1) Department of Cell Biology, University of Camerino
62032 Camerino, ITALY
2) Agriculture Industrial Development, SpA
Research Center, Z.I. Blocco Palma 1°
95100 Catania, ITALY

ABSTRACT In order to develop a quantitative assay for a serious fungal wilt pathogen (*Phoma tracheiphila*), causal agent of the "mal secco" disease of lemon, we made a library of *Phoma tr*. DNA in pBR322. The clones were then tested for specificity to the fungus. In this manner, we obtained a probe suitable for studying many parameters of the host/pathogen relationship.

INTRODUCTION

The fungal disease "mal secco" seriously affects lemon culture in the mediterranean countries (1). The causative agent is the wilt pathogen, *Phoma tracheiphila*, generally restricted to the xylem, making detection difficult (2). Entry into the canopy via germinating conidi is through natural wounds and leaf scars, but not through stomata (3). It is believed that the vegetative hyphae reach vascular tissue by digestion of cell walls and middle lamella material, using exocellular enzymes (4).

The lack of a convenient, quantitative assay of the pathogen has limited field and laboratory studies on this host:pathogen system. We decided therefore to develop a quantitative detection method for Phoma tr. with four objectives in mind: 1) determine the effectiveness of fungicidal applications, 2) track the inocula in the field and under controlled greenhouse conditions, 3) survey the extent of infected plants in orchards and nurseries, 4) study compatible: incompatible reactions in the laboratory.

Toward these ends, we have developed a spot hybridization assay that is sensitive and specific. This method has also been applied by others to plant virus detection (5).

MATERIALS AND METHODS

Plants and fungi. Lemon seedlings, cv Femminello, were grown axenically on M&S agar medium (6) in the dark at 24°C for approximately 12 days. The pathogenic fungus, Phoma tr., was grown axenically in vitro (7).

Nucleic acid extraction. Roots and hypocotyls from lemon seedlings were frozen in liquid nitrogen and ground to a dust in the Mikro-Dismembrator (Braun). The powder was mixed with a solution containing 50 mM Tris-HCl pH 8.3, 3% sodium dodecylsulfate and 5% phenol. The resulting suspension was extracted three times with phenol saturated with 50 mM Tris-HCl pH 8.3 and containing 0.1% each 8-hydroxyquinoline and m-cresol. The aqueous phase was precipitated in ethanol, vacuum desiccated and resuspended in sterile distilled water or in 10 mM Tris-HCl pH 7.6, 1mM EDTA (TE). Total fungal DNA was obtained essentially according to Garber and Yoder (8). Plasmid DNA was purified from transformed E.coli cells by the alkaline lysis method (9).

Labeling of the probes. DNA was labeled with ^{32}P by nick-translation (10).

Cloning strategy. DNA isolated from Phoma tr. as indicated above was partially digested with Sau 3AI (fig 1). The resulting fragments were ligated with Bam HI cut pBR322 plasmid and the total mixture was used to transform E.coli (HB 101) cells (fig 2). Ampicillin resistant-tetracycline sensitive colonies were propagated on liquid medium and screened for the presence of recombinant plasmids.

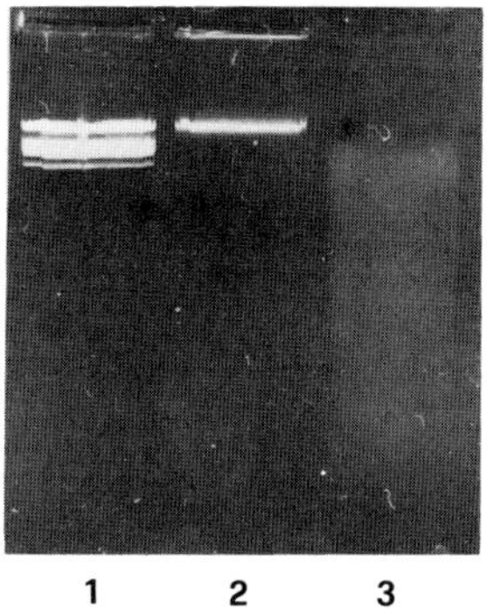

FIGURE 1. Agarose gel electrophoresis of a partial digest of total Phoma tr. DNA with Sau 3AI (lane 3). Lane 1: phage lambda DNA digested with Eco RI; lane 2: undigested Phoma tr. DNA.

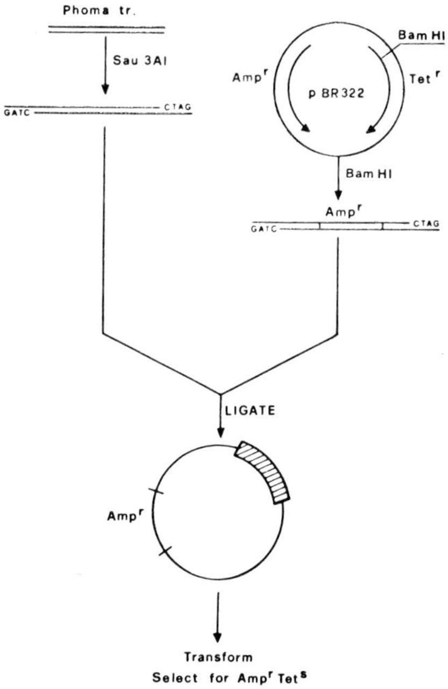

FIGURE 2. Cloning strategy of Phoma tr. DNA in E.coli.

Spot-hybridization conditions. Nitrocellulose filters carrying DNA were preincubated in 3x SSC, 0.08% bovine serum albumin, 0.08% Ficoll, 0.08% polyvinylpyrrolidone and about 50 ug/ml trout sperm DNA at 68°C for 3 hours. Hybridizations were carried out in the same buffer containing about 10^7 cpm of ^{32}P-labeled, nick-translated plasmid DNA for 16 hours. After hybridization, the filters were washed 3 times in 2x SSC, 0.1% SDS at 68°C. Exposure to film was at -80°C for up to 7 days.

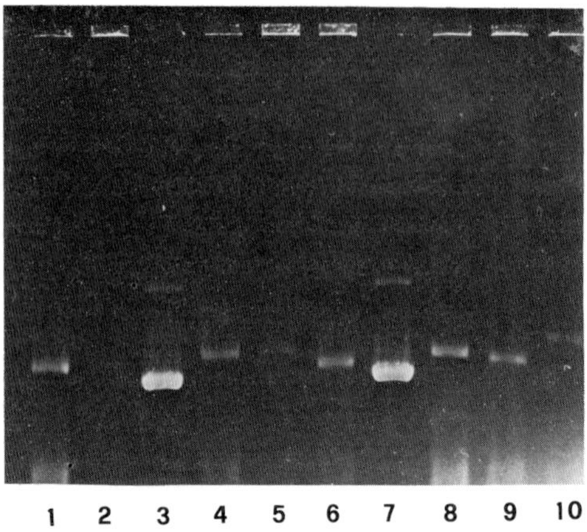

FIGURE 3. Agarose gel electrophoresis of plasmid DNA from amp^r - tet^s E.coli colonies; lanes 3 and 7: pB322.

RESULTS AND DISCUSSION

Preliminary investigations with nick-translated Phoma tr. DNA, used as a probe in spot-hybridization experiments, showed us that fungal DNA hybridizes to lemon DNA to a certain extent. In order to circumvent this problem and obtain an unambiguous probe for Phoma tr., we prepared a library of fungal DNA in pB322 as described in Materials and Methods. Sampling at random, the recombinant plasmids obtained were found to contain inserts with an average length of about 1 Kbp (fig 3); however, much larger inserts were occasionally found as in the case of pPho7 (fig 4).

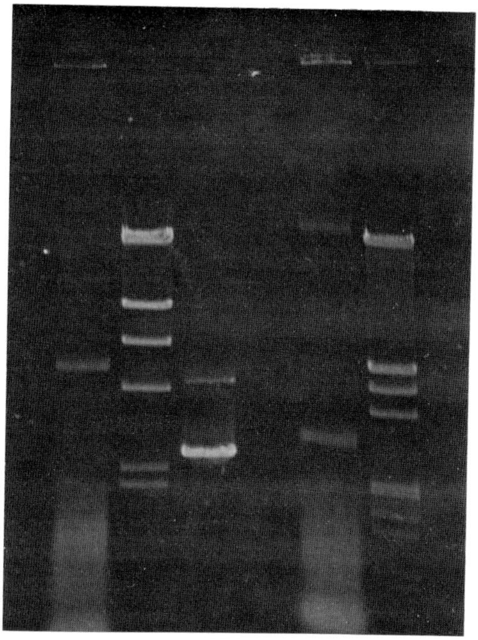

FIGURE 4. Agarose gel electrophoresis of two clones of Phoma tr. DNA in pBR322 showing the range of sizes of the recombinant plasmids obtained. Lane 1: pPho71; lane 4: pPho166; lane 2,3 and 5 are respectively: phage lambda DNA digested with Hind III, pBR322, phage lambda DNA digested with Hind III and Eco RI.

The recombinant plasmids were then screened in order to select those which gave a strong signal when hybridized with Phoma tr. DNA but negligible cross-hybridization with lemon DNA. An example is shown (fig 5). In addition, we are screening recombinant plasmids for nonhybridization to DNA obtained from other commonly encountered fungal and bacterial organisms found under field conditions in Sicilian citrus orchards.

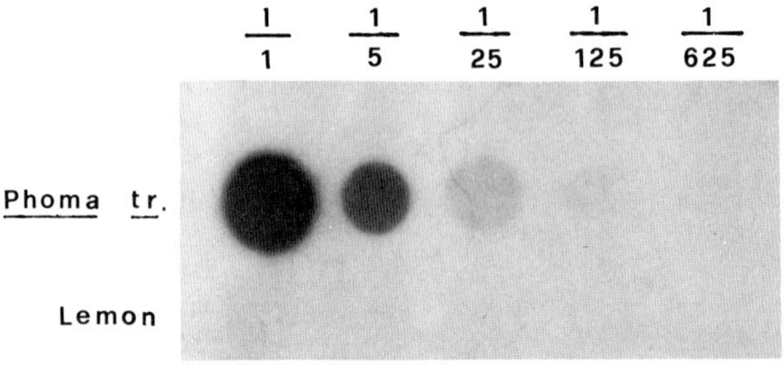

FIGURE 5. Spot hybridization of *Phoma tr.* and lemon DNA's. The DNA isolated from lemon seedlings and *Phoma tr.* cultures were serially diluted before spotting on to nitrocellulose and probed with nick-translated pPho14 DNA. The first spots are about 250 ng DNA.

The present probe is now suitable for studyıng many parameters of the host-pathogen relationship in the laboratory and greenhouse. However, we are screening further in order to maximize sensitivity and specificity for etiological field studies.

ACKNOWLEDGMENTS

We wish to express our gratitude to Augusto Amici, Anna Gilletti, Angelo La Marca, and Antonella Maucioni for assistance in this work.

REFERENCES

1. Salerno M, Cutuli G, and Somma V (1976). The mal secco disease in citrus orchards of Italy. Brief references to past history, present research, and trends for the future (in Italian). Italia Agric. 1:96.
2. Petri L (1930). Further research on the morphology, biology, and parasitism of *Deuterophoma tracheiphila* (in Italian). Boll. Staz. Patol. Veg. 10:191.
3. Zucker WV, Catara A (1985). Observations in the scanning electron microscope on the foliar penetration of *Phoma*

tracheiphila, in press.
4. Graniti A (1969). Host-parasite relations in citrus disease as exemplified by Phytopthera gummosis and Deuterophoma "mal secco". In Chapman HD (ed): Proc. 1st Internat. Citrus Sympos. Riverside: Univ. Calif., p 1187.
5. Maule AJ, Hull R, Donson J (1983). The application of spot hybridization to the detection of DNA and RNA viruses in plant tissues. J Virol Meth 6:215.
6. Murashige T, Skoog F (1962). A revised medium for rapid growth and bioassays with tobacco tissue cultures. Physiologia Plantarum 15:473.
7. Nachmias A, Barash I, Buchner V, Solel Z and Strobel GA (1979). Isolation of a phytotoxic glycopeptide from lemon leaves infected with Phoma tracheiphila. Physiol Plant Pathol 14.135.
8. Garber RC, Yoder OC (1983). Isolation of DNA from filamentous fungi and separation into nuclear, mitochondrial, ribosomal and plasmid components. Anal Biochem 135:416.
9. Birnboim HC, Doly J (1979). A rapid alkaline extraction procedure for screening recombinant plasmid DNA. Nucleic Acids Res 7:1513.
10. Rigby PWJ, Dieckman M, Rhodes C, Berg P (1977). Labeling deoxyribonucleic acid to high specific activity by nick-translation with DNA polymerase I. J Mol Biol 113:237.

Index